Thin Films for Energy Harvesting, Conversion, and Storage

Thin Films for Energy Harvesting, Conversion, and Storage

Special Issue Editors

Zhong Chen
Yuxin Tang
Xin Zhao

MDPI • Basel • Beijing • Wuhan • Barcelona • Belgrade

MDPI

Special Issue Editors
Zhong Chen
Nanyang Technological University
Singapore

Yuxin Tang
University of Macau
China

Xin Zhao
Nanyang Technological University
Singapore

Editorial Office
MDPI
St. Alban-Anlage 66
4052 Basel, Switzerland

This is a reprint of articles from the Special Issue published online in the open access journal *Coatings* (ISSN 2079-6412) from 2018 to 2019 (available at: https://www.mdpi.com/journal/coatings/special_issues/film_energy)

For citation purposes, cite each article independently as indicated on the article page online and as indicated below:

LastName, A.A.; LastName, B.B.; LastName, C.C. Article Title. *Journal Name* **Year**, *Article Number*, Page Range.

ISBN 978-3-03921-724-3 (Pbk)
ISBN 978-3-03921-725-0 (PDF)

Contents

About the Special Issue Editors

Zhong Chen (Professor) received his PhD from the University of Reading, UK. After completing his studies in early 1997, he joined the newly established Institute of Materials Research and Engineering, a national research institute funded by Singapore government. In March 2000, he moved to Nanyang Technological University (NTU) where he served as Assistant Professor, and has since been promoted to Associate Professor and then Professor in the School of Materials Science and Engineering. Prof. Chen's research interests include 1) surface engineering, thin films and nanostructured materials, and 2) mechanical behavior of materials. He is author of over 350 peer-reviewed journal papers, 5 book chapters, and 6 granted patents. According to Google Scholar, his journal articles have received over 13,000 citations with an h-index of 59. Since joining NTU, Prof. Chen has supervised 30 PhD and 5 MEng graduates. Prof. Chen has served as Editor or on the editorial board for 8 academic journals. He has also served as a reviewer for more than 70 academic journals and a number of research funding agencies, including the European Research Council.

Yuxin Tang (Assistant Professor) is Assistant Professor in the Institute of Applied Physics and Materials Engineering (IAPME) at University of Macau. He graduated from Nanyang Technological University (NTU, Singapore) with a PhD in Materials Science (2013), where he developed functional materials for environmental protection. Follow this, he continued as a Postdoc at NTU, where he worked on advanced functional materials for energy conversion and storage application. He has received prestigious awards and honors for this work, including the Emerging Investigators for *Journal of Materials Chemistry A* (2017), World Future Foundation PhD Prize Award for best PhD Thesis (2013), and Chinese Government Award for Outstanding Self-financed Students Studying Abroad (2013). To date, he has published over 80 scientific articles in such journals as *Chemical Society Reviews, JACS, Advanced Materials, Angewandte Chemie, Advanced Energy Materials, Advanced Functional Materials and Small*.

Xin Zhao received his PhD degree from Nanjing University (China) in 2014. He has been a Research Fellow at the School of Materials Science and Engineering of Nanyang Technological University since his appointment in September 2014. His research interests include photoelectrochemical water splitting using solar energy for hydrogen generation, electrochemical hydrogen energy generation, and photoluminescence materials for white LEDs. To date, he has published 39 scientific articles in such journals as *Advanced Energy Materials, Applied Catalysis B: Environmental, Advanced Functional Materials, Joule, Energy & Environmental Science, Journal of Materials Chemistry A* and *Angewandte Chemie International Edition*.

Preface to "Thin Films for Energy Harvesting, Conversion, and Storage"

Thin film-based energy harvesting, conversion, and storage devices have attracted great attention due to their attractive potential for improved efficiency, manufacturability, and production costs. Nowadays, thin film electrodes have been used in a wide range of fields, such as photovoltaics, fuel cells, supercapacitors, flow batteries, and rechargeable metal ion batteries. In order to enhance the efficiency of energy conversion, rational thin film design strategies, novel thin film materials, and the fundamental understanding of structure–property correlation have been systematically studied. For examples, thin film materials can be fabricated by various methods, including thermal evaporation method, electrochemical deposition, atomic layer deposition, chemical vapor deposition, pulsed laser deposition, and molecular beam epitaxy. By controlling these thin film fabrication techniques, thin film electrodes with desirable properties are obtained and can be used towards improving the device performance via fundamental understanding of the structure–property–performance correlations.

This Special Issue contains 9 research articles and 2 review articles. In their research article, Hu et al. present a DFT-based model on the adsorption behavior of H_2O, H^+, Cl^-, and OH^- on clean and Cr-doped Fe (110) planes, and the surface energy study suggests that the Cr-doped Fe(110) surface is more stable than Fe(110) and Cr(110) facets upon adsorption of these four typical adsorbates. This study could provide guidance for the design of corrosion-resistant devices. Bonomo et al. investigate the effect of sensitization on the electrochemical properties of nanostructured NiO, and the photoelectrochemical cells displaying the highest efficiencies of solar conversion were those that employed sensitized NiO electrodes with the lowest values of charge transfer resistance through the dye/NiO junction in the absence of illumination. This finding indicates that the electronic communication between the NiO substrate and the dye sensitizer is the most important factor in the electrochemical and photoelectrochemical processes occurring at this type of modified semiconductor. Mohammed et al. fabricated a few-layer graphene nano-flake thin film by an affordable vacuum kinetic spray method at room temperature and modest low vacuum conditions. Meantime, the proposed affordable supercapacitors show a high areal capacitance and a small equivalent series resistance. Gao et al. synthesized $Cu_2ZnSn(S,Se)_4$ (CZTSSe) and $Cu_2Zn(Sn,Ge)(S,Se)_4$ (CZTGSSe) thin films using a non-vacuum solution method. Based on the CZTSSe and CZTGSSe films, solar cells were prepared. The as-fabricated CZTGSSe solar cells exhibited a lower diode ideality factor and lower reverse saturation current density. She et al. report the fabrication of the mixed nickel–cobalt–molybdenum metal oxide nanosheet arrays for hybrid supercapacitor applications. In their paper, Yu et al. construct a $LaFeO_3$ perovskite nanoparticle-modified TiO_2 nanotube array, and the fabricated sample displays excellent photocatalytic performance. Zhu et al. have demonstrated facile formation of ultrathin Al_2O_3-coated $LiNi_{0.8}Co_{0.1}Mn_{0.1}O_2$ cathode material by an atomic layer deposition (ALD) method. Enhanced electrochemical performance was obtained by optimizing the thickness of the Al_2O_3 layer. In their brief report, Li et al. propose a metal–dielectric metal structure based on a Fabry–Pérot cavity, and the as-prepared narrow-band absorber can be easily fabricated by the mature thin film technology independent of any nanostructure, which makes it an appropriate candidate for photodetectors, sensing, and spectroscopy. Chen et al. performed an in situ investigation of the early-stage $CH_3NH_3PbI_3$ (MAPbI$_3$) and $CH(NH_2)_2PbI_3$ (FAPbI$_3$) degradation under high water vapor pressure. Their experimental results highlight the importance of the compositional and morphological changes in early stage degradation in perovskite materials. In one of the two review articles, Shin and Choi summarize the recent studies of semitransparent solar cells and discuss the major problems to be overcome towards commercialization of these solar cells. Hu et al. present a review of the latest processes for designing anode materials to improve the efficiency of photoelectrochemical water splitting. This review is helpful for researchers who are working in or are considering entering the field to better appreciate the state of the art, and to make a better choice when they embark on new research in photocatalytic water splitting materials.

Zhong Chen, Yuxin Tang, Xin Zhao
Special Issue Editors

coatings

MDPI

Editorial

Special Issue: "Thin Films for Energy Harvesting, Conversion, and Storage"

Zhong Chen [1,*], Xin Zhao [1,*] and Yuxin Tang [2,*]

1   School of Materials Science and Engineering, Nanyang Technological University, 50 Nanyang Avenue, Singapore 639798, Singapore
2   Institute of Applied Physics and Materials Engineering, University of Macau, Macau, China
*   Correspondence: ASZChen@ntu.edu.sg (Z.C.); xinzhao@ntu.edu.sg (X.Z.); yxtang@um.edu.mo (Y.T.)

Received: 23 September 2019; Accepted: 23 September 2019; Published: 25 September 2019

Abstract: Efficient clean energy harvesting, conversion, and storage technologies are of immense importance for the sustainable development of human society. To this end, scientists have made significant advances in recent years regarding new materials and devices for improving the energy conversion efficiency for photovoltaics, thermoelectric generation, photoelectrochemical/electrolytic hydrogen generation, and rechargeable metal ion batteries. The aim of this Special Issue is to provide a platform for research scientists and engineers in these areas to demonstrate and exchange their latest research findings. This thematic topic undoubtedly represents an extremely important technological direction, covering materials processing, characterization, simulation, and performance evaluation of thin films used in energy harvesting, conversion, and storage.

Keywords: thin films; synthesis; characterization; energy harvesting; energy conversion; energy storage

Thin film-based energy harvesting, conversion, and storage devices have attracted great attention due to their attractive potential for improved efficiency, manufacturability, and production costs. Nowadays, thin film electrodes have been used in a wide range of fields, such as photovoltaics, fuel cells, supercapacitors, flow batteries, and rechargeable metal ion batteries [1,2]. In order to enhance the efficiency of energy conversion, rational thin film design strategies, novel thin film materials, and the fundamental understanding of structure–property correlation have been systematically studied [3,4]. For examples, thin film materials can be fabricated by various methods, including thermal evaporation method, electrochemical deposition, atomic layer deposition, chemical vapor deposition, pulsed laser deposition, and molecular beam epitaxy. By controlling these thin film fabrication techniques, thin film electrodes with desirable properties are obtained and can be used towards improving the device performance via fundamental understanding of the structure–property–performance correlations.

This Special Issue contains 9 research articles and 2 review articles. In their research article, Hu et al. [5] present a DFT-based model on the adsorption behavior of H_2O, H^+, Cl^-, and OH^- on clean and Cr-doped Fe (110) planes, and the surface energy study suggests that the Cr-doped Fe(110) surface is more stable than Fe(110) and Cr(110) facets upon adsorption of these four typical adsorbates. This study could provide guidance for the design of corrosion-resistant devices. Bonomo et al. [6] investigate the effect of sensitization on the electrochemical properties of nanostructured NiO, and the photoelectrochemical cells displaying the highest efficiencies of solar conversion were those that employed sensitized NiO electrodes with the lowest values of charge transfer resistance through the dye/NiO junction in the absence of illumination. This finding indicates that the electronic communication between the NiO substrate and the dye sensitizer is the most important factor in the electrochemical and photoelectrochemical processes occurring at this type of modified semiconductor. Mohammed et al. [7] fabricated a few-layer graphene nano-flake thin film by an affordable vacuum kinetic spray method at room temperature and modest low vacuum conditions. Meantime, the proposed affordable supercapacitors show a high areal capacitance and a small equivalent series resistance. Gao et al. [8] synthesized $Cu_2ZnSn(S,Se)_4$ (CZTSSe) and $Cu_2Zn(Sn,Ge)(S,Se)_4$ (CZTGSSe) thin films using a non-vacuum solution method. Based on the CZTSSe and CZTGSSe films, solar cells were prepared. The as-fabricated CZTGSSe solar cells exhibited a lower diode ideality factor and lower reverse saturation current density. She et al. [9] report the fabrication of the mixed nickel–cobalt–molybdenum metal oxide nanosheet arrays for hybrid supercapacitor applications. In their paper, Yu et al. [10] construct a $LaFeO_3$ perovskite nanoparticle-modified TiO_2 nanotube array, and the fabricated sample displays excellent photocatalytic performance. Zhu et al. [11] have demonstrated facile formation of ultrathin Al_2O_3-coated $LiNi_{0.8}Co_{0.1}Mn_{0.1}O_2$ cathode material by an atomic layer deposition (ALD) method. Enhanced electrochemical performance was obtained by optimizing the thickness of the Al_2O_3 layer. In their brief report, Li et al. [12] propose a metal–dielectric metal structure based on a Fabry–Pérot cavity, and the as-prepared narrow-band absorber can be easily fabricated by the mature thin film technology independent of any nanostructure, which makes it an appropriate candidate for photodetectors, sensing, and spectroscopy. Chen et al. [13] performed an in situ investigation of the early-stage $CH_3NH_3PbI_3$ (MAPbI$_3$) and $CH(NH_2)_2PbI_3$ (FAPbI$_3$) degradation under high water vapor pressure. Their experimental results highlight the importance of the compositional and morphological changes in early stage degradation in perovskite materials. In one of the two review articles, Shin and Choi [14] summarize the recent studies of semitransparent solar cells and discuss the major problems to be overcome towards commercialization of these solar cells. Hu et al. [15] present a review of the latest processes for designing anode materials to improve the efficiency of photoelectrochemical water splitting. This review is helpful for researchers who are working in or are considering entering the field to better appreciate the state of the art, and to make a better choice when they embark on new research in photocatalytic water splitting materials.

Conflicts of Interest: The author declares no conflict of interest.

References

1. Tang, Y.; Zhang, Y.; Li, W.; Ma, B.; Chen, X. Rational material design for ultrafast rechargeable lithium-ion batteries. *Chem. Soc. Rev.* **2015**, *44*, 5926–5940. [CrossRef] [PubMed]
2. Tang, Y.; Jiang, Z.; Xing, G.; Li, A.; Kanhere, P.D.; Zhang, Y.; Sum, T.C.; Li, S.; Chen, X.; Dong, Z.; et al. Efficient Ag@ AgCl Cubic Cage Photocatalysts Profit from Ultrafast Plasmon-Induced Electron Transfer Processes. *Adv. Funct. Mater.* **2013**, *23*, 2932–2940. [CrossRef]
3. Lee, T.D.; Ebong, A.U. A review of thin film solar cell technologies and challenges. *Renew. Sustain. Energy Rev.* **2017**, *70*, 1286–1297. [CrossRef]
4. Choudhary, S.; Upadhyay, S.; Kumar, P.; Singh, N.; Satsangi, V.R.; Shrivastav, R.; Dass, S. Nanostructured bilayered thin films in photoelectrochemical water splitting–A review. *Int. J. Hydrogen Energy* **2012**, *37*, 18713–18730. [CrossRef]
5. Hu, J.; Wang, C.M.; He, S.J.; Zhu, J.B.; Wei, L.P.; Zheng, S.L. A DFT-Based Model on the Adsorption Behavior of H_2O, H^+, Cl^-, and OH^- on Clean and Cr-Doped Fe(110) Planes. *Coatings* **2018**, *8*, 51. [CrossRef]
6. Bonomo, M.; Gatti, D.; Barolo, C.; Dini, D. Effect of Sensitization on the Electrochemical Properties of Nanostructured NiO. *Coatings* **2018**, *8*, 232. [CrossRef]
7. Mohammed, M.M.M.; Chun, D.-M. Electrochemical Performance of Few-Layer Graphene Nano-Flake Supercapacitors Prepared by the Vacuum Kinetic Spray method. *Coatings* **2018**, *8*, 302. [CrossRef]
8. Gao, C.; Sun, Y.L.; Yu, W. Influence of Ge Incorporation from $GeSe_2$ Vapor on the Properties of $Cu_2ZnSn(S,Se)_4$ Material and Solar Cells. *Coatings* **2018**, *8*, 304. [CrossRef]
9. She, Y.; Tang, B.; Li, D.L.; Tang, X.S.; Qiu, J.; Shang, Z.G.; Hu, W. Mixed Nickel-Cobalt-Molybdenum Metal Oxide Nanosheet Arrays for Hybrid Supercapacitor Applications. *Coatings* **2018**, *8*, 340. [CrossRef]
10. Yu, J.D.; Xiang, S.W.; Ge, M.Z.; Zhang, Z.Y.; Huang, J.Y.; Tang, Y.X.; Sun, L.; Lin, C.J.; Lai, Y.K. Rational Construction of $LaFeO_3$ Perovskite Nanoparticle-Modified TiO_2 Nanotube Arrays for Visible-Light Driven Photocatalytic Activity. *Coatings* **2018**, *8*, 374. [CrossRef]

11. Zhu, W.C.; Huang, X.; Liu, T.T.; Xie, Z.Q.; Wang, Y.; Tian, K.; Bu, L.M.; Wang, H.B.; Gao, L.J.; Zhao, J.Q. Ultrathin Al_2O_3 Coating on $LiNi_{0.8}Co_{0.1}Mn_{0.1}O_2$ Cathode Material for Enhanced Cycleability at Extended Voltage Ranges. *Coatings* **2019**, *9*, 92. [CrossRef]
12. Li, Q.; Li, Z.Z.; Xiang, X.J.; Wang, T.T.; Yang, H.G.; Wang, X.Y.; Gong, Y.; Gao, J.S. Tunable Perfect Narrow-Band Absorber Based on a Metal-Dielectric-Metal Structure. *Coatings* **2019**, *9*, 393. [CrossRef]
13. Chen, S.; Solanki, A.; Pan, J.S.; Sun, T.C. Compositional and Morphological Changes in Water-Induced Early-Stage Degradation in Lead Halide Perovskites. *Coatings* **2019**, *9*, 535. [CrossRef]
14. Shin, D.H.; Choi, S.-H. Recent Studies of Semitransparent Solar Cell. *Coatings* **2018**, *8*, 329. [CrossRef]
15. Hu, J.; Zhao, S.; Zhao, X.; Chen, Z. Strategies of Anode Materials Design towards Improved Photoelectrochemical Water Splitting Efficiency. *Coatings* **2019**, *9*, 309. [CrossRef]

coatings

MDPI

Article

Compositional and Morphological Changes in Water-Induced Early-Stage Degradation in Lead Halide Perovskites

Shi Chen [1,*], Ankur Solanki [2,3], Jisheng Pan [4] and Tze Chein Sum [2]

1   Institute of Applied Physics and Materials Engineering, University of Macau, Macau, SAR, China
2   Division of Physics and Applied Physics, School of Physical and Mathematical Sciences, Nanyang
    Technological University, 21 Nanyang Link, Singapore 637371, Singapore
3   Department of Science, School of Technology, Pandit Deendayal Petroleum University,
    Gandhiagar 382007, India
4   Institute of Materials Research and Engineering, A*STAR (Agency for Science, Technology and Research), 2
    Fusionopolis Way, Innovis #08-03, Singapore 138634, Singapore
*   Correspondence: shichen@um.edu.mo; Tel.: +852-8822-4294

Received: 20 July 2019; Accepted: 14 August 2019; Published: 22 August 2019

Abstract: With tremendous improvements in lead halide perovskite-based optoelectronic devices ranging from photovoltaics to light-emitting diodes, the instability problem stands as the primary challenge in their development. Among all factors, water is considered as one of the major culprits to the degradation of halide perovskite materials. For example, $CH_3NH_3PbI_3$ (MAPbI$_3$) and $CH(NH_2)_2PbI_3$ (FAPbI$_3$) decompose into PbI$_2$ in days under ambient conditions. However, the intermediate changes of this degradation process are still not fully understood, especially the changes in early stage. Here we perform an in-situ investigation of the early-stage MAPbI$_3$ and FAPbI$_3$ degradation under high water vapor pressure. By probing the surface and bulk of perovskite samples using near-ambient pressure X-ray photoelectron spectroscopy (NAP-XPS) and XRD, our findings clearly show that PbI$_2$ formation surprisingly initiates below the top surface or at grain boundaries, thus offering no protection as a water-blocking layer on surface or grain boundaries to slow down the degradation process. Meanwhile, significant morphological changes are observed in both samples after water vapor exposure. In comparison, the integrity of MAPbI$_3$ film degrades much faster than the FAPbI$_3$ film against water vapor. Pinholes and large voids are found in MAPbI$_3$ film while only small number of pinholes can be found in FAPbI$_3$ film. However, the FAPbI$_3$ film suffers from its phase instability, showing a fast α-to-δ phase transition. Our results highlight the importance of the compositional and morphological changes in the early stage degradation in perovskite materials.

Keywords: halide perovskite; degradation; water; PbI$_2$ formation; morphology

1. Introduction

Halide perovskite materials with excellent optoelectronic properties show extensive potential in applications including photovoltaics [1], photodetectors [2], light-emitting diodes, and so on [3]. However, the inherent instability prevents these materials from long-term usage in devices [4]. The two widely used perovskite materials, methylammonium lead triiodide (MAPbI$_3$) and formamidinium lead triiodide (FAPbI$_3$), are susceptible to degradation by multiple factors, including water, oxygen, UV light, electrical field, and heating [4,5]. Synergistic degradation by the combination of multiple factors were also seen. Recent reports found the degradation process is greatly accelerated when perovskite is exposed to water, oxygen, and light together [6,7]. However, the excess water is still considered as the one of the major culprits causing the degradation of perovskite materials. Without any protection, both MAPbI$_3$ and FAPbI$_3$ films cannot last more than a few days in ambient air, quickly decomposed into PbI$_2$ [8].

Based on these observations, two general strategies are proposed to improve device stability. The first strategy is to prevent perovskite layers from making any contact with water. Methods associated with this strategy include encapsulation [9,10] and surface passivation [11]. The second strategy is to enhance the durability of perovskite against water. Methods associated with this strategy include organic and inorganic doping [12–14] and film quality improvement [15]. These efforts successfully extend the lifetime of PSCs to thousands of hours, but still far from fulfilling commercialization requirements [4]. To further extend the lifetime of perovskite materials, new strategy based on thorough understanding of perovskite degradation is needed.

It is clear that MAPbI$_3$ and FAPbI$_3$ decompose into PbI$_2$ at the end, but there are still uncertainties in the process of the degradation. For example, many studies suggest an indirect degradation pathway with sequential formation of two hydrated intermediates, monohydrate (CH$_3$NH$_3$PbI$_3$·H$_2$O) and dihydrate ((CH$_3$NH$_3$)$_4$PbI$_6$·2H$_2$O), during ingress of water [16]. The monohydrate forms first and is reversible, while dihydrate forms only after monohydrate. Prolonged exposure dihydrate with water leads to final decomposition [17]. However, this indirect pathway may not be always valid. Direct PbI$_2$ formation without steps of monohydrate and dihydrate formation was shown in some rigorous studies, raising doubts on the validity of the indirect degradation pathway. Schlipf et al. reported an earlier PbI$_2$ formation than the appearance of monohydrate in their in-situ XRD measurements [18]. Recent near ambient pressure X-ray photoelectron spectroscopy (NAPXPS) study also claimed the surface of a perovskite thin film prepared by thermal deposition quickly decomposed into PbI$_2$ at only 30% of relative humidity level [19]. Therefore, a direct degradation pathway leading to PbI$_2$ may exist, causing premature deterioration of device performance.

Meanwhile, it is also uncertain the role of PbI$_2$ in degradation. Excessive PbI$_2$ appeared beneficial to the device performance, though it may affect the long-term photostability [20,21]. Previous studies usually assumed the PbI$_2$ is formed from the surface, their data usually suggest a linear degradation speed, implying no water-blocking effect from PbI$_2$ layer on surface [22]. Studies using in-vivo XRD measurements confirmed PbI$_2$ formation in sample but was unable to determine whether PbI$_2$ is on the top of surface or not [8,23]. PbI$_2$ layer was observed by surface sensitive techniques such as XPS, but in these studies, perovskite films were usually measured after complete decomposition and missed the critical PbI$_2$ formation period [22,24].

Another uncertainty is the effect of morphological change in the early stage of degradation. Morphology is considered as one of the important factors that affects the performance and lifetime of devices. Interestingly, it can be beneficial if a small fraction of water is introduced during fabrication [25]. Improved performance and lifetime are witnessed in these devices. These improvements are believed to be due to increased grain size as well as trap passivation [26,27]. Morphological change may be detrimental if grains are grown too large and cause film disintegration. However, the process at which the film loses its integrity is largely ignored. Previous studies focused on non-morphological changes such as trap passivation and grain boundary variation at early degradation stage [17,28]. Others studied much later degradation stage, when the perovskite is completely reverted to PbI$_2$ [8,29]. Therefore, the morphology evolution in degradation needs further investigation.

Here we report a study focusing on the critical changes in the early stage of MAPbI$_3$ and FAPbI$_3$ degradation in a precise and controlled condition. Multiple experimental techniques are applied to study water-induced compositional and morphological evolutions under high water vapor pressure at room temperature. Surprisingly, the surface of degraded samples remained stoichiometric to the pristine phase while PbI$_2$ is found in bulk. This finding excludes the possibility of self-passivation by PbI$_2$ and supports the coexistence of both degradation pathways under high water partial pressure condition. Meanwhile, prominent morphological changes such as pinholes and large voids are observed in MAPbI$_3$ sample, revealing loss of film integrity initiated in the early stage degradation. In comparison, FAPbI$_3$ film remains largely intact with much less pinholes, indicating a much slower morphology evolution. However, the FAPbI$_3$ sample suffers from water-induced phase change. XRD data show the dominant δ phase in the degraded FAPbI$_3$ sample, indicating a fast α to δ phase transition in the early stage of the degradation. Our results successfully clarify compositional and morphological changes in the early-stage degradation of perovskite thin films.

2. Materials and Methods

The MAPbI$_3$ sample and FAPbI$_3$ sample are prepared by standard solution preparation methods with anti-solvent treatment. All organic cation salts were purchased from Dyesol (Queanbeyan, Australia) while lead iodide was bought from Acros Organics (Geel, Belgium). MAPbI$_3$ and FAPbI$_3$ perovskite solutions were prepared by mixing precursors in stoichiometric ratio (1 M concentration) in anhydrous dimethylformamide (DMF from Sigma Aldrich, St. Louis, MO, USA). The perovskite thin films were spun-coated on cleaned indium-tin-oxide (ITO) substrates at 5000 rpm for 12 s. The anti-solvent treatment was performed by dripping 100 μL toluene on the spinning substrates 9 s prior to the end of spinning. Subsequently, MAPbI$_3$ samples were annealed at 100 °C for 30 min while FAPbI$_3$ samples were annealed at 160 °C for 10 min. Prepared samples were transferred from glovebox to XPS system through an air-tight container to minimize contact with external atmosphere. XPS measurements were conducted in-situ before and after water vapor dosing in the high-pressure gas cell. SEM and XRD were ex-situ measured before and after water vapor dosing in near-ambient pressure X-ray photoelectron spectroscopy (NAP-XPS) system.

The film crystallinity was measured using Bruker D8 discover X-ray diffractometer (Billerica, MA, USA) with Cu Kα radiation (λ = 1.54 Å). The morphology is measured by JEOL FESEM 6700 (Akishima, Japan). The surface composition was measured by PREVAC NAP-XPS system (Rogow, Poland) with monochromatic Al Kα X-ray source (*hν* = 1486.7 eV).

3. Results and Discussion

To investigate the degradation on the perovskite surface, it is important to control the hydration level precisely. Many previous works used ambient condition with different relative humidity (RH) values to probe the degradation process [23,30,31]. However, RH is not an absolute unit and can largely vary due to temperature change. Therefore, RH value is not an accurate parameter to measure the content of water. Furthermore, gases in the air such as oxygen may also have implications on the perovskite degradation, making the process more complicated [28,32]. Here we use the gas cell inside a NAPXPS system to study the interaction of perovskite samples to water vapor. The in-situ environment excludes any other external factors and the measured water partial pressure is more accurate than RH value.

MAPbI$_3$ and FAPbI$_3$ samples are exposed to 23 mbar of water vapor pressure in NAPXPS cells separately for two times with 1 h each. The water vapor partial pressure is about 80% RH for the measured cell temperature of 23 °C. The lower exposure pressure at 18 mbar or below shows no distinguished changes. Extending the exposure time to 6 h resulted in further decreased XPS intensities as well as I/Pb, N/Pb, and C/Pb ratios. Therefore, we only discuss the 2 h exposure in detail. To monitor potential surface composition changes, XPS high-resolution spectra from MAPbI$_3$ and FAPbI$_3$ before and after water vapor exposure were acquired (Figure 1). It should be highlighted that measurements were done at an ultrahigh vacuum condition after the system was fully recovered from water exposure. Therefore, signals from monohydrate and dihyrate are not expected. For MAPbI$_3$, the spectra of I $3d_{5/2}$, Pb $4f$, and N $1s$ contain single peaks at 619.6, 138.8, and 402.7 eV, respectively. All of them originated from MAPbI$_3$ [33]. In the spectrum of C $1s$, two peaks at 286.7 and 285.5 eV can be distinguished. The higher binding energy peak (C1) is from C–N bonding in MA cation. The lower binding energy peak (C2) is attributed to the adventitious carbon and it is not related to perovskite itself. The binding energies of carbon peaks are consistent with previous reports [33,34]. After water vapor exposure, peak intensities of I $3d_{5/2}$, Pb $4f$, N $1s$, and C1 gradually decrease. Only the C2 peak shows a slight increase. Meanwhile, the peak position remains unchanged. The drop in peak intensities are also observed for the FAPbI$_3$ sample (Figure 1e–g). However, the magnitude of the drop is smaller. The relative intensities of different elements are summarized in Figure 2. These intensities in FAPbI$_3$ drop about 10% to 20%, while in MAPbI$_3$, the intensities drop more than 30% for iodine and close to 50% for nitrogen. To clarify if this decrease is related to the decomposition, the normalized atomic ratios between iodine, lead, nitrogen, and C1 are compared, as listed in Table 1. For MAPbI$_3$, the I/Pb ratio is 3.1 in the pristine sample, indicating a slightly iodine rich on the surface. After the first and second water vapor exposure, this ratio further increased to 3.5 and 3.7, respectively. Meanwhile, the N/Pb ratio and C/Pb ratio are also maintained above 1, indicating no sign of organic cation deficiency at surface. Therefore, the atomic ratio change of Pb/I only suggests enrichment of organic cations or deficiency of lead atoms. None of these changes support PbI$_2$ formation. For FAPbI$_3$, the composition change is even smaller. The I/Pb ratio is between 3.1–3.6 and N/Pb ratio is between 2.1–2.2. The C/Pb ratio is slightly lower than 1, but no systematic decrease after water vapor exposure was observed. From the atomic ratio data, it can be concluded that there is no sign of PbI$_2$ formation. It appears to contradict a previous NAPXPS study, in which the perovskite surface is completely decomposed at 9 mbar of water partial pressure [19]. This contradiction can be justified by the difference in sample fabrication. Unlike the solution process method, the perovskite films prepared by vacuum deposition usually have smaller domain sizes and poorer crystalline quality. A lower-quality film may result in much faster degradation. Instead, our results are consistent with studies using solution-processed perovskite samples, where no significant surface degradation were reported [24,35].

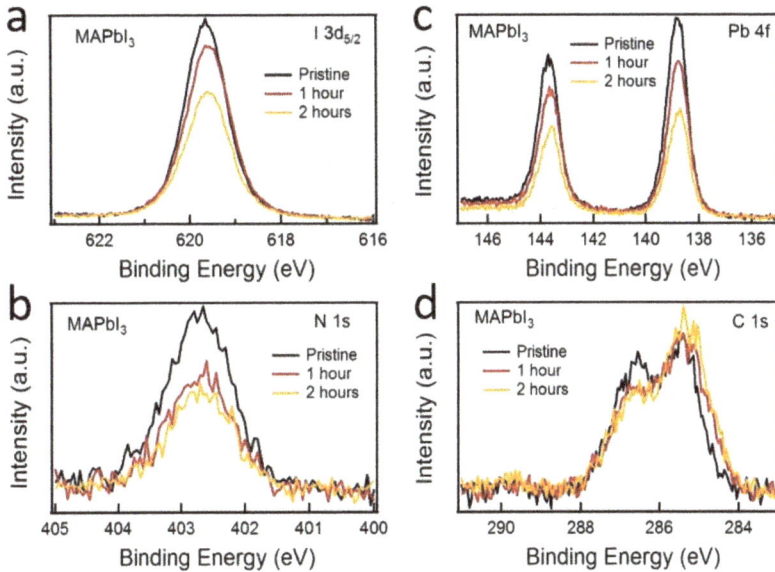

Figure 1. *Cont.*

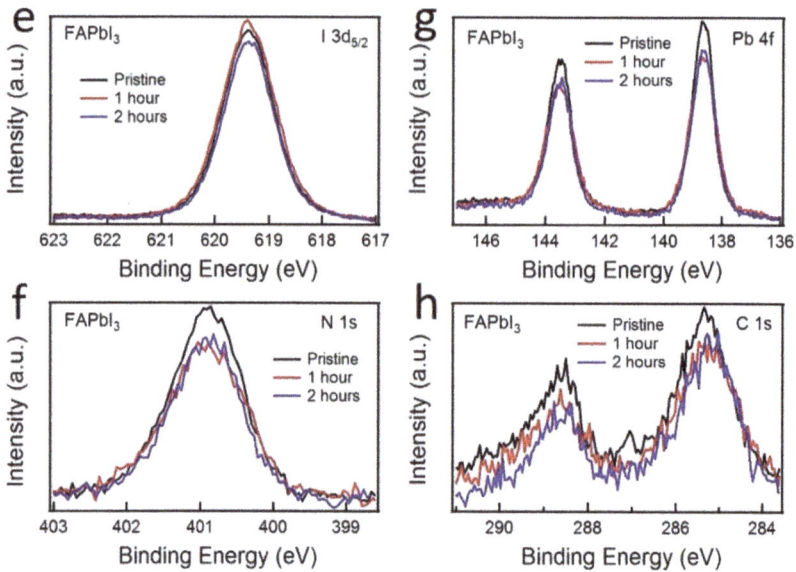

Figure 1. High-resolution spectra of I $3d_{5/2}$, N $1s$, Pb $4f$ and C $1s$ obtained from MAPbI$_3$ (**a–d**) and FAPbI$_3$ (**e–h**) samples before and after water vapor exposure to different periods, respectively.

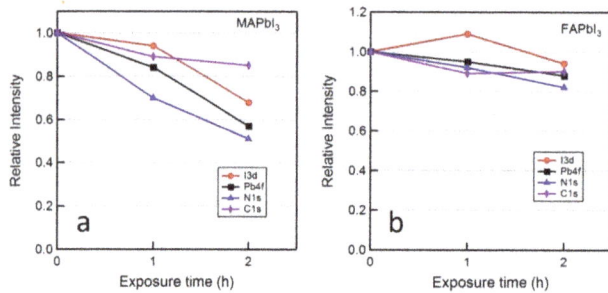

Figure 2. The relative intensities change after different water vapor exposure period. (**a**) Elemental intensity changes of MAPbI$_3$ sample. (**b**) Elemental intensity changes of FAPbI$_3$ sample.

Table 1. Atomic ratio of different elements obtained by XPS from MAPbI$_3$ and FAPbI$_3$ before and after water vapor exposure.

Exposure Time	MAPbI$_3$			FAPbI$_3$		
	I/Pb	N/Pb	C/Pb	I/Pb	N/Pb	C/Pb
0 h	3.1	1.6	1.5	3.1	2.2	0.8
1 h	3.5	1.0	1.5	3.6	2.1	0.7
2 h	3.7	1.5	2.4	3.3	2.1	0.9

Since there is no trace of PbI$_2$ formation on the surface, the decrease in intensities is probably due to reduced film coverage. This is supported by the observation in XPS wide scans (Figure 3). After water vapor exposure, new peaks originated from In $3d$ and O $1s$, indicating the partial exposure of ITO substrates. This is clear evidence that large voids form in MAPbI$_3$ thin film. The In $3d$ signal is much stronger in the MAPbI$_3$ sample, suggesting greater film area shrinkage. Therefore, it can be concluded that though the surface composition barely changes, the film coverage greatly reduced. Large voids compromise the device integrity and is probably the determining factor to device lifetime.

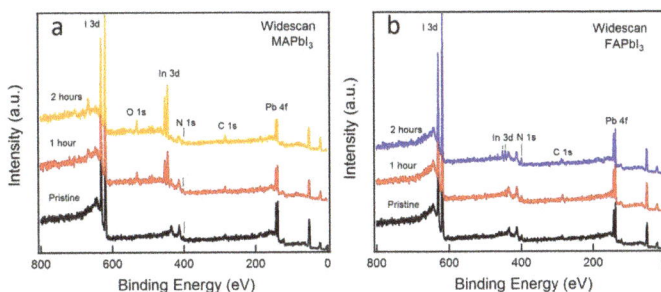

Figure 3. XPS wide scans obtained from MAPbI$_3$ and FAPbI$_3$ samples before and after water vapor exposure. (**a**) MAPbI$_3$ sample; (**b**) FAPbI$_3$ sample.

Besides surface sensitive XPS measurements, XRD is used as a complementary technique to evaluate the changes in the bulk. In Figure 4, the XRD spectra of MAPbI$_3$ and FAPbI$_3$ are shown before and after water vapor exposure. The spectra of both pristine samples only contain peaks from perovskite itself. No PbI$_2$ peak at 12.7° is observed [36]. After water vapor exposure, a small PbI$_2$ peak is observed in both samples. Since there is no trace of PbI$_2$ formation on surface, the signal must come from the site below the very top surface. It appears counterintuitive that PbI$_2$ formation starts from the bulk instead of surface. One probable explanation is due to the fast diffusion of water molecules and organic cations [37,38]. The water diffusion into the bulk may cause decomposition both inside and on the surface. However, when volatile organic cation escapes via the surface, it may partially compensate the loss of cations at the surface region, delaying PbI$_2$ formation on the surface. It is also possible that part of PbI$_2$ crystals formed along the grain boundaries without much exposure to the top surface. Without surface PbI$_2$ formation, it may be beneficial to keep the interface unchanged, but it also means that slowing down the degradation by PbI$_2$ passivation will not occur. In the FAPbI$_3$ sample, the diffraction peak (001) at 13.8° disappeared and a peak at 11.8° emerged. Other high-order peaks also emerged at new two theta degrees. This change is consistent with the α-to-δ phase transition observed in FAPbI$_3$ previously [35]. Therefore, FAPbI$_3$ is more sensitive to water-induced phase change.

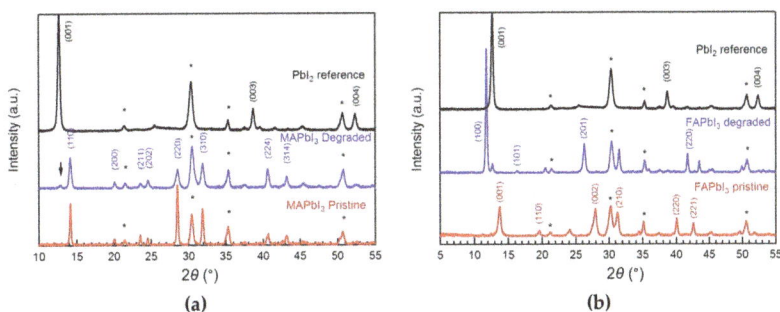

Figure 4. XRD data of MAPbI$_3$ and FAPbI$_3$ sample before and after water vapor exposure. PbI$_2$ spectra are shown as a reference. Peaks marked with * is from indium-tin-oxide (ITO) substrate. (**a**) XRD spectra of the MAPbI$_3$ sample before and after water exposure. (**b**) XRD spectra of the FAPbI$_3$ sample before and after water exposure.

The XPS data suggest film coverage is reduced after water vapor exposure. To understand how severe the reduction in surface coverage is, the surface of both samples are ex-situ measured by SEM

(Figure 5). The pristine MAPbI$_3$ film is compact with no observable pinholes, while the FAPbI$_3$ film only contains a few. The FAPbI$_3$ film shows similar grain size with small bright grains on the surface. The formation of these bright grains is probably due to the annealing process to convert the film from δ phase to α phase. However, during annealing, grain recrystallization also occurs, leaving small grains scattered on the surface. This phenomenon was observed in the previous study as well [39]. After water vapor exposure, the number of pinholes is greatly increased on the MAPbI$_3$ surface. These pinholes are formed along the grain boundaries and do not affect remaining grains. Meanwhile, a grain-coarsening effect is observed. The average grain size has increased from about 0.13 to 0.23 μm^2. All these observations can be attributed to water-induced recrystallization [26]. On the FAPbI$_3$ surface, the increase in pinholes is far less. Instead, needle-like crystals scatter on the surface, probably related to the δ phase of FAPbI$_3$ [40]. In summary, more pinholes are found on the MAPbI$_3$ surface, indicating a poorer morphology stability against water.

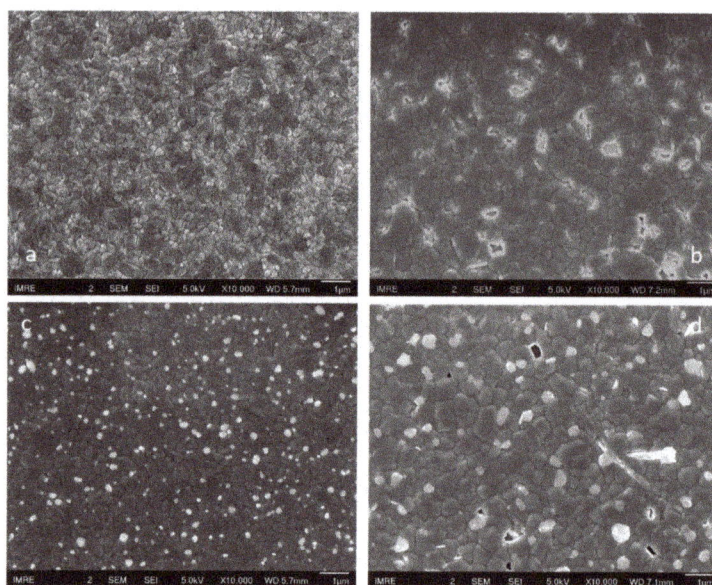

Figure 5. SEM images of MAPbI$_3$ and FAPbI$_3$ samples before and after water vapor exposure for 2 h. (**a**) Pristine MAPbI$_3$ sample; (**b**) MAPbI$_3$ sample after degradation; (**c**) pristine FAPbI$_3$ sample; (**d**) FAPbI$_3$ sample after degradation.

Except small pinholes, a much larger-scale reconstruction pattern is also observed on the MAPbI$_3$ surface (Figure 6). This reconstruction is not seen on the FAPbI$_3$ surface. The pattern has a spherulite-like feature with sub-millimeter scale. This pattern was shown in previous works, but its appearance was not discussed [17,41]. This spherulite-like pattern is highly symmetric, where each half assembles an alluvial fan. Inside the pattern, the film has been completely reconstructed, forming long crystals with fused boundary pointing towards the joint of two alluvial fans. Meanwhile, voids are seen between reconstructed patterns or between reconstruction pattern and nearby film. The symmetry of this pattern implies the film reconstruction initiates from the center and then spreads out. The formation of these reconstruction center could be due to either localized surface hydrophilic spots, which make water soaking easy or temporary, and localized humidity level variation, which may cause micro water drops formation on the surface, or both. Interestingly, the relative two-fold symmetric alluvial fan-like structure also indicates non-uniform diffusion on the surface. Along preferential water diffusion direction, adsorbed water diffuses much faster, allowing quick bilateral movement of water to two opposite ends. At the direction perpendicular to the preferential diffusion direction, water diffuses much slower and does not spread out. It is possible that both monohydrate and dihydrate were formed during recrystallization. However, the film is always measured after water exposure and at ultrahigh vacuum condition. Therefore, from the morphology shown in SEM images, perovskite crystals reversed back from hydrates. The void created by this pattern is much larger than the pinholes formed at grain boundaries. Therefore, it poses a much more severe threat to integrity of devices, and may cause premature device failure. The SEM results are consistent with the XPS data in which nearly 30% and 10% of the film area disappeared after water exposure of MAPbI$_3$ and FAPbI$_3$, respectively.

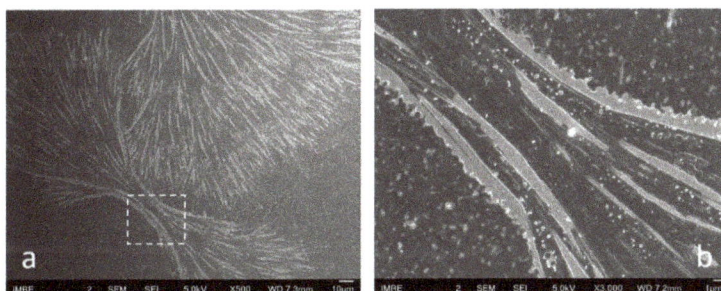

Figure 6. Large-scale morphology changes on the MAPbI$_3$ sample after water exposure. (**a**) Spherulite-like pattern on the MAPbI$_3$ surface. (**b**) Zoomed image of the center of spherulite-like pattern.

4. Conclusions

In conclusion, we find the degradation of perovskite materials involves multiple changes. For MAPbI$_3$, the initial degradation is dominated by water-induced recrystallization, which causes drastic changes in film morphology. For FAPbI$_3$, the initial change mainly occurs due to the phase transition. For both perovskite films, we reveal the PbI$_2$ formation occurring in the bulk rather than on the surface, probably mainly at grain boundaries. Therefore, PbI$_2$ in the degraded film does not act as a water-blocking barrier to slow down the degradation process. The tendency of PbI$_2$ formation at grain boundaries rather than on the surface is also observed in non-stoichiometric sample with excess PbI$_2$ or stoichiometric samples degraded by heat. Therefore, it can be generally concluded that degradation process of perovskite film is not slowed by the emergence of PbI$_2$. Our results highlight the importance of the external water-blocking layer because PbI$_2$ does not form a continuous film on the surface. Also, a more robust device structure with self-supporting layers may have better resistance to morphology change at the earlier stage of degradation.

Author Contributions: Conceptualization, S.C.; investigation, S.C., A.S., and J.P.; writing—original draft preparation, S.C.; writing—review and editing, S.C., A.S., J.P., and T.C.S., funding acquisition, S.C. and T.C.S.

Funding: This research was funded by Ministry of Education AcRF Tier 2 (MOE2016-T2-1-034), the Singapore National Research Foundation Investigatorship Programme (NRF-NRFI-2018-04) and the University of Macau startup fund (SRG2018-00140-IAPME).

Acknowledgments: Authors acknowledge technical support in experiments from Lim Poh Chong and Bussolotti Fabio in IMRE.

Conflicts of Interest: The authors declare no conflict of interest.

References

1. Yang, W.S.; Park, B.-W.; Jung, E.H.; Jeon, N.J.; Kim, Y.C.; Lee, D.U.; Shin, S.S.; Seo, J.; Kim, E.K.; Noh, J.H.; et al. Iodide management in formamidinium-lead-halide–based perovskite layers for efficient solar cells. *Science* **2017**, *356*, 1376–1379. [CrossRef] [PubMed]
2. Dou, L.T.; Yang, Y.; You, J.B.; Hong, Z.R.; Chang, W.H.; Li, G.; Yang, Y. Solution-processed hybrid perovskite photodetectors with high detectivity. *Nat. Commun.* **2014**, *5*, 5404. [CrossRef] [PubMed]

3. Du, B.; Xia, Y.; Wei, Q.; Xing, G.; Chen, Y.; Huang, W. All-inorganic perovskite nanocrystals-based light emitting diodes and solar cells. *Chemnanomat* **2019**, *5*, 266–277. [CrossRef]

4. Leijtens, T.; Eperon, G.E.; Noel, N.K.; Habisreutinger, S.N.; Petrozza, A.; Snaith, H.J. Stability of metal halide perovskite solar cells. *Adv. Energy Mater.* **2015**, *5*, 1500963. [CrossRef]

5. Li, B.; Li, Y.; Zheng, C.; Gao, D.; Huang, W. Advancements in the stability of perovskite solar cells: Degradation mechanisms and improvement approaches. *RSC Adv.* **2016**, *6*, 38079–38091. [CrossRef]

6. Aristidou, N.; Eames, C.; Islam, M.S.; Haque, S.A. Insights into the increased degradation rate of $CH_3NH_3PbI_3$ solar cells in combined water and O_2 environment. *J. Mater. Chem. A* **2017**, *5*, 25469–25475. [CrossRef]

7. Norbert, H.N.; Felix, L.; Brus, V.V.; Shargaieva, O.; Rappich, J. Unraveling the light-induced degradation mechanisms of $CH_3NH_3PbI_3$ perovskite film. *Adv. Electron. Mater.* **2017**, *3*, 1700158. [CrossRef]

8. Habisreutinger, S.N.; Leijtens, T.; Eperon, G.E.; Stranks, S.D.; Nicholas, R.J.; Snaith, H.J. Carbon nanotube/polymer composites as a highly stable hole collection layer in perovskite solar cells. *Nano Lett.* **2014**, *14*, 5561–5568. [CrossRef]

9. Wei, Z.; Zheng, X.; Chen, H.; Long, X.; Wang, Z.; Yang, S. A multifunctional C + epoxy/Ag-paint cathode enables efficient and stable operation of perovskite solar cells in watery environments. *J. Mater. Chem. A* **2015**, *3*, 16430–16434. [CrossRef]

10. Bella, F.; Griffini, G.; Correa-Baena, J.-P.; Saracco, G.; Gratzel, M.; Hagfeldt, A.; Turri, S.; Gerbaldi, C. Improving efficiency and stability of perovskite solar cells with photocurable fluoropolymers. *Science* **2016**, *354*, 203–206. [CrossRef]

11. Yang, S.; Wang, Y.; Liu, P.; Cheng, Y.; Zhao, H.; Yang, H. Functionalization of perovskite thin films with moisture-tolerant molecules. *Nat. Energy* **2016**, *1*, 15016. [CrossRef]

12. Hu, Y.; Qiu, T.; Bai, F.; Miao, X.; Zhang, S. Enhancing moisture-tolerance and photovoltaic performances of FAPbI3 by bismuth incorporation. *J. Mater. Chem. A* **2017**, *5*, 25258–25265. [CrossRef]

13. Wang, Z.; Lin, Q.; Chmiel, F.P.; Sakai, N.; Herz, L.M.; Snaith, H.J. Efficient ambient-air-stable solar cells with 2D-3D heterostructured butylammonium-caesium-formamidinium lead halide perovskites. *Nat. Energy* **2017**, *2*, 17135. [CrossRef]

14. Saidaminov, M.I.; Kim, J.; Jain, A.; Quintero-Bermudez, R.; Tan, H.; Long, G.; Tan, F.; Johnston, A.; Zhao, Y.; Voznyy, O.; et al. Suppression of atomic vacancies via incorporation of isovalent small ions to increase the stability of halide perovskite solar cells in ambient air. *Nat. Energy* **2018**, *3*, 648–654. [CrossRef]

15. Kim, J.H.; Williams, S.T.; Cho, N.; Chueh, C.-C.; Jen, A.K.-Y. Enhanced environmental stability of planar heterojunction perovskite solar cells based on blade-coating. *Adv. Energy Mater.* **2015**, *5*, 1401229. [CrossRef]

16. Song, Z.; Abate, A.; Watthage, S.C.; Liyanage, G.K.; Phillips, A.B.; Steiner, U.; Graetzel, M.; Heben, M.J. Perovskite solar cell stability in humid air: Partially reversible phase transitions in the PbI_2–CH_3NH_3I–H_2O system. *Adv. Energy Mater.* **2016**, *6*, 1600846. [CrossRef]

17. Leguy, A.M.A.; Hu, Y.; Campoy-Quiles, M.; Alonso, M.I.; Weber, O.J.; Azarhoosh, P.; van Schilfgaarde, M.; Weller, M.T.; Bein, T.; Nelson, J.; et al. Reversible hydration of $CH_3NH_3PbI_3$ in films, single crystals, and solar cells. *Chem. Mater.* **2015**, *27*, 3397–3407. [CrossRef]

18. Schlipf, J.; Hu, Y.H.; Pratap, S.; Bießmann, L.; Hohn, N.; Porcar, L.; Bein, T.; Docampo, P.; Muller-Buschbaum, P. Shedding light on the moisture stability of 3D/2D hybrid perovskite heterojunction thin films. *ACS Appl. Energy Mater.* **2019**, *2*, 1011–1018. [CrossRef]

19. Ke, J.C.-R.; Walton, A.S.; Lewis, D.J.; Tedstone, A.; O'Brien, P.; Thomas, A.G.; Flavell, W.R. In situ investigation of degradation at organometal halide perovskite surfaces by X-ray photoelectron spectroscopy at realistic water vapour pressure. *Chem. Commun.* **2017**, *53*, 5231–5234. [CrossRef]

20. Bi, D.; Tress, W.; Dar, M.I.; Gao, P.; Luo, J.; Renevier, C.; Schenk, K.; Abate, A.; Giordano, F.; Baena, J.-P.C.; et al. Efficient luminescent solar cells based on tailored mixed-cation perovskites. *Sci. Adv.* **2016**, *2*, e1501170. [CrossRef]

21. Jesper, J.; Juan-Pablo, C.B.; Elham, H.A.; Bertrand, P.; Samuel, D.S.; Marine, E.F.B.; Wolfgang, T.; Kurt, S.; Joel, T.; Jacques, E.M.; et al. Unreacted PbI_2 as a double-edge sword for enhancing the performance of perovskite solar cells. *J. Am. Chem. Soc.* **2016**, *138*, 10331–10343. [CrossRef]

22. Wang, C.C.; Gao, Y.L. Stability of perovskites at the surface analytic level. *J. Phys. Chem. Lett.* **2018**, *9*, 4657–4666. [CrossRef] [PubMed]

23. Wei, W.; Hu, Y.H. Catalytic role of H_2O in degradation of inorganic-organic perovskite ($CH_3NH_3PbI_3$) in air. *Int. J. Energy Res.* **2017**, *41*, 1063–1069. [CrossRef]

24. Philippe, B.; Park, B.-W.; Lindblad, R.; Oscarsson, J.; Ahmadi, S.; Johansson, E.M.J.; Rensmo, H. Chemical and electronic structure characterization of lead halide perovskites and stability behavior under different exposures—A photoelectron spectroscopy investigation. *Chem. Mater.* **2015**, *27*, 1720–1731. [CrossRef]
25. You, J.; Yang, Y.; Hong, Z.; Song, T.B.; Meng, L.; Liu, Y.; Jiang, C.; Zhou, H.; Chang, W.H.; Li, G.; et al. Moisture assisted perovskite film growth for high performance solar cells. *Appl. Phys. Lett.* **2014**, *105*, 183902. [CrossRef]
26. Zhang, W.; Xiong, J.; Li, J.; Daoud, W.A. Mechanism of water effect on enhancing the photovoltaic performance of triple-cation hybrid perovskite solar cells. *ACS Appl. Mater. Interfaces* **2019**, *11*, 12699–12708. [CrossRef]
27. Solanki, A.; Lim, S.S.; Mhaisalkar, S.; Sum, T.C. Role of water in suppressing recombination pathways in $CH_3NH_3PbI_3$ perovskite solar cells. *ACS Appl. Mater. Interfaces* **2019**, *11*, 25474–25482. [CrossRef]
28. Yang, J.M.; Yuan, Z.C.; Liu, X.J.; Braun, S.; Li, Y.Q.; Tang, J.X.; Gao, F.; Duan, C.G.; Fahlman, M.; Bao, Q.Y. Oxygen- and water-induced energetics degradation in organometal halide perovskites. *ACS Appl. Mater. Interfaces* **2018**, *10*, 16225–16230. [CrossRef]
29. Christians, J.A.; Miranda Herrera, P.A.; Kamat, P.V. Transformation of the excited state and photovoltaic efficiency of $CH_3NH_3PbI_3$ perovskite upon controlled exposure to humidified Air. *J. Am. Chem. Soc.* **2015**, *137*, 1530–1538. [CrossRef]
30. Shirayama, M.; Kato, M.; Miyadera, T.; Sugita, T.; Fujiseki, T.; Hara, S.; Kadowaki, H.; Murata, D.; Chikamatsu, M.; Fujiwara, H.; et al. Degradation mechanism of $CH_3NH_3PbI_3$ perovskite materials upon exposure to humid air. *J. Appl. Phys.* **2016**, *119*, 115501. [CrossRef]
31. Li, D.; Bretschneider, S.A.; Bergmann, V.W.; Hermes, I.M.; Mars, J.; Klasen, A.; Lu, H.; Tremel, W.; Mezger, M.; Butt, H.J.; et al. Humidity-induced grain boundaries in MAPbI3 perovskite films. *J. Phys. Chem. C* **2016**, *120*, 6363–6368. [CrossRef]
32. Ralaiarisoa, M.; Salzmann, I.; Zu, F.S.; Koch, N. Effect of water, oxygen, and air exposure on $CH_3NH_3PbI_{3-x}Cl_x$ perovskite surface electronic properties. *Adv. Electron. Mater.* **2018**, *4*, 1800307. [CrossRef]
33. Chen, S.; Goh, T.W.; Sabba, D.; Chua, J.; Mathews, N.; Huan, C.H.A.; Sum, T.C. Energy level alignment at the methylammonium lead iodide/copper phthalocyanine interface. *APL Mater.* **2014**, *2*, 081512. [CrossRef]
34. Hawash, Z.; Raga, S.R.; Son, D.Y.; Ono, L.K.; Park, N.G.; Qi, Y.B. Interfacial modification of perovskite solar cells using an ultrathin MAI layer leads to enhanced energy level alignment, efficiencies, and reproducibility. *J. Phys. Chem. Lett.* **2017**, *8*, 3947–3953. [CrossRef] [PubMed]
35. Yamanaka, S.; Hayakawa, K.; Cojocaru, L.; Tsuruta, R.; Sato, T.; Mase, K.; Uchida, S.; Nakayama, Y. Electronic structures and chemical states of methylammonium lead triiodide thin films and the impact of annealing and moisture exposure. *J. Appl. Phys.* **2018**, *123*, 165501. [CrossRef]
36. Smecca, E.; Numata, Y.; Deretzis, I.; Pellegrino, G.; Boninelli, S.; Miyasaka, T.; La Magna, A.; Alberti, A. Stability of solution-processed MAPbI3 and FAPbI3 layers. *Phys. Chem. Chem. Phys.* **2016**, *18*, 13413–13422. [CrossRef] [PubMed]
37. Sun, P.P.; Chi, W.J.; Li, Z.S. Effects of water molecules on the chemical stability of MAGeI3 perovskite explored from a theoretical viewpoint. *Phys. Chem. Chem. Phys.* **2016**, *18*, 24526–24536. [CrossRef] [PubMed]
38. Tong, C.; Geng, W.; Tang, Z.; Yam, C.; Fan, X.; Liu, J.; Lau, W.; Liu, L. Uncovering the veil of the degradation in perovskite $CH_3NH_3PbI_3$ upon humidity exposure: A first-principles study. *J. Phys. Chem. Lett.* **2015**, *6*, 3289–3295. [CrossRef]
39. Liu, T.; Zong, Y.; Zhou, Y.; Yang, M.; Li, Z.; Game, O.S.; Zhu, K.; Zhu, R.; Gong, Q.; Padture, N.P. High-performance formamidinium-based perovskite solar cells via microstructure-mediated δ-to-α phase transformation. *Chem. Mater.* **2017**, *29*, 3246–3250. [CrossRef]
40. Tseng, W.S.; Jao, M.H.; Hsu, C.C.; Huang, J.S.; Wu, C.I.; Yeh, N.C. Stabilization of hybrid perovskite $CH_3NH_3PbI_3$ thin films by graphene passivation. *Nanoscale* **2017**, *9*, 19227–19235. [CrossRef]
41. Yang, J.L.; Fransishyn, K.M.; Kelly, T.L. Comparing the effect of mesoporous and planar metal oxides on the stability of methylammonium lead iodide thin films. *Chem. Mater.* **2016**, *28*, 7344–7352. [CrossRef]

coatings

MDPI

Article

Tunable Perfect Narrow-Band Absorber Based on a Metal-Dielectric-Metal Structure

Qiang Li [1], Zizheng Li [1,*], Xiangjun Xiang [2], Tongtong Wang [1], Haigui Yang [1], Xiaoyi Wang [1], Yan Gong [3,4,*] and Jinsong Gao [1,4]

[1] Key Laboratory of Optical System Advanced Manufacturing Technology, Changchun Institute of Optics, Fine Mechanics and Physics, Chinese Academy of Sciences, Changchun 130033, China; liqiang@ciomp.ac.cn (Q.L.); wangtt@ciomp.ac.cn (T.W.); yanghg@ciomp.ac.cn (H.Y.); wangxiaoyi@ciomp.ac.cn (X.W.); gaojs@ciomp.ac.cn (J.G.)

[2] Research Center of Laser Fusion, China Academy of Engineering Physics, Mianyang 621900, China; dennis55555@163.com

[3] Jiangsu Key Laboratory of Medical Optics, Suzhou Institute of Biomedical Engineering and Technology, Chinese Academy of Sciences, Suzhou 215163, China

[4] College of Da Heng, University of Chinese Academy of Sciences, Beijing 100049, China

[*] Correspondence: lizizheng@ciomp.ac.cn (Z.L.); gongy@sklao.ac.cn (Y.G.)

Received: 17 May 2019; Accepted: 17 June 2019; Published: 18 June 2019

Abstract: In this paper, a metal-dielectric-metal structure based on a Fabry–Perot cavity was proposed, which can provide near 100% perfect narrow-band absorption. The lossy ultrathin silver film was used as the top layer spaced by a lossless silicon oxide layer from the bottom silver mirror. We demonstrated a narrow bandwidth of 20 nm with 99.37% maximum absorption and the absorption peaks can be tuned by altering the thickness of the middle SiO_2 layer. In addition, we established a deep understanding of the physics mechanism, which provides a new perspective in designing such a narrow-band perfect absorber. The proposed absorber can be easily fabricated by the mature thin film technology independent of any nano structure, which make it an appropriate candidate for photodetectors, sensing, and spectroscopy.

Keywords: thin film; coatings; metal-dielectric-metal structure; Fabry–Perot cavity; perfect absorption

1. Introduction

Perfect absorbers are of great interest with respect to both fundamental theory and practical applications in many fields, such as solar cells [1–5], sensing [6–10], photo-detection [11,12], and thermal emitting [13,14]. Perfect absorbers can absorb all incident electromagnetic radiation at the desired wavelength, which means reflection and transmission are efficiently suppressed. In recent years, plasmonic structures have attracted much attention due to their excellent ability to achieve the light and matter interaction in infinitesimal space [15–20], by which specific reflection, transmission, and absorption can be achieved. Different kinds of nano structures, such as nanoparticles [21–23], nanocones [24,25], nanohole arrays [26–30], and nano gratings [31–33] were proposed, which can excite surface plasmons resulting in the enhancement of absorption. Many perfect absorbers using nano structures have demonstrated ultra-high absorption and their absorption spectra can be engineered by adjusting the geometry, size, and periodicity of the structures at the same time. However, the fabrication of these nanostructures usually involves costly precise nano fabrication processing steps, such as focused ion beam (FIB) and electron beam lithography (EBL), which turn out to be a challenge for applications in large areas of these patterned absorbers. This is one of the main factors severely limiting their application.

Compared to plasmonic structures, unpatterned thin films made up of one or more films of dielectric or metallic materials are widely used as color filters [34,35], antireflective coatings [36,37], and reflector mirrors [38,39]. Many traditional optical coatings rely on the Fabry–Perot interference effect to achieve a specific function. As early as in 1952, a triple layer consisting of a metal substrate, a dielectric layer, and a thin top metal layer, was proposed as a perfect absorber for radar waves [40]. Recently, Capasso et al. proposed a simple thin film as a perfect absorber, which consists of an ultrathin lossy semiconductor film and a gold reflective mirror [41]. This structure can achieve the interference effect in an ultrathin absorbing film with a thickness of a few nanometers. Unlike optical coatings, the transmission and reflection phase change at the interface are not zero or π anymore. The imaginary part of the complex refractive index has a critical impact on the phase change leading to the strong Fabry–Perot-type interference effect, which can form a dip in reflection spectrum. Moreover, thin films can be easily fabricated by mature thin film technology instead of costly nanofabrication techniques, which makes it a candidate in solar cells, photodetectors, and filters. This work has had important research significance and since then many studies have achieved perfect absorption using ultrathin films [42–46]. However, they usually focus on broadband perfect absorption.

In this paper, we propose a Ag-SiO$_2$-Ag (Metal-Dielectric-Metal, MDM) triple layer structure, which can exhibit perfect absorption in the visible and near infrared region with a narrow band. This kind of triple layer structure forms a FP cavity, and its optical spectrum is controlled easily by the thickness of the layers rather than the sub-wavelength size of the nanostructures, which makes it a more compatible candidate in many applications. Moreover, a deep comprehension of the physical mechanism is also studied by the theoretical analysis and finite-different time-domain (FDTD) algorithm, which paves the way to design this kind of perfect absorber. Therefore, the proposed perfect narrow band absorber, with the advantages of low-cost, large area, and a simple fabrication process, is promising for photodetectors, photovoltaics, sensing, and spectroscopy.

2. Simulation and Analysis

The proposed triple layer structure is schematically shown in Figure 1a and consists of a bottom Ag layer, a middle SiO$_2$ layer, and a top ultrathin silver (Ag) layer in sequence. In this metal-dielectric-metal (MDM) structure, Ag is chosen for its high reflection in the visible and near infrared region and low material loss. SiO$_2$ is selected for its stable characteristic and appreciable transparency in the specific band. We choose glass as the substrate and the thickness of the bottom Ag layer is 100 nm so that no incident light can pass through. Therefore, the absorption A is equal to $1 - R$ where R is the reflectance. The thickness of the middle SiO$_2$ layer and the top ultrathin Ag layer are set to d and t. The 3D finite-difference time-domain (FDTD) algorithm is used to simulate the optical properties of the proposed MDM structure, where the perfectly matched layers (PML) are applied in the Z axis and the periodic boundary conditions are used for a unit cell in the X-Y plane. At the same time a 2 nm $\times$ 2 nm $\times$ 2 nm discrete mesh is used for the simulation region. The Ag and SiO$_2$ permittivity in our simulation are from the Palik data [47]. The incident light is a plane wave propagating along the negative Z direction with the wavelength ranging from 400 to 2000 nm.

As we have described before, a FP cavity consisting of two Ag layers separated by a dielectric SiO$_2$ spacer can form resonance when the Bragg principle is met. The thickness of the SiO$_2$ has a significant influence for controlling the absorption peaks at the desired wavelength. The FDTD simulation results in Figure 1b show that high order absorption peaks will appear (the value of m is the resonance mode order) because of the FP resonance as the SiO$_2$ thickness increases. There are six resonance modes corresponding to six narrow absorption bands when the thickness of SiO$_2$ increases from 50 to 900 nm. In the simulated results shown in Figure 1c four absorption peaks are observed at wavelengths of 480.0, 633.0, 941.0, and 1864.0 nm, with a full width at half maximum (FWHM) of 8, 12, 7, and 14 nm when the thickness of the top Ag and SiO$_2$ layers are 30 and 600 nm, respectively. At the same time, the corresponding near electric field maps in Figure 1d confirmed that m order FP resonance was generated in the SiO$_2$ cavity at the wavelength of the absorption peaks in Figure 1b. Moreover, the largest electric field enhancement for modes from $m = 1$ to $m = 6$ are 5, 6, 14, 22, 12, and 8, respectively, while the electric field enhancement in the air above the top ultrathin Ag layer is much weaker than it in the SiO_2 cavity. The proposed MDM structure can achieve a narrowband absorption and the absorption peaks can shift rapidly when the SiO_2 thickness changes, and this characteristic benefits a large number of potential applications, such as color filters, bolometers, and sensors.

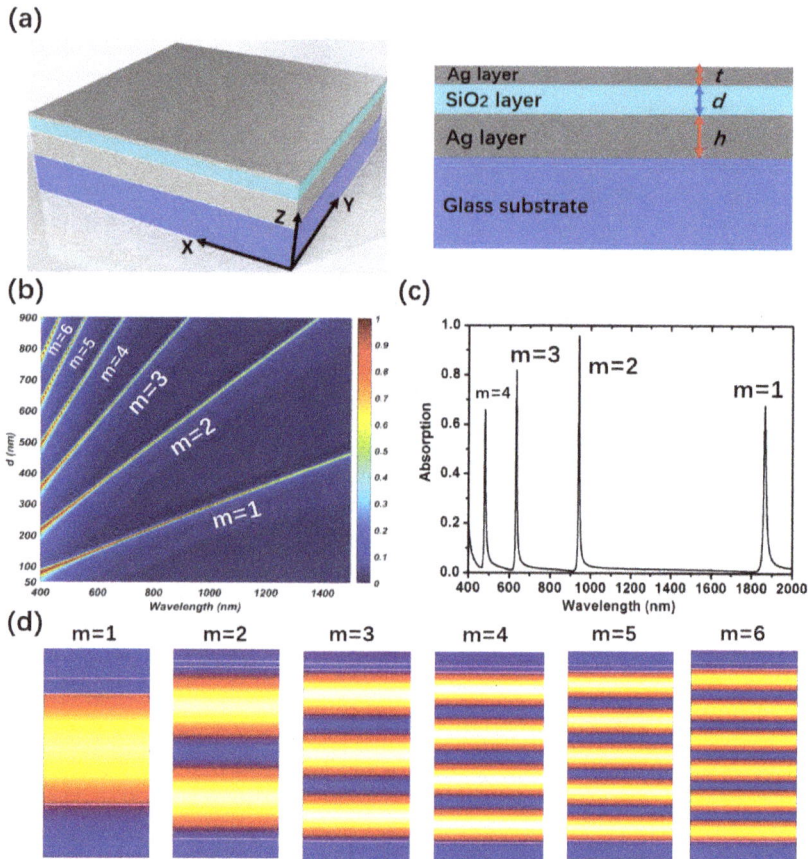

Figure 1. (**a**) Schematic diagram of the MDM structure perfect absorber. The thickness of the three layers are *t*, *d*, and *h*, respectively; (**b**) The relationship between the absorption and the SiO_2 thickness *d* by FDTD simulation when the thickness of the top Ag layer is 30 nm; (**c**) Simulated absorption spectra extracted from Figure 1b when *d* is 600 nm; (**d**) Electric field distribution in the SiO_2 cavity at the absorption peaks from $m = 1$ to $m = 6$.

In the previous section, we proposed a three-layer MDM structure forming an asymmetric Fabry−Perot cavity which can achieve multi-narrowband perfect absorption. Now we take the first order of FP resonance ($m = 1$) for example to analyze the optical properties of the triple layer structure and the physical mechanism of the perfect absorption with an ultra-narrow band. Firstly, the transfer matrix method (TMM) is carried out to study the effect of the middle SiO_2 thickness on the absorption spectrum. As discussed before, the bottom and top Ag thin films can form a FP resonator cavity and the thickness of the SiO_2 spacing layer determines the resonance wavelength. The reflection of the triple MDM structure was calculated using the TMM algorithm with varying SiO_2 thickness *d* from 80 to 200 nm in a 5 nm interval when the thickness of the top ultrathin Ag layer *t* was fixed at

30 nm, as shown in Figure 2a. The calculated absorption using $A = 1 - R$ is shown in Figure 2b, which exhibits a tunable narrow band perfect absorption in the visible region. In addition, we identified the relationship of the resonance wavelength and the SiO_2 cavity length by the minimum value of the calculated reflection curves, as shown in Figure 2c, in which we can see there is a linear relationship between the resonance wavelength and thickness d. We choose five different SiO_2 thicknesses—90, 110, 130, 160, and 180—and simulated the absorption spectrum by the FDTD algorithm, as shown in Figure 2d. The absorption spectra calculated by TMM in Figure 2c and the FDTD algorithm in Figure 2d agree well with each other, which can verify the correctness of the calculation and simulation.

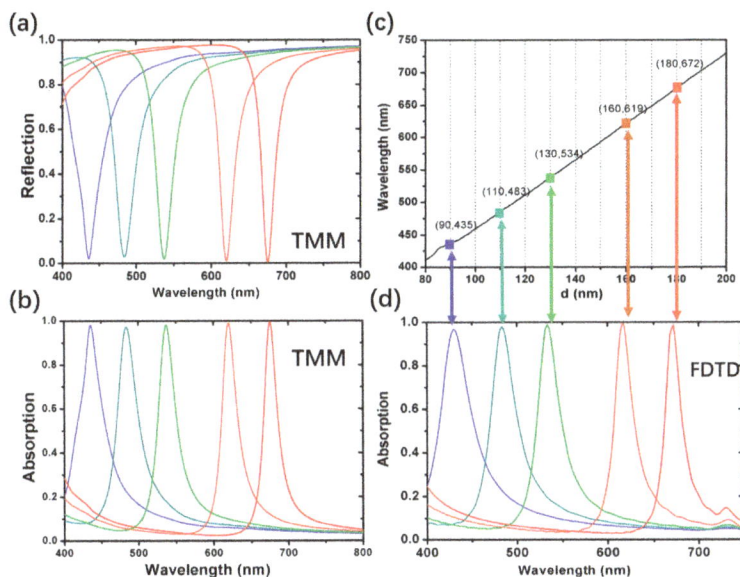

Figure 2. (**a,b**) The calculated reflection and absorption of five SiO_2 thickness d (90, 110, 130, 160, and 180 nm) by the TMM algorithm. (**c**) The relationship of resonance wavelength and the SiO_2 thickness d calculated by the TMM algorithm; (**d**) The five colorful solid lines represent the simulated absorption spectra by the FDTD algorithm when the thickness of the SiO_2 d are 90, 110, 130, 160, and 180 nm, respectively.

In order to clarify the physical mechanism of the narrowband perfect absorber, the effect of the different top Ag layer thickness t on the absorption properties is discussed. The FDTD algorithm was used to simulate the reflection and transmittance of a single Ag layer by changing the thickness from 10 to 70 nm on a glass substrate, and the calculated absorption is shown in Figure 3a. As we can see, the absorption mainly concentrates in the short wavelength region, which is determined by the property of the Ag material. Figure 3b shows the absorption curves when the thickness of the Ag layer is 10, 30, and 50 nm, respectively. The absorption is relatively low so as no resonant behavior is observed. Compared with the single Ag layer, the MDM triple layer structure shows obviously resonant absorption, as shown in Figure 3c (here the middle SiO_2 thickness is fixed to be 130 nm). If there is no top Ag silver film, the structure turns out to be an Ag reflection mirror coated by a SiO_2 layer, which is why the reflection is ultrahigh leading to almost zero absorption when t is 0 nm. As the thickness of the top Ag layer increases, an obvious resonant absorption band can be generated due to the FP cavity. For a 30 nm thickness top Ag layer the resonance wavelength is approximately 534 nm and the absorption is higher than other thicknesses, as shown in Figure 3d. Although the bandwidth is narrower when the thickness t is 50 nm, its absorption declines dramatically. Therefore, the ultrathin top Ag layer in our work is optimized at 30 nm for perfect absorption.

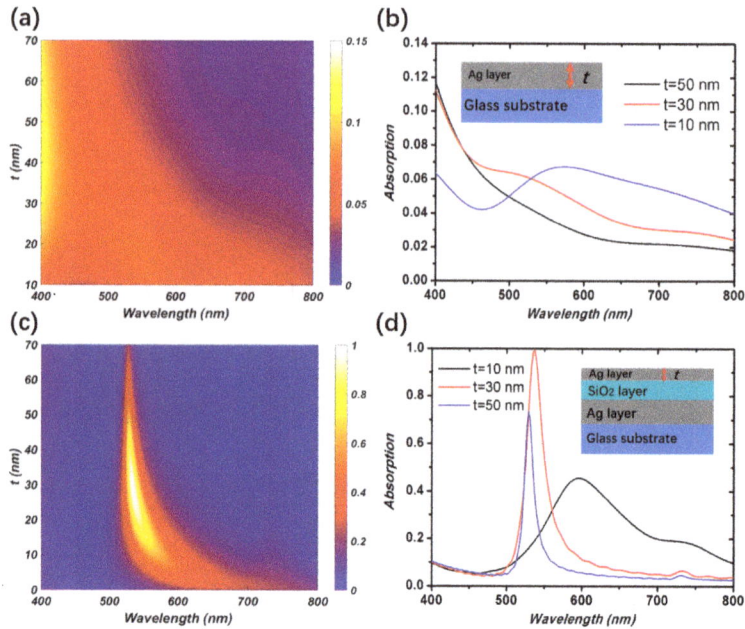

Figure 3. The simulated absorption by the FDTD algorithm. (**a**) The absorption curves when a single Ag layer coated on the glass substrate with varying thickness from 0–70 nm; (**b**) The absorption curves taken from Figure 3a when the thickness of the top Ag layer is 10, 30 and 50 nm; (**c**) The absorption as a function of wavelength of the MDM structure with a top Ag thickness from 0–70 nm when the middle SiO_2 thickness is fixed at 130 nm; (**d**) The absorption curves taken from Figure 3c when the thickness of the top Ag layer is 10, 30 and 50 nm.

In addition, the distributions of the FDTD simulated electric field (|E|) at wavelengths of 534 nm (FP resonance) and 700 nm (no FP resonance) for the triple layer MDM structure with 30 nm top Ag and 130 nm middle SiO_2 layers are shown in Figure 4a,b. As we can see, the electric field is enhanced in the air region above the ultrathin top Ag layer and in the middle SiO_2 cavity there is a small E-field at the non-resonance wavelength (700 nm) of the FP cavity, while for the resonance wavelength of 534 nm, we can observe that the E-field is mainly concentrated in the SiO_2 cavity because of the FP resonance, which leads to a four times enhancement in the E-field intensity. Figure 4c shows the distributions of the E-field in the visible region. According with Figure 4b, the E-field intensity is obviously enhanced at the resonance wavelength of 534 nm and the E-field enhancement band is narrow leading to a narrow absorption band. The FDTD simulation results indicate that the proposed triple MDM structure realizes the narrow band absorption via the FP cavity, and the perfect absorption has a relation with the E-field enhancement in the middle SiO_2 cavity.

In order to clarify the relation of perfect absorption and the E-field enhancement in the cavity, we calculated the absorption of three layers, respectively, using the FDTD algorithm for the MDM structure with a 30 nm top Ag layer and a 130 nm thickness middle SiO_2 layer when the FP resonance occurs in the cavity at the wavelength of 534 nm. The power monitors were set from the bottom Ag layer to the ultrathin top layer Ag layer with a 1 nm interval to record the power intensity if the incident light passed through them. The difference values between two neighboring monitors can show the absorption in this 1 nm region. The simulated result shows in Figure 5a that incident light is completely absorbed by two Ag layers because the SiO_2 material has no absorption property. The incident light power is mainly absorbed by the top ultrathin Ag layer. We also obtained the absorption in the bottom and top Ag layers as about 62.75% and 36.28%, respectively. The remaining 1% of incident light is reflected back, which is in accordance with the simulated absorption curve in Figure 2b.

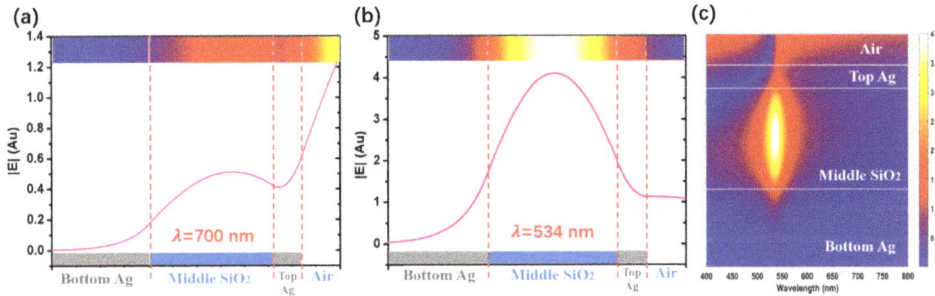

Figure 4. (**a**,**b**) Simulated E-field intensity distributions by the FDTD algorithm at the resonance wavelength of 534 nm and non-resonance wavelength of 700 nm, respectively. The three layers of the MDM absorber consist of a 30 nm top Ag layer, a 130 nm middle SiO$_2$ layer, and a 100 nm bottom Ag layer. (**c**) Simulated E-field distributions as a function of wavelength in the visible region.

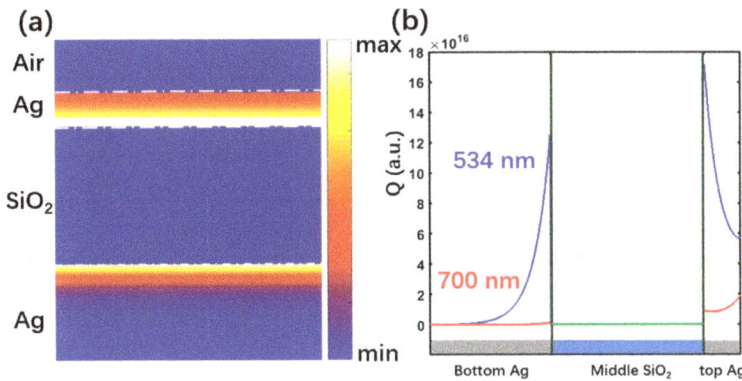

Figure 5. (**a**) Simulated energy distribution at the 534 nm resonance wavelength when the middle SiO$_2$ layer is 130 nm; and (**b**) the Ohmic loss calculated by Equation (1) at the resonance wavelength and non-resonance wavelength.

Furthermore, the absorption distribution in the proposed triple layers can be directly calculated by the local Ohmic loss formula [48]:

$$A(r,w) = \frac{1}{2}\varepsilon_0 w Im\varepsilon(w)|E|^2 \tag{1}$$

in which w is the angular frequency, $Im\varepsilon(w)$ is the imaginary part of the dielectric permittivity, and E is the simulated E-field by the FDTD algorithm, as shown in Figure 4. In this way, we calculated the absorption distribution in the triple layer stack using Equation (1), as plotted in Figure 5b. As we can see, the calculated Ohmic loss at the resonance wavelength and non-resonance wavelength is completely different. The blue line indicated the Ohmic loss at resonance wavelength is much larger than the red line which is indicated the Ohmic loss at non-resonance wavelength. The Ohmic loss totally occurs in both bottom and top metal Ag layers, and the ultrathin top Ag layer plays a dominating role, which is in accordance with the simulated results in Figure 5a. Obviously, the proposed triple layer perfect absorber is based on the FP cavity, and the thickness of the middle SiO$_2$ determines the wavelength of resonance. When the FP resonance formed in the cavity, the selected incident light can be reflected constantly by the two Ag layers, and some incident energy is transformed to ohmic loss each time light is reflected until all the power is consumed. The stronger E-field in the metal, the more power it can consume, which lays the foundation of the proposed perfect absorber.

3. Experiment Results and Discussion

We used the method of magnetron sputtering to fabricate the triple layer structure. Ag and SiO$_2$ layers were deposited alternately on a glass substrate via DC and RF sputtering in a vacuum chamber (SKY-450, Sky Technology Development, Shenyang, China). The deposition rate of Ag and SiO$_2$ are 4.28 and 2.63 Å/s, respectively. After research of the fabrication technology, the detailed deposition parameters are shown as Table 1 below.

Table 1. Coating process parameters.

Material	Power (W)	Ar (Sccm)	O$_2$ (Sccm)	Vacuum Degree (Pa)	Deposition Rate (Å/s)
Ag	150	80	0	1.0	4.28
SiO$_2$	200	80	20	1.5	2.63

The optical character of a single SiO$_2$ layer has an important influence on our proposed perfect absorber. We use a spectroscopic ellipsometer to measure the optical constant of the SiO$_2$ and Ag thin film fabricated by the magnetron sputtering method. The measured results are shown in Figure 6, which illustrates that the deposition parameters we chose are appropriate.

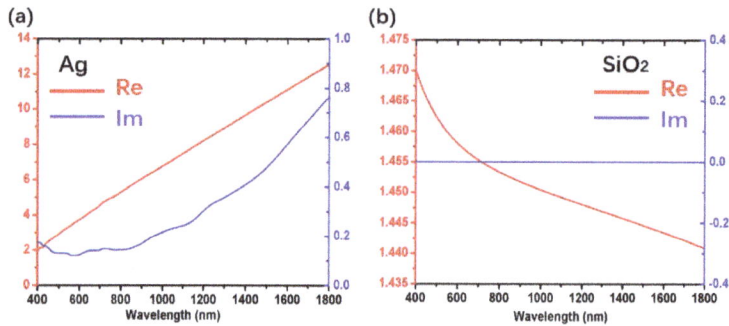

Figure 6. The optical constant of (**a**) Ag and (**b**) SiO$_2$ from the results measured by a spectroscopic ellipsometer.

After studying the process, a sample with a 100 nm bottom Ag layer, a 600 nm middle SiO$_2$ layer, and a 30 nm ultrathin top layer was fabricated. Its reflection was measured by a spectrometer (Perkin Elmer Lambda 900, Waltham, MA, USA), and the absorption calculated by $A = 1 - R$, as shown in Figure 7. Three resonance absorption peaks corresponding to 4, 3, and 2 order FP cavities can be observed at the wavelengths of 480.0, 633.0, and 941.0 nm. The absorption bands are narrow and agree perfectly with the FDTD simulation results, which can verify the correctness of the FDTD simulation.

In addition, we achieved narrow-band perfect absorption in the visible region using the first order of FP resonance. The fabricated samples with five different middle SiO$_2$ thicknesses present vivid colors, from left to right, as Figure 8a shows. The scanning electron microscope (SEM, JMS-6510, JEOL, Tokyo, Japan) image shows that the SiO$_2$ and Ag layers are clearly seen in Figure 8b. The reflection and absorption spectra obtained by FDTD simulation (the solid lines) and experiment (the dotted lines) are shown in Figure 8c, from which we can see the absorption can reach above 99.37% and their full width at half maximum (FWHM) are about 20–50 nm leading to a highest quality factor of 35.7. The results obtained by experiment compared with FDTD simulation are shown in detail in Table 2.

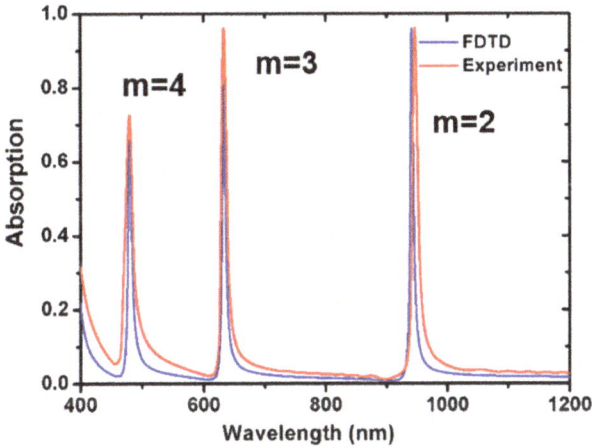

Figure 7. Simulated and experimental absorption for the 100 nm bottom Ag layer, 600 nm middle SiO₂ layer, and 30 nm top Ag layer.

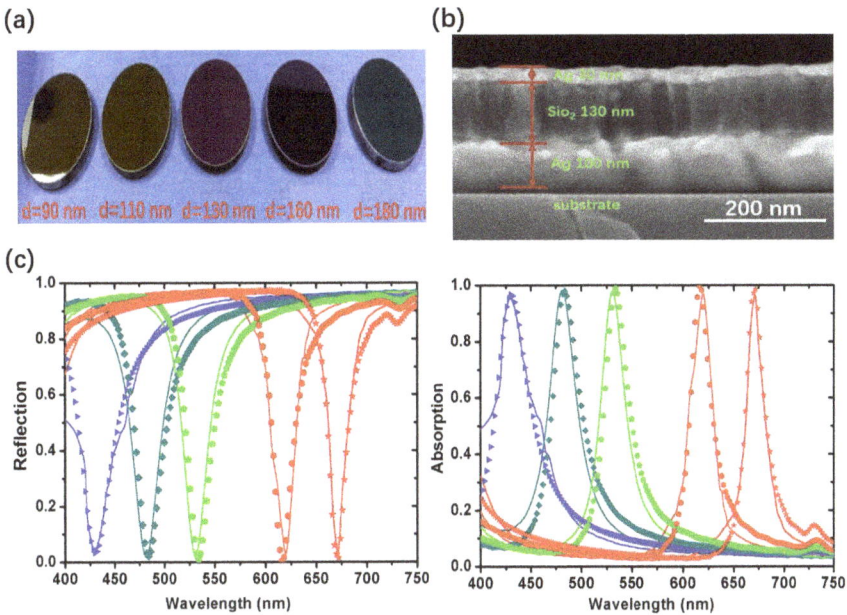

Figure 8. (**a**) Five samples with different SiO₂ thickness: 90, 110, 130, 160, and 180 nm, from left to right; (**b**) SEM image of the structure showing the Ag and SiO₂ layers; (**c**) Simulated and experimental reflection and absorption of five samples; The solid lines are the FDTD simulation results, while the dotted lines are the experimental results.

Table 2. The experiment and simulation results at resonance wavelengths for five different thicknesses.

d (nm)	Simulated Resonance Wavelength (nm)	Simulated Max Absorption	Experimental Resonance Wavelength (nm)	Experimental Max Absorption	FWHM (nm)
90	436	0.9779	440	0.9868	54
110	485	0.9887	482	0.9865	30
130	534	0.9899	530	0.9937	26
160	617	0.9966	620	0.9925	24
180	671	0.9984	675	0.9917	20

4. Conclusions

In conclusion, we propose a Ag-SiO$_2$-Ag triple layer structure, which can exhibit perfect absorption in the visible and near infrared region with a narrow band based on the Fabry−Perot cavity. Its optical spectrum is controlled easily by the thickness of the layers and the highest experimental absorption can reach about 99.37% with a narrow bandwidth of 20–50 nm. Moreover, a deep comprehension of the physical mechanism is also studied by theoretical analysis and the FDTD algorithm, which paves the way to design this kind of triple structure. The function of each layer and the physical mechanism in the proposed MDM triple layers are clarified using FDTD simulation. This kind of perfect absorber can be fabricated using mature deposition film technology with the advantages of low-cost, large area, and a simple fabrication process, which make it a promising solution for photodetectors, photovoltaics, sensing, and spectroscopy.

Author Contributions: Conceptualization: Q.L. and Z.L.; Methodology: Q.L. and Z.L.; Validation: Q.L. and X.W.; Formal Analysis: Q.L. and H.Y.; Investigation: T.W., X.X. and X.W.; Data Curation: Q.L., Y.G. and J.G.; Writing—Original Draft Preparation: Q.L.; Writing—Review and Editing: H.Y., Q.L. and Z.L.

Funding: This research was funded by the National Natural Science Foundation of China (Nos. 61705226 and 61875193), the Changchun Science and Technology Innovation "Shuangshi Project" Major Scientific and Technological Project (No. 19SS004), and the Science and Technology Innovation Project of Jilin Province (No. 20190201126JC).

Conflicts of Interest: The authors declare no conflict of interest.

References

1. Zhang, G.; Finefrock, S.; Liang, D.; Yadav, G.G.; Yang, H.; Fang, H.; Wu, Y. Semiconductor nanostructure-based photovoltaic solar cells. *Nanoscale* **2011**, *3*, 2430. [CrossRef] [PubMed]
2. Guo, C.F.; Sun, T.; Cao, F.; Liu, Q.; Ren, Z. Metallic nanostructures for light trapping in energy-harvesting devices. *Light Sci. Appl.* **2014**, *3*, e161.
3. Su, Y.-H.; Ke, Y.-F.; Cai, S.-L.; Yao, Q.-Y. Surface plasmon resonance of layer-by-layer gold nanoparticles induced photoelectric current in environmentally-friendly plasmon-sensitized solar cell. *Light Sci. Appl.* **2012**, *1*, e14. [CrossRef]
4. Yu, M.; Long, Y.Z.; Sun, B.; Fan, Z. Recent advances in solar cells based on one-dimensional nanostructure arrays. *Nanoscale* **2012**, *4*, 2783–2796.
5. Azad, A.K.; Kort-Kamp, W.J.M.; Sýkora, M.; Weisse-Bernstein, N.R.; Luk, T.S.; Taylor, A.J.; Dalvit, D.A.R.; Chen, H.-T. Metasurface Broadband Solar Absorber. *Sci. Rep.* **2016**, *6*, 20347. [CrossRef]
6. Zheng, J.; Yang, W.; Wang, J.; Zhu, J.; Qian, L.; Yang, Z. An ultranarrow SPR linewidth in the UV region for plasmonic sensing. *Nanoscale* **2019**, *11*, 4061–4066. [CrossRef] [PubMed]
7. Yong, Z.; Zhang, S.; Gong, C.; He, S. Narrow band perfect absorber for maximum localized magnetic and electric field enhancement and sensing applications. *Sci. Rep.* **2016**, *6*, 24063. [CrossRef] [PubMed]
8. Park, B.; Yun, S.H.; Cho, C.Y.; Kim, Y.C.; Shin, J.C.; Jeon, H.G.; Huh, Y.H.; Hwang, I.; Baik, K.Y.; Lee, Y.I.; et al. Surface plasmon excitation in semitransparent inverted polymer photovoltaic devices and their applications as label-free optical sensors. *Light Sci. Appl.* **2014**, *3*, e222. [CrossRef]

9.   Lu, X.; Zhang, T.; Wan, R.; Xu, Y.; Zhao, C.; Guo, S. Numerical investigation of narrowband infrared absorber and sensor based on dielectric-metal metasurface. *Opt. Express* **2018**, *26*, 10179–10187. [CrossRef]

10. Allsop, T.; Arif, R.; Neal, R.; Kalli, K.; Kundrát, V.; Rozhin, A.; Culverhouse, P.; Webb, D.J. Photonic gas sensors exploiting directly the optical properties of hybrid carbon nanotube localized surface plasmon structures. *Light Sci. Appl.* **2016**, *5*, e16036. [CrossRef]

11. Li, W.; Valentine, J. Metamaterial Perfect Absorber Based Hot Electron Photodetection. *Nano Lett.* **2014**, *14*, 3510–3514. [CrossRef] [PubMed]

12. Chalabi, H.; Schoen, D.; Brongersma, M.L. Hot-Electron Photodetection with a Plasmonic Nanostripe Antenna. *Nano Lett.* **2014**, *14*, 1374–1380. [CrossRef] [PubMed]

13. Costantini, D.; Lefebvre, A.; Coutrot, A.-L.; Moldovan-Doyen, I.; Hugonin, J.-P.; Boutami, S.; Marquier, F.; Benisty, H.; Greffet, J.-J. Plasmonic Metasurface for Directional and Frequency-Selective Thermal Emission. *Phys. Rev. Appl.* **2015**, *4*, 014023. [CrossRef]

14. Liu, X.; Tyler, T.; Starr, T.; Starr, A.F.; Jokerst, N.M.; Padilla, W.J. Taming the Blackbody with Infrared Metamaterials as Selective Thermal Emitters. *Phys. Rev. Lett.* **2011**, *107*, 045901. [CrossRef] [PubMed]

15. Ameling, R.; Dregely, D.; Giessen, H. Strong coupling of localized and surface plasmons to microcavity modes. *Opt. Lett.* **2011**, *36*, 2218–2220. [CrossRef]

16. Li, Q.; Gao, J.; Yang, H.; Liu, H.; Wang, X.; Li, Z.; Guo, X. Tunable plasmonic absorber based on propagating and localized surface plasmons using metal-dielectric-metal structure. *Plasmonics* **2017**, *12*, 1037–1043. [CrossRef]

17. Yu, P.; Li, J.; Tang, C.; Cheng, H.; Liu, Z.; Li, Z.; Liu, Z.; Gu, C.; Li, J.; Chen, S.; et al. Controllable optical activity with non-chiral plasmonic metasurfaces. *Light Sci. Appl.* **2016**, *5*, e16096. [CrossRef]

18. Huang, Y.; Zhang, X.; Li, J.; Ma, L.; Zhang, Z. Analytical plasmon dispersion in subwavelength closely spaced Au nanorod arrays from planar metal–insulator–metal waveguides. *J. Mater. Chem. C* **2017**, *5*, 6079–6085. [CrossRef]

19. Sannomiya, T.; Saito, H.; Junesch, J.; Yamamoto, N. Coupling of plasmonic nanopore pairs: Facing dipoles attract each other. *Light Sci. Appl.* **2016**, *5*, e16146. [CrossRef]

20. Nelson, J.W.; Knefelkamp, G.R.; Brolo, A.G.; Lindquist, N.C. Digital plasmonic holography. *Light Sci. Appl.* **2018**, *7*, 52. [CrossRef]

21. Ding, T.; Sigle, D.; Zhang, L.; Mertens, J.; De Nijs, B.; Baumberg, J. Controllable Tuning Plasmonic Coupling with Nanoscale Oxidation. *ACS Nano* **2015**, *9*, 6110–6118. [CrossRef]

22. Nazirzadeh, M.A.; Atar, F.B.; Turgut, B.B.; Okyay, A.K. Random sized plasmonic nanoantennas on Silicon for low-cost broad-band near-infrared photodetection. *Sci. Rep.* **2014**, *4*, 7103. [CrossRef] [PubMed]

23. Liu, Z.; Liu, X.; Huang, S.; Pan, P.; Chen, J.; Liu, G.; Gu, G. Automatically Acquired Broadband Plasmonic-Metamaterial Black Absorber during the Metallic Film-Formation. *ACS Appl. Mater. Interfaces* **2015**, *7*, 4962–4968. [CrossRef] [PubMed]

24. Li, Q.; Gao, J.; Li, Z.; Yang, H.; Liu, H.; Wang, X.; Li, Y. Absorption enhancement in nanostructured silicon fabricated by self-assembled nanosphere lithography. *Opt. Mater.* **2017**, *70*, 165–170. [CrossRef]

25. He, X.; Yan, S.; Lu, G.; Zhang, Q.; Wu, F.; Jiang, J. An ultra-broadband polarization-independent perfect absorber for the solar spectrum. *RSC Adv.* **2015**, *5*, 61955–61959.

26. Liu, H.; Wang, X.; Zhao, J.; Li, Z.; Yang, H.; Gao, J.; Li, Q. Novel aluminum plasmonic absorber enhanced by extraordinary optical transmission. *Opt. Express* **2016**, *24*, 25885. [CrossRef]

27. Siddique, R.H.; Mertens, J.; Hölscher, H.; Vignolini, S. Scalable and controlled self-assembly of aluminum-based random plasmonic metasurfaces. *Light Sci. Appl.* **2017**, *6*, e17015. [CrossRef]

28. Li, Q.; Li, Z.; Wang, X.; Wang, T.; Liu, H.; Yang, H.; Gong, Y.; Gao, J. Structurally tunable plasmonic absorption bands in a self-assembled nano-hole array. *Nanoscale* **2018**, *10*, 19117–19124. [CrossRef]

29. Sain, B.; Kaner, R.; Prior, Y. Phase-controlled propagation of surface plasmons. *Light Sci. Appl.* **2017**, *6*, e17072. [CrossRef]

30. Lu, J.Y.; Nam, S.H.; Wilke, K.; Raza, A.; Lee, Y.E.; Alghaferi, A.; Fang, N.X.; Zhang, T. Localized Surface Plasmon-Enhanced Ultrathin Film Broadband Nanoporous Absorbers. *Adv. Opt. Mater.* **2016**, *4*, 1255–1264. [CrossRef]

31. Shan, H.; Yu, Y.; Wang, X.; Luo, Y.; Zu, S.; Du, B.; Han, T.; Li, B.; Li, Y.; Wu, J.; et al. Direct observation of ultrafast plasmonic hot electron transfer in the strong coupling regime. *Light Sci. Appl.* **2019**, *8*, 9. [CrossRef]

32. Li, G.; Shen, Y.; Xiao, G.; Jin, C. Double-layered metal grating for high-performance refractive index sensing. *Opt. Express* **2015**, *23*, 8995. [CrossRef] [PubMed]

33. Floess, D.; Chin, J.Y.; Kawatani, A.; Dregely, D.; Habermeier, H.-U.; Weiss, T.; Giessen, H. Tunable and switchable polarization rotation with non-reciprocal plasmonic thin films at designated wavelengths. *Light Sci. Appl.* **2015**, *4*, e284. [CrossRef]

34. Willey, R. Achieving narrow bandpass filters which meet the requirements for DWDM. *Thin Solid Films* **2001**, *398*, 1–9. [CrossRef]

35. Tang, H.; Gao, J.; Zhang, J.; Wang, X.; Fu, X. Preparation and Spectrum Characterization of a High Quality Linear Variable Filter. *Coatings* **2018**, *8*, 308. [CrossRef]

36. Loh, J.Y.Y.; Kherani, N. Design of Nano-Porous Multilayer Antireflective Coatings. *Coatings* **2017**, *7*, 134. [CrossRef]

37. Matsuoka, Y.; Mathonnèire, S.; Peters, S.; Masselink, W.T. Broadband multilayer anti-reflection coating for mid-infrared range from 7 μm to 12 μm. *Appl. Opt.* **2018**, *57*, 1645–1649. [CrossRef]

38. Li, Z.; Li, Q.; Quan, X.; Zhang, X.; Song, C.; Yang, H.; Wang, X.; Gao, J. Broadband High-Reflection Dielectric PVD Coating with Low Stress and High Adhesion on PMMA. *Coatings* **2019**, *9*, 237. [CrossRef]

39. Hu, C.; Liu, J.; Wang, J.; Gu, Z.; Li, C.; Li, Q.; Li, Y.; Zhang, S.; Bi, C.; Fan, X.; et al. New design for highly durable infrared-reflective coatings. *Light Sci. Appl.* **2018**, *7*, 17175. [CrossRef]

40. Salisbury, W.W. Absorbent Body for Electromagnetic Waves. U.S. Patent 2599944, 10 June 1952.

41. Mikhail, A.K.; Romain, B.; Patrice, G.; Federico, C. Nanometre optical coatings based on strong interference effects in highly absorbing media. *Nat. Mater.* **2013**, *12*, 20–24. [CrossRef]

42. Park, J.; Kang, J.-H.; Vasudev, A.P.; Schoen, D.T.; Kim, H.; Hasman, E.; Brongersma, M.L. Omnidirectional Near-Unity Absorption in an Ultrathin Planar Semiconductor Layer on a Metal Substrate. *ACS Photonics* **2014**, *1*, 812–821. [CrossRef]

43. Pu, M.; Feng, Q.; Wang, M.; Hu, C.; Huang, C.; Ma, X.; Zhao, Z.; Wang, C.; Luo, X. Ultrathin broadband nearly perfect absorber with symmetrical coherent illumination. *Opt. Express* **2012**, *20*, 2246. [CrossRef] [PubMed]

44. Li, Z.; Palacios, E.; Butun, S.; Kocer, H.; Aydin, K. Omnidirectional, broadband light absorption using large-area, ultrathin lossy metallic film coatings. *Sci. Rep.* **2015**, *5*, 15137. [CrossRef] [PubMed]

45. Kats, M.A.; Capasso, F. Ultra-thin optical interference coatings on rough and flexible substrates. *Appl. Phys. Lett.* **2014**, *105*, 131108. [CrossRef]

46. Kats, M.A.; Sharma, D.; Lin, J.; Genevet, P.; Blanchard, R.; Yang, Z.; Qazilbash, M.M.; Basov, D.N.; Ramanathan, S.; Capasso, F. Ultra-thin perfect absorber employing a tunable phase change material. *Appl. Phys. Lett.* **2012**, *101*, 221101. [CrossRef]

47. Palik, E.D. *Handbook of Optical Constants of Solids*; Academic Press: New York, NY, USA, 1985.

48. Hao, J.; Zhou, L.; Qiu, M. Nearly total absorption of light and heat generation by plasmonic metamaterials. *Phys. Rev. B* **2011**, *83*, 165107. [CrossRef]

coatings

MDPI

Article

Ultrathin Al_2O_3 Coating on $LiNi_{0.8}Co_{0.1}Mn_{0.1}O_2$ Cathode Material for Enhanced Cycleability at Extended Voltage Ranges

Wenchang Zhu [1,2], **Xue Huang** [1,2], **Tingting Liu** [3], **Zhiqiang Xie** [4], **Ying Wang** [4], **Kai Tian** [1,2], **Liangming Bu** [1,2], **Haibo Wang** [1,2,5], **Lijun Gao** [1,2,*] **and Jianqing Zhao** [1,2,*]

[1] College of Energy, Soochow Institute for Energy and Materials InnovationS, Soochow University, Suzhou 215006, China; 20164208119@stu.suda.edu.cn (W.Z.); 20174208074@stu.suda.edu.cn (X.H.); 20174208072@stu.suda.edu.cn (K.T.); 20164208112@stu.suda.edu.cn (L.B.); wanghb@suda.edu.cn (H.W.)
[2] Key Laboratory of Advanced Carbon Materials and Wearable Energy Technologies of Jiangsu Province, Soochow University, Suzhou 215006, China
[3] Suzhou University of Science and Technology & Jiangsu Key Laboratory of Environmental Science and Engineering, Suzhou 215001, China; liutt@mail.usts.edu.cn
[4] Department of Mechanical & Industrial Engineering, Louisiana State University, Baton Rouge, LA 70803, USA; zxie5@lsu.edu (Z.X.); ywang@lsu.edu (Y.W.)
[5] Institute of Chemical Power Sources, Soochow University, Suzhou 215600, China
[*] Correspondence: gaolijun@suda.edu.cn (L.G.); jqzhao@suda.edu.cn (J.Z.);
Tel.: +86-512-6522-9905 (L.G. & J.Z.)

Received: 20 December 2018; Accepted: 31 January 2019; Published: 3 February 2019

Abstract: Ni-rich $LiNi_{0.8}Co_{0.1}Mn_{0.1}O_2$ oxide has been modified by ultrathin Al_2O_3 coatings via atomic layer deposition (ALD) at a growth rate of 1.12 Å/cycle. All characterizations results including TEM, SEM, XRD and XPS together confirm high conformality and uniformity of the resultant Al_2O_3 layer on the surface of $LiNi_{0.8}Co_{0.1}Mn_{0.1}O_2$ particles. Coating thickness of the Al_2O_3 layer is optimized at ~2 nm, corresponding to 20 ALD cycles to enhance the electrochemical performance of Ni-rich cathode materials at extended voltage ranges. As a result, 20 Al_2O_3 ALD-coated $LiNi_{0.8}Co_{0.1}Mn_{0.1}O_2$ cathode material can deliver an initial discharge capacity of 212.8 mAh/g, and an associated coulombic efficiency of 84.0% at 0.1 C in a broad voltage range of 2.7–4.6 V vs. Li^+/Li in the first cycle, which were both higher than 198.2 mAh/g and 76.1% of the pristine $LiNi_{0.8}Co_{0.1}Mn_{0.1}O_2$ without the Al_2O_3 protection. Comparative differential capacity (dQ/dV) profiles and electrochemical impedance spectra (EIS) recorded in the first and 100th cycles indicated significant Al_2O_3 ALD coating effects on suppressing phase transitions and electrochemical polarity of the Ni-rich $LiNi_{0.8}Co_{0.1}Mn_{0.1}O_2$ core during reversible lithiation/delithiation. This work offers oxide-based surface modifications with precise thickness control at an atomic level for enhanced electrochemical performance of Ni-rich cathode materials at extended voltage ranges.

Keywords: Al_2O_3 oxide; atomic layered deposition; $LiNi_{0.8}Co_{0.1}Mn_{0.1}O_2$; Ni-rich cathode material; lithium ion battery

1. Introduction

Lithium ion batteries have been dominantly applied in hybrid electric vehicles (HEVs) and pure electric vehicles (EVs) [1,2], which significantly enhances our living environment. In particular for electric transportation systems and future large-scale applications of stationary grid energy storage, lithium ion batteries are required to provide higher energy and power densities, together with excellent long-term service life and acceptable cost [3]. Performance of lithium ion batteries critically depends on the electrochemical properties and behaviors of electrode materials both in cathodes and anodes [4–6]. Because of the rapid development of practical Si–C composite anode materials, developing high-performance cathode materials with high specific capacity, high working potential, excellent cycle life, reliable safety, and low cost has become an urgent requirement [7–9]. Li-excess and Ni-rich layered oxides have been extensively investigated as high-energy cathode materials for superior lithium ion batteries [10,11]. Despite delivering an attractive discharge capacity of ~250 mAh/g, Li-excess layered cathode materials have been suffering from two crucial issues. One is the layered-to-spinel phase transition during prolonged electrochemical cycles, leading to continuous voltage fading. The other is low coulombic efficiency in the first cycle that results from the initial electrochemical activation of the Li^+-inactive Li_2MnO_3 component [10,12,13]. These two problems impede practical applications of Li-excess cathode materials. Alternatively, Ni-rich layered oxides have been considered as feasible cathode materials by offering a reversible capacity of ~200 mAh/g and a stable working voltage at ~3.8 V vs. Li^+/Li.

It is well-known that the resultant electrochemical performance of Ni-rich cathode materials is essentially determined on the specific composition of Ni, Co, and Mn elements within these ternary transition metal oxides, $LiNi_{1-x-y}Co_xMn_yO_2$ ($x + y < 0.5$) [11,14,15]. Higher Ni content results in higher lithium storage capacity of $LiNi_{1-x-y}Co_xMn_yO_2$, but leads to the undesired formation of Li^+/Ni^{2+} cation mixing in the layered structure; the expense of safety in material usage and more complexity in the material preparation [16,17]. Mn and Co dopants have been demonstrated to enhance the structural robustness and cationic ordering for $LiNi_{1-x-y}Co_xMn_yO_2$ variants, respectively. However, the specific capacity has to be sacrificed, when Mn and Co contents are increased. In order to rapidly promote practical applications of Ni-rich cathode materials for superior lithium ion batteries, poor cycling and thermal stabilities of $LiNi_{1-x-y}Co_xMn_yO_2$ should be urgently addressed, especially for materials with high Ni content. Capacity fading can be exacerbated, when these Ni-rich $LiNi_{1-x-y}Co_xMn_yO_2$ cathodes are cycled at broad voltage ranges with a high upper cut-off voltage above 4.3 V vs. Li^+/Li. One reason is that the intensive oxidation capability of $Ni^{3+/4+}$ redox could aggravate parasitic side reactions during lithiation/delithiation of $LiNi_{1-x-y}Co_xMn_yO_2$ cathode materials, leading to undesired electrolyte decomposition and solid electrolyte interphase (SEI) formation at the surface of the cathode. The other reason is the induced phase transition from the initial layered structure to spinel-like and/or NiO-like rock-salt phases at the surface of $LiNi_{1-x-y}Co_xMn_yO_2$ particles, because the formation of Li^+ vacancies in high concentration at deep charge destabilizes the initial layered structure of $LiNi_{1-x-y}Co_xMn_yO_2$ [18]. These unfavorable structural transformations are consecutive, and gradually overspread from the surface to bulk Ni-rich layered oxides [17,19]. In addition, the formation of impure phases definitely brings about structural degradation and increased kinetic barrier for lithium ion diffusion, causing capacity fading and inferior rate capability of $LiNi_{1-x-y}Co_xMn_yO_2$ cathode materials [20,21].

Surface modifications, either on particles of active cathode materials or at the surface of the entire cathode, have been demonstrated to improve electrochemical performance for various cathode materials even in harsh operation conditions, such as elevated working temperatures and broad voltage ranges [22,23]. Oxides are considered common coating materials, because of their low material cost, simple synthetic procedure, and high effectiveness in protecting internal electrode materials. Various examples have been reported in the literature [24–28], involving Al_2O_3, ZrO_2, ZnO, TiO_2, Cr_2O_3, SnO_2, etc. However, associated electrochemical performance of modified cathode materials is sensitively determined by qualities of the coated oxides in terms of thickness, conformality, uniformity, and integrity. Oxide coatings prepared by atomic layer deposition (ALD) have been demonstrated to considerably enhance electrochemical performance for various cathode materials, involving layered $LiCoO_2$ [29,30], spinel $LiMn_2O_4$, and $LiMn_{1.5}Ni_{0.5}O_4$ [31,32]; Ni-rich $LiNi_{1-x-y}Co_xMn_yO_2$ and $LiNi_{0.8}Co_{0.15}Al_{0.05}O_2$ [33,34]; and Li-excess $Li[Li_{1-x}M_x]O_2$ oxides [35]. The ALD technique can result in highly conformational and uniform oxide coating layers at the entire surface of the underlying substrates, and thus has been considered as an effective method for surface modifications of different cathode materials. Ni-rich cathode materials have been dominantly used for lithium ion batteries to power currently-developed electric automobiles, but still suffer from poor service life, especially in harsh operation conditions, such as at elevated working temperatures and/or in extended voltage ranges. As reported in the literature [33,34], Al_2O_3 ALD coatings contribute to enhanced cycling stabilities of $LiNi_{0.5}Co_{0.2}Mn_{0.3}O_2$, $LiNi_{0.8}Co_{0.1}Mn_{0.1}O_2$, and $LiNi_{0.8}Co_{0.15}Al_{0.05}O_2$ by protecting coated Ni-rich oxide particles from parasitic side reactions at the cathode's surface. However, the insulating property of Al_2O_3 material makes it is necessary to precisely control the coating's thickness, in order to maximize the electrochemical performance of Ni-rich cathode materials. In addition, it is also urgent to understand structural features and chemical environments between oxide coatings and Ni-rich oxide particles.

The ALD technique was employed in this work to coat ultrathin Al_2O_3 onto Ni-rich $LiNi_{0.8}Co_{0.1}Mn_{0.1}O_2$ target material (marked as n Al_2O_3 ALD NCM811, then responses to the number of ALD cycles related to the thickness control of Al_2O_3 coating layer). TEM observations were used to examine the structural characteristics of resultant Al_2O_3 coatings and determine the ALD growth rate of Al_2O_3 oxide. XPS spectra were used to study chemical environments of different n Al_2O_3 ALD NCM811 (n = 10, 20 and 50) materials. Electrochemical performance of three n Al_2O_3 ALD NCM811 cathode materials were evaluated and compared with the bare NCM811 cathode in a broad voltage range of 2.7–4.6 V vs. Li^+/Li. This work offers a high-quality oxide coating for Ni-rich cathode materials by using the ALD method, which significantly contributes to improved electrochemical performance in extended voltage ranges for superior lithium ion batteries.

2. Materials and Methods

2.1. Synthesis of Ni-Rich Layered $LiNi_{0.8}Co_{0.1}Mn_{0.1}O_2$ Material

In a typical sol (solution)-gel route, stoichiometric amounts of $CH_3COOLi \cdot 2H_2O$, $Ni(CH_3COO)_2 \cdot 4H_2O$, $Co(CH_3COO)_2 \cdot 4H_2O$, and $Mn(CH_3COO)_2 \cdot 4H_2O$ were dissolved in 50 mL of distilled water. An extra 3 mol.% of the lithium source was used to compensate for the volatilization of lithium at high heating temperatures. The green sol was obtained after the solvent was completely evaporated, and was then transferred to dry in vacuum at 120 °C for 12 h. Dried gel heat treatments were carried out at 850 °C for 5 h, under pure O_2 flow at a temperature ramp of 5 °C/min. The resultant black $LiNi_{0.8}Mn_{0.1}Co_{0.1}O_2$ (marked as NCM811) powder was collected after cooling to room temperature.

2.2. Al_2O_3 Coating on $LiNi_{0.8}Co_{0.1}Mn_{0.1}O_2$ Particles Through Atomic Layer Deposition

The atomic layer deposition of Al_2O_3 coating on $LiNi_{0.8}Co_{0.1}Mn_{0.1}O_2$ particles was performed in a Savannah 100 ALD system (Cambridge NanoTech Inc., Waltham, MA, USA) at 120 °C, by applying $Al(CH_3)_3$ (Trimethylalumium, TMA) and H_2O as precursors with exposure time of 0.01 s, waiting time of 5 s, and purge time of 40 s, respectively. Nitrogen gas was used as the flow gas, and the flow rate was set at 20 sccm. The vacuum condition in the reaction chamber was controlled at 10^{-3}–10^{-2} torr, and the overpressure threshold for the coating processes was set at 250 torr. Two self-terminating reactions involved in the ALD growth of Al_2O_3 layer are described as follows:

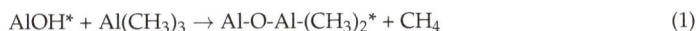

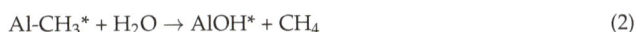

$$AlOH^* + Al(CH_3)_3 \rightarrow Al\text{-}O\text{-}Al\text{-}(CH_3)_2{}^* + CH_4 \tag{1}$$

$$Al\text{-}CH_3{}^* + H_2O \rightarrow AlOH^* + CH_4 \tag{2}$$

$LiNi_{0.8}Co_{0.1}Mn_{0.1}O_2$ particles coated with the Al_2O_3 layer under different ALD cycles were marked as n Al_2O_3 ALD NCM811 (the n corresponds to the number of ALD cycles).

2.3. Characterizations

Crystallographic structures of bare NCM811 and Al_2O_3-coated NCM811 materials were examined by X-ray diffraction (XRD), using a Bruker D8 Advance automatic diffractometer (Bruker, Billerica, MA, USA) with Cu Kα radiation. Morphologies and energy dispersive spectroscopic (EDS) mappings of different samples were observed by field emission scanning electron microscopy (FESEM) on a Hitachi SU-8010 (Hitachi High-Tech Corp., Tokyo, Japan) and FEI Quanta FEG 250 microscopies (FEI, Hillsboro, OR, USA), respectively, with an acceleration voltage at 15/20 kV in a secondary electrons (SE) detection mode. The Everhart Thornley Detector (ETD) was used to collected electron signals. Detailed structural features of different specimens were captured by transmission electron microscopy (TEM) and high-resolution transmission electron microscopy (HRTEM) coupled with selected area electron diffraction (SAED) on a FEI Tecnai G2 T20 equipment (FEI, Hillsboro, OR, USA), at an acceleration voltage of 200 kV. Chemical environments and valent states of cations within different materials were characterized by X-ray photoelectron spectroscopic (XPS) measurements on an ESCALAB 250Xi XPS device (Thermo Fisher Scientific, Waltham, MA USA). All XPS spectra were calibrated according to the binding energy of the C 1s peak at 284.6 eV.

2.4. Electrochemical Measurements

The working electrodes were composed of 80 wt.% active cathode materials, 10 wt.% super p as the conductive carbon, and 10 wt.% polyvinylidenefluoride (PVDF) as the binder. These cathodes were assembled into two-electrode CR2025-type coin cells (Shenzhen Kejing Star Technology Co., LTD., Shenzhen, China), in an Ar-filled glove box (MBraun, Garching, Germany) for electrochemical measurements, with the metallic lithium foil as the reference, and counter electrodes with the glass microfiber (Whatman, Grade GF/B, Little Chalfont, UK) as the separator. The electrolyte was a 1 M $LiPF_6$ solution dissolved in ethylene carbonate (EC) and dimethyl carbonate (DMC) at a volumetric ratio of 1:1. The real loading of active material was ~3.5 mg in the cathode with a diameter of 12 mm and a thickness of ~50 mm for all electrode materials. The electrolyte volume used during the coin cell assembly was ~0.5 mL per unit. Galvanostatic charge and discharge were performed at different current densities in a voltage range of 2.7–4.6 V vs. Li^+/Li on a LAND battery testing system (Jinnuo, Wuhan, China). The current density corresponding to 1 C was 200 mA/g. Electrochemical impedance spectroscopy (EIS) was conducted in a frequency range of 10 mHz–100 kHz with an AC amplitude of 5 mV on the Autolab electrochemical workstation (PGSTAT302N, Metrohm Autolab, Utrecht, The Netherlands).

3. Results and Discussion

3.1. Characteristics of Al$_2$O$_3$ Layer Deposited on NCM811 Particles via ALD Coating

As shown in the schematic illustrations in Figure 1a, the ALD growth of Al_2O_3 coating at the surface of NCM811 particles was subjected to a sequence of chemisorption and self-terminating surface reactions (Equations (1) and (2) in Section 2.2). Detailed deposition mechanisms for the different oxides, involving Al_2O_3, ZnO and ZrO_2, can be found in our previous work [12,36–40]. As a result, ultrathin oxide coatings under ALD growth always reveal desirable conformality, uniformity, robustness, and pinhole-free features. Furthermore, thickness of the resultant oxide coatings can be precisely controlled at the Angstrom or atomic monolayer level, which is determined by the number of ALD cycles. Accordingly, morphologic and structural differences of pristine NCM811, 20 Al_2O_3 ALD NCM811, and 50 Al_2O_3 ALD NCM811 materials are compared in Figure 1b–j. The pristine NCM811 material (Figure 1b) reveals a distinct aggregation of tremendous primary nanoparticles, which have smooth facets and sharp edges with an average particle size of ~500 nm. Such morphologic features generally indicate the high crystallinity and purity of the prepared NCM811 particles. By contrast, FESEM images of the 20 Al_2O_3 ALD NCM811 particles showed apparent roughness at the entire surface after coating the Al_2O_3 layer in 20 ALD cycles (Figure 1c), implying the conformal and uniform surface modification of the NCM811 substrate. The elemental energy dispersive spectroscopic (EDS) mapping was captured on the selected 20 Al_2O_3 ALD NCM811 particles (Figure 1d), which revealed even distributions of Al, Ni, Co, Mn, and O elements, and thus again demonstrated desirable conformality and uniformity of Al_2O_3 coatings at the surface of NCM811 particles. In order to comprehensively understand structural characteristics of the resultant Al_2O_3 layer, 50 Al_2O_3 ALD NCM811 particles with increased coating thickness were examined by TEM observations, and compared to bare NCM811 and the 20 Al_2O_3 ALD NCM811 materials. As shown in Figure 1e, TEM images of NCM811 particles showed consistent contrast throughout the whole observation area, revealing a single solid structure without any surface modifications. HRTEM images of the selected region in Figure 1e, marked by a blue square, showed perfect lattice fringes with a d-space of 0.47 nm (Figure 1h), corresponding to (003) planes in the layered structure of NCM811 oxide. By contrast, TEM images of the 20 Al_2O_3 ALD NCM811 and 50 Al_2O_3 ALD NCM811 particles captured in Figure 1f,g gave rise to distinct contrast differences between the dark NCM811 core and light Al_2O_3 coating shells at the surface, respectively. It was noticeable that the ALD coating contributed to satisfactory uniformity and conformality of the resultant Al_2O_3 layer, which was in accordance with that coated at the surface of different cathode materials reported in the literature [36,38]. The precise thickness was 5.6 nm as shown in the HRTEM image of the 50 Al_2O_3 ALD NCM811 (Figure 1j), corresponding to an Al_2O_3 ALD growth rate of 1.12 Å/cycle. Accordingly, 20 ALD cycles resulted in an Al_2O_3 coating with 2.2 nm thickness as shown in Figure 1i for the 20 Al_2O_3 ALD NCM811 materials. Furthermore, these two HRTEM observations indicated an amorphous feature of the Al_2O_3 coating layer, which was distinguishable from the internal crystalline NCM811 particles. As a result, SEM and TEM images demonstrated the conformational and uniform coating of the Al_2O_3 layer for the surface modification of Ni-rich NCM811 particles through ALD processes, together with thickness control at the sub-nano level.

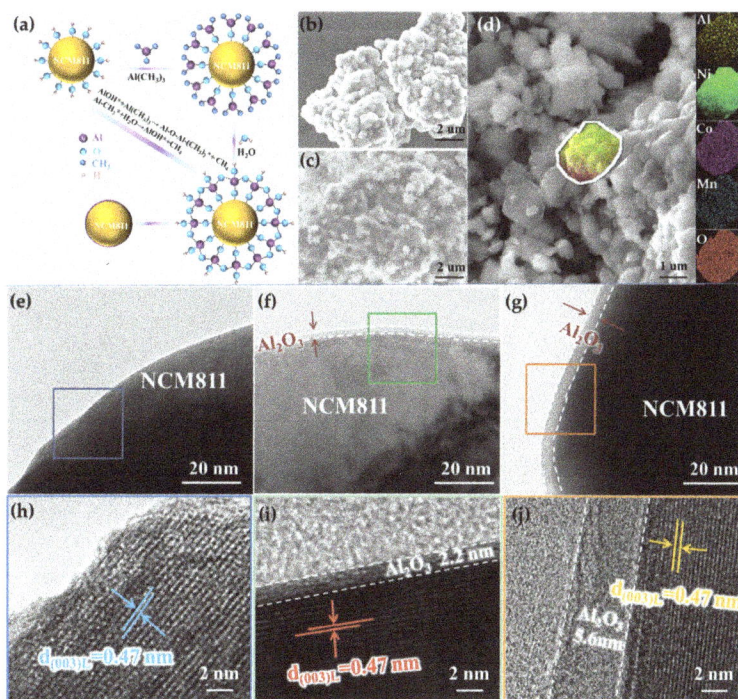

Figure 1. (**a**) Schematics illustration of the Al_2O_3 ALD growth at the surface of NCM811 particles; FESEM images of (**b**) bare NCM811, and (**c**) 20 Al_2O_3 ALD NCM811; EDS mapping of (**d**) 20 Al_2O_3 ALD NCM811; TEM images of (**e**) bare NCM811, (**f**) 20 Al_2O_3 ALD NCM811, and (**g**) 50 Al_2O_3 ALD NCM811; and HRTEM images of (**h**) bare NCM811, (**i**) 20 Al_2O_3 ALD NCM811, and (**j**) 50 Al_2O_3 ALD NCM811.

Al$_2$O$_3$ coating effects on changing the crystal structure of Ni-rich NCM811 core were evaluated using XRD patterns of different coated materials, coupled with quantitative analyses from Rietveld refinements. As shown in Figure 2a, the XRD pattern of pristine NCM811 material was well indexed to a layered α-NaFeO$_2$ structure with a R-3m space group (JCPDS No. 74-0919). It also showed a desirable intensity ratio of characteristic 003 over 104 reflections, i.e., I_{003}/I_{104} = 1.48 and distinct peak splitting of two 006/102 and 108/110 doublets, indicating outstanding structural ordering of the layered NCM811 material. The Rietveld refinement on the XRD pattern of pristine NCM811 resulted in lattice parameters of a = 2.8746 Å, c = 14.2102 Å, and V = 101.72 Å^3, together with a low estimated Li$^+$/Ni^{2+} cation mixing of 3.68% based on a low reliability-factor of R_{wp} = 6.62% (Table 1). These XRD features all indicate an excellent layered structure of the prepared Ni-rich NCM811 material, in accordance with HRTEM images (Figure 1h). XRD patterns and the associated refinement results of NCM811 particles coated with 10, 20, and 50 Al$_2$O$_3$ ALD layers are displayed in Figure 2b–d, respectively. The calculated lattice parameters and cation mixings of these three coated materials, with gradually-increasing coating thickness of Al$_2$O$_3$, are summarized in Table 1 compared to bare NCM811, together with values of standard deviations (s.d.) for the calculations. The Al$_2$O$_3$ phase was not detectable even in the 50 Al$_2$O$_3$ ALD NCM811 samples, owing to the amorphous phase. However, the Al$_2$O$_3$ coating slightly suppressed the crystal structural of the internal NCM811 host. Coatings thicker than 2 nm, i.e., 20 ALD cycles (20 cycle* 1.12 Å/cycle = 2.24 nm), resulted in a detectable decrease in a, c, and V values, according to the refinement results of the 20 Al$_2$O$_3$ ALD NCM811 sample, while the 10 Al$_2$O$_3$ ALD NCM811 showed little change in lattice parameters as compared with that in the pristine NCM811 (Table 1). Although the thick Al$_2$O$_3$ coating reduced a and c values of the 50 Al$_2$O$_3$ ALD NCM811 material, its c/a ratio was almost the same with the NCM811, again indicating high conformality and uniformity of surface modifications via oxide ALD coatings. On the other hand, XRD patterns of the three Al$_2$O$_3$-coated NCM811 materials revealed I_{003}/I_{104} values equal to 1.54, 1.46, and 1.45 after 10, 20, and 50 ALD cycles (Figure 2b–d), respectively, indicating little effects of Al$_2$O$_3$ ALD coating on the structural ordering of internal NCM811 material. The estimated cation mixings within the layered structures of these three samples wre all in an acceptable range of 3%–4%, consistent with that of the pristine NCM811 material (Table 1).

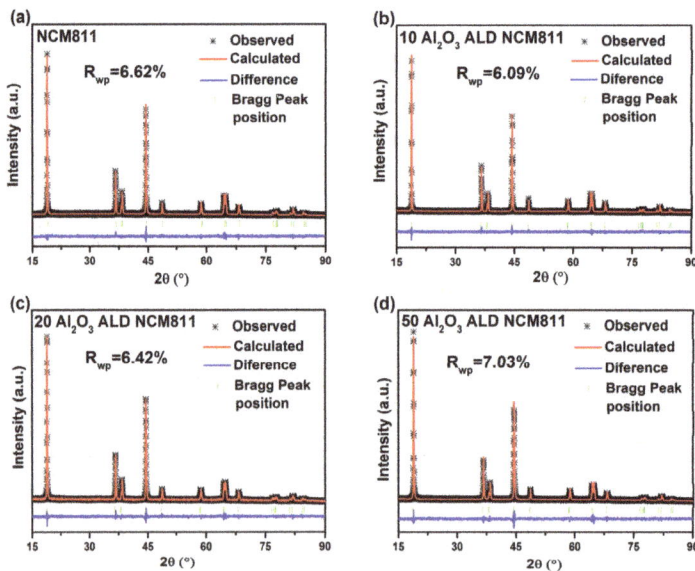

Figure 2. XRD patterns of (**a**) bare NCM8111; (**b**) 10 Al$_2$O$_3$ ALD NCM8111; (**c**) 20 Al$_2$O$_3$ ALD NCM8111; and (**d**) 50 Al$_2$O$_3$ ALD NCM8111 materials, coupled with rietveld refinement results.

Table 1. Crystal parameters of different Al_2O_3-coated NCM811 materials in comparison to bare NCM811, according to XRD patterns and refinement results as shown in Figure 2.

Samples	a (Å)	s.d.	c (Å)	s.d.	Volume (Å³)	s.d.	c/a	Li⁺ at 3b Sites	s.d.	R_{wp} (%)	R_p (%)
NCM811	2.8746	3×10^{-5}	14.2102	1.8×10^{-3}	101.72	2.87×10^{-3}	4.9432	0.9632	2.203×10^{-3}	6.62	4.81
10 Al_2O_3 ALD NCM811	2.8745	3×10^{-5}	14.2095	3.1×10^{-3}	101.71	3.1×10^{-3}	4.9433	0.9634	2.351×10^{-3}	6.09	5.12
20 Al_2O_3 ALD NCM811	2.8743	2×10^{-5}	14.2094	3.6×10^{-3}	101.65	3.05×10^{-3}	4.9436	0.9697	1.936×10^{-3}	6.42	5.18
50 Al_2O_3 ALD NCM811	2.8739	4×10^{-5}	14.2071	5.2×10^{-3}	101.49	7.93×10^{-3}	4.9435	0.9666	7.725×10^{-3}	7.03	5.58

In order to confirm Al_2O_3 coating formation and its possible impacts on the chemical environments of transition metal cations at the surface of NCM811 particles, characteristic Al 2p, Ni 2p, Co 2p, and Mn 2p XPS peaks of bare NCM811 and three Al_2O_3-coated NCM811 materials were measured and compared. As shown in Figure 3a, 10 Al_2O_3 ALD NCM811, 20 Al_2O_3 ALD NCM811, and 50 Al_2O_3 ALD NCM811 samples all showed sharp Al 2p XPS peaks, verifying successful Al_2O_3 coating at the surface of NCM811 particles. These collected Al 2p peaks slightly shifted to higher binding energies at 75.4, 75.6, and 75.7 eV, respectively, compared to 74.6 eV of pure Al_2O_3 oxide reported in NIST XPS Database, implying possible chemical bonding between the Al_2O_3 coating layer and NCM811 particles. As reported in the literature [36–40], the oxide coating and substrate in a typical ALD growth can result in a favorable chemical bond of substrate–O–M_xO_y (M = Al, Zn, Zr, Sn, etc.), this process is illustrated in Figure 1a. However, it is unexpected to see that all the characteristic of Ni 2p, Co 2p, and Mn 2p XPS peaks captured on three Al_2O_3-coated NCM811 particles also gave rise to higher bonding energies compared to those of pristine NCM811 particles, when the coating thickness of Al_2O_3 was gradually increased (Figure 3b–d). These interesting XPS results may be attributed to the tremendous –OH groups that were left at the surface of Al_2O_3 coating at the end of ALD growth (Figure 1a), which are highly electronegative, and thus probably induce reduced electron cloud densities of Al, Ni, Co, and Mn cations. The thicker Al_2O_3 coating resulted in increased surface area of Al_2O_3-coated NCM811 particles as well as the quantity of residual –OH groups at the most outside surface; hence, the inductive effect was enhanced by the increased number of –OH groups, leading to binding energy shifts of different cations to higher values. XPS results indicate a probable chemical bonding between Al_2O_3 and NCM811, which may facilitate charge transfer and lithium ion diffusion through the oxide coating layer.

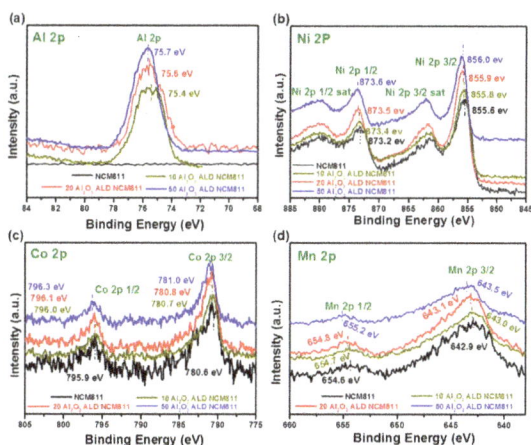

Figure 3. Characteristics XPS spectra captured on the 10 Al_2O_3 ALD NCM811, 20 Al_2O_3 ALD NCM811, and 50 Al_2O_3 ALD NCM811 materials compared to the bare NCM811: (**a**) Al 2p, (**b**) Ni 2p, (**c**) Co 2p, and (**d**) Mn 2p peaks.

3.2. Enhanced Electrochemical Performance of Ni-Rich NCM811 Cathode Materials by Al$_2$O$_3$ ALD Coating

Effects of Al$_2$O$_3$ ALD coating on improving electrochemical performance of Ni-rich NCM811 cathode material are evaluated in Figure 4, where different cathode materials were subjected to a broad voltage range of 2.7–4.6 V vs. Li$^+$/Li. Figure 4a compares resultant cycling performance of the Al$_2$O$_3$-coated NCM811 cycled at 0.1 C with bare NCM811 material. The coating thickness of Al$_2$O$_3$ layer was optimized at ~2 nm, corresponding to 20 ALD cycles. As a result, the 20 Al$_2$O$_3$ ALD NCM811 cathode delivered an impressive discharge capacity of 212.8 mAh/g and an associated initial columbic efficiency of 84.0% in the first cycle, which were both higher than 198.2 mAh/g and 76.1% of the pristine NCM811 without Al$_2$O$_3$ protection. Accordingly, the charge/discharge curves of these two cathode materials cycled at 1st and 100th cycles are shown in Figure 4b. The improved electrochemical reversibility with increased specific capacity of the 20 Al$_2$O$_3$ ALD NCM811 in the first cycle can be attributed to the conformal surface modification from the Al$_2$O$_3$ coating (Figure 1a), which effectively suppressed detrimental side reactions even when the NCM811 cathode material was cycled in an extended voltage range up to 4.6 V. In addition to protecting the internal Ni-rich core material from the HF attack in electrolytes as reported in the literature [29,41], the conformational and uniform Al$_2$O$_3$ coating also reduced the electrolyte decomposition induced by high oxidation of Ni^{3+}/Ni^{4+} redox, especially under the deeply-charged condition at a high cut-off voltage of 4.6 V. As shown in Figure 4a,b, the pristine NCM811 cathode material suffered from distinct capacity decay to 127.5 mAh/g after 100 electrochemical cycles, while the 20 Al$_2$O$_3$ ALD NCM811 retained a much higher capacity of 157.2 mAh/g in the 100th cycle. Furthermore, the NCM811 also encountered an obvious working voltage fading from 3.80 initially to 3.76 V in the 100th cycle. By contrast, electrochemical polarity was effectively inhibited in the 20 Al$_2$O$_3$ ALD NCM811 cathode material, resulting in a well-preserved working voltage from 3.84 V in the first cycle to 3.83 V after 100 cycles. The 10 Al$_2$O$_3$ ALD NCM811 delivered moderate cycling performance as shown in Figure 4a, between the NCM811 and 20 Al$_2$O$_3$ ALD NCM811, which was attributed to the thinner Al$_2$O$_3$ coating around 1 nm. On the other hand, the much thicker coating of Al$_2$O$_3$ layer (>5 nm) after 50 ALD cycles resulted in satisfactory cycling stability, but the lithium storage capacity had to be sacrificed (Figure 4a), owing to the inferior conductivity of insulating Al$_2$O$_3$ oxide [42,43]. According to reports in the literature [13,44], Ni-rich cathode materials undergo inevitable phase transitions among three hexagonal (H1, H2, and H3) and one monoclinic (M) phases in their layered structures during lithiation/delithiation, which further aggravate undesired capacity fading in long-term electrochemical cycles. In order to examine possible phase transitions, related differential capacity (dQ/dV) profiles are shown in Figure 4c, which were transferred from charge/discharge curves in Figure 4b. The initial anodic peak of pristine NCM811 showed the much higher intensity by contrast with 20 Al$_2$O$_3$ ALD NCM811, indicating severe parasitic side reactions that occurred during the first phase transformation from the hexagonal H1 to monoclinic M phases. Subsequent M to H2, and H2 to H3 took place around at 4.0 and 4.2 V as marked in Figure 4c, respectively, when the Ni-rich cathode materials were charged to a high cutoff voltage. In successive discharge processes, phase transitions from H3 to H2 to M were reduced, resulting in dominant cathodic peaks around 3.75 V both in NCM811 and 20 Al$_2$O$_3$ ALD NCM811 cathodes. However, two redox pairs at higher voltages around 4.0 and 4.2 V were detected in the 100th cycle of pristine NCM811 cathode (peak voltages of detected anodic and cathodic peaks are marked by underlines), while the 20 Al$_2$O$_3$ ALD NCM811 gave rise to the smoother dQ/dV curves above 3.9 V, revealing significantly suppressed phase transitions for its enhanced cycling stability after the optimal Al$_2$O$_3$ ALD coating. Furthermore, total resistance changes of these two cathodes after 100 cycles were determined by EIS spectra as shown in Figure 4d. These measurements were conducted on their fully-discharged states after prolonged electrochemical cycles. EIS spectra of fresh cells were also recorded at open circuit voltage states for the comparison. It is worth noting that two fresh cells of NCM811 and the 20 Al$_2$O$_3$ ALD NCM811 cathodes both showed small resistances around 25 Ω. However, two Nyquist circles with larger diameters were captured in both cathodes after 100 cycles. The selected equivalent circuit, as the inset, was used to calculate the associated resistances based on fitting solid lines. For the pristine NCM811, one Nyquist circle in high frequency may be attributed to impure phases, such as spinel-like and/or rock-salt phases formed during prolonged cycles as reported in the literature [45,46], and others in low frequency were assigned to the solid electrolyte interphase (SEI) film. By contrast, two Nyquist circles in 20 Al_2O_3 ALD NCM811 resulted from the Al_2O_3 coating and SEI film at the surface of cycled cathode material, respectively. It is noticeable that introduction of 2 nm Al_2O_3 coating led to acceptable resistance at 29 Ω, which was smaller than the 35 Ω caused by the secondary impure structure that formed at the surface of NCM811 without surface modification. Moreover, the 20 Al_2O_3 ALD NCM811 resulted in considerable decrease (160 Ω) corresponding to SEI film resistance, while the pristine NCM811 brought about larger SEI resistance (250 Ω). Electrochemical performance demonstrated significant effects of the optimized Al_2O_3 ALD coating on reducing electrochemical polarity and improving reversibility of Ni-rich cathode materials for superior lithium ion batteries.

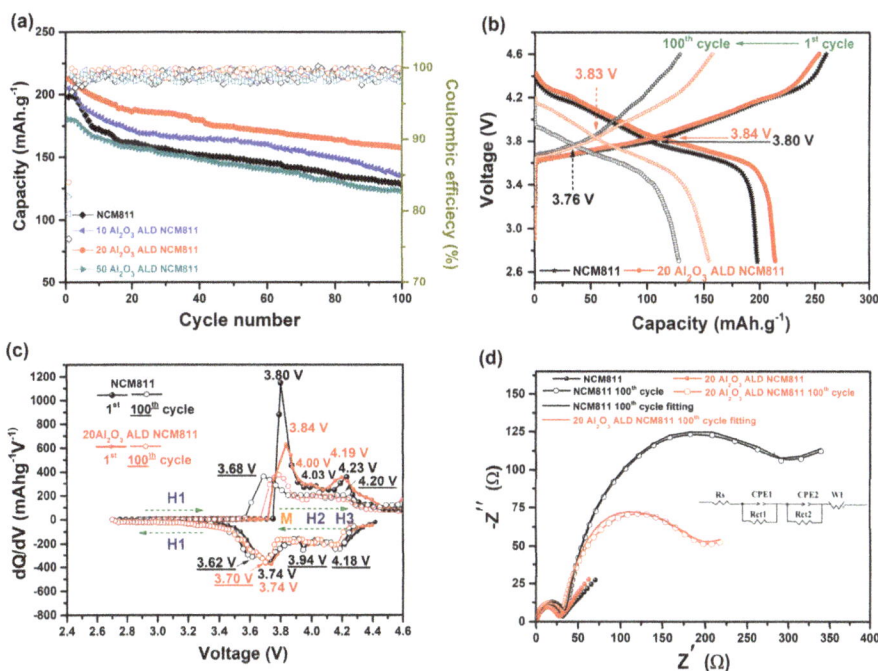

Figure 4. Electrochemical performance of different cathode materials in a broad voltage range of 2.7–4.6 V vs. Li^+/Li: (**a**) cycling performance at 0.1 C for 100 cycles; (**b**) charge/discharge curves of bare NCM811 and 20 Al_2O_3 ALD NCM811 in the first and 100th cycles at 0.1 C; (**c**) dQ/dV profiles of bare NCM811 and 20 Al_2O_3 ALD NCM811 in the first and 100th cycles; and (**d**) EIS spectra captured on fresh cells (solid patterns) at open circuit voltage states and after 100 cycles at fully-discharged states (hollow patterns).

4. Conclusions

Enhanced cycling stability of Ni-rich $LiNi_{0.8}Co_{0.1}Mn_{0.1}O_2$ cathode material in an extended voltage range has been realized by surface modification of ultrathin Al_2O_3 coating through atomic layer deposition. The oxide ALD growth leads to impressive conformality and uniformity of resultant Al_2O_3 layer, and the coating thickness can be precisely controlled at 1.12 Å per ALD cycle. The optimal thickness of the Al_2O_3 coating for $LiNi_{0.8}Co_{0.1}Mn_{0.1}O_2$ cathode material is in 2 nm level through performing 20 ALD cycles. Either the thicker or thinner coating thickness leads to decreased lithium storage capacity together with inferior cycling performance of different Al_2O_3

coated $LiNi_{0.8}Co_{0.1}Mn_{0.1}O_2$ cathodes. The optimized 20 Al_2O_3 ALD modified $LiNi_{0.8}Co_{0.1}Mn_{0.1}O_2$ cathode material distinctly shows increased discharge capacity and columbic efficiency in the first cycle, as well as improved capacity retention after successive 100 cycles in comparison with the pristine material. Electrochemical features in the first and 100th cycle of cathode materials, with and without Al_2O_3 protection were compared, involving charge/discharge curves, differential capacity profiles, and electrochemical impedance spectra. The Al_2O_3 coating has been demonstrated to suppress parasitic side reactions and phase transitions during prolonged lithiation/delithiation of the $LiNi_{0.8}Co_{0.1}Mn_{0.1}O_2$ cathode, resulting in reduced electrochemical polarity and enhanced cycling stability for Ni-rich cathode materials cycled in broad voltage ranges.

Author Contributions: Conceptualization, J.Z.; Methodology, W.Z., T.L., Z.X., and X.H.; Software, L.B.; Data Curation, W.Z., X.H., L.B., and K.T.; Writing—Original Draft Preparation, W.Z.; Writing—Review and Editing, Y.W., H.W., J.Z., and L.G.; Project Administration, L.G. and J.Z.; Funding Acquisition, L.G. and J.Z.

Funding: This research was funded by the National Natural Science Foundation of China (Nos. 21703147 and U1401248); the Jiangsu Provincial Natural Science Foundations for the Young Scientist (No. BK20170338); the General Financial Grant from the China Postdoctoral Science Foundation (No. 2016M601876); and the Open Fund of Jiangsu Key Laboratory of Materials and Technology for Energy Conversion (No. MTEC-2017M01).

Acknowledgments: The authors acknowledge Suzhou Key Laboratory for the Advanced Carbon Materials and Wearable Energy Technologies, Suzhou 215006, China.

Conflicts of Interest: The authors declare no conflicts of interest.

References

1. Poizot, P.; Laruelle, S.; Grugeon, S.; Grugeon, L.; Tarascon, J.M. Nano-sized transition-metal oxides as negative-electrode materials for Lithium-ion batteries. *Nature* **2000**, *407*, 496–499. [CrossRef]
2. Mizushima, K.; Jones, P.C.; Wiseman, P.J.; Goodenough, J.B. Li_xCoO_2 ($0 < x < -1$): A new cathode material for batteries of high energy density. *Mater. Res. Bull.* **1980**, *15*, 783–789. [CrossRef]
3. Thackeray, M.M.; Wolverton, C.; Isaacs, E.D. Electrical energy storage for transportation-approaching the limits of, and going beyond, Lithium-ion batteries. *Energy Environ. Sci.* **2012**, *5*, 7854–7863. [CrossRef]
4. Tang, Y.; Deng, J.; Li, W.; Malyi, O.I.; Zhang, Y.; Zhou, X.; Pan, S.; Wei, J.; Cai, Y.; Chen, Z. Water-soluble sericin protein enabling stable solid-electrolyte interphase for fast charging high voltage battery electrode. *Adv. Mater.* **2017**, *29*, 1701828. [CrossRef]
5. Tang, Y.; Zhang, Y.; Malyi, O.I.; Bucher, N.; Xia, H.; Xi, S.; Zhu, Z.; Lv, Z.; Li, W.; Wei, J. Identifying the origin and contribution of surface storage in $TiO_2(B)$ nanotube electrode by in situ dynamic valence state monitoring. *Adv. Mater.* **2018**, *36*, 1802200. [CrossRef]
6. Zhang, Y.; Malyi, O.I.; Tang, Y.; Wei, J.; Zhu, Z.; Xia, H.; Li, W.; Guo, J.; Zhou, X.; Chen, Z. Reducing the charge carrier transport barrier in functionally layer-graded electrodes. *Angew. Chem.* **2017**, *56*, 14847. [CrossRef]
7. Yi, R.; Dai, F.; Gordin, M.L.; Chen, S.; Wang, D. Lithium-ion batteries: micro-sized Si–C composite with interconnected nanoscale building blocks as high-performance anodes for practical application in Lithium-ion batteries. *Adv. Energy Mater.* **2013**, *3*, 273. [CrossRef]
8. Fergus, J.W. Recent developments in cathode materials for Lithium ion batteries. *J. Power Sources* **2010**, *195*, 939–954. [CrossRef]
9. Lee, J.; Kitchaev, D.A.; Kwon, D.H.; Lee, C.W.; Papp, J.K.; Liu, Y.S.; Lun, Z.; Clément, R.J.; Shi, T.; Mccloskey, B.D. Reversible Mn^{2+}/Mn^{4+} double redox in Lithium-excess cathode materials. *Nature* **2018**, *556*, 185–190. [CrossRef]
10. Jiang, K.C.; Wu, X.L.; Yin, Y.X.; Lee, J.S.; Kim, J.; Guo, Y.G. Superior hybrid cathode material containing Lithium-excess layered material and graphene for Lithium-ion Batteries. *ACS Appl. Mater. Interfaces* **2012**, *4*, 4858–4863. [CrossRef]
11. Zhao, J.; Zhang, W.; Huq, A.; Misture, S.T.; Zhang, B.; Guo, S.; Wu, L.; Zhu, Y.; Chen, Z.; Amine, K. In situ probing and synthetic control of cationic ordering in Ni-Rich layered oxide cathodes. *Adv. Energy Mater.* **2017**, *7*, 1601266. [CrossRef]
12. Zhao, J.; Aziz, S.; Wang, Y. Hierarchical functional layers on high-capacity Lithium-excess cathodes for superior Lithium ion batteries. *J. Power Sources* **2014**, *247*, 95–104. [CrossRef]

13. Zhao, J.; Huang, R.; Gao, W.; Zuo, J.M.; Zhang, X.F.; Misture, S.T.; Chen, Y.; Lockard, J.V.; Zhang, B.; Guo, S. An Ion-exchange promoted phase transition in a Li-excess layered cathode material for high-performance Lithium ion batteries. *Adv. Energy Mater.* **2015**, *5*, 1401937. [CrossRef]

14. Martha, S.K.; Sclar, H.; Framowitz, Z.S.; Kovacheva, D.; Saliyski, N.; Gofer, Y.; Sharon, P.; Golik, E.; Markovsky, B.; Aurbach, D. A comparative study of electrodes comprising nanometric and submicron particles of $LiNi_{0.5}Mn_{0.5}O_2$, $LiNi_{0.33}Mn_{0.33}Co_{0.33}O_2$, and $LiNi_{0.4}Mn_{0.4}Co_{0.2}O_2$ layered compounds. *J. Power Sources* **2009**, *189*, 248–255. [CrossRef]

15. Li, X.; Cheng, F.; Guo, B.; Chen, J. Template-synthesized $LiCoO_2$, $LiMn_2O_4$, and $LiNi_{0.8}Co_{0.2}O_2$ nanotubes as the cathode materials of Lithium ion batteries. *J. Phys. Chem. B* **2005**, *109*, 14017–14024. [CrossRef]

16. Liu, W.; Oh, P.; Liu, X.; Lee, M.J.; Cho, W.; Chae, S.; Kim, Y.; Cho, J. Nickel-rich layered Lithium transition-metal oxide for high-energy Lithium-ion batteries. *Angew. Chem.* **2015**, *54*, 4440–4457. [CrossRef]

17. Zhong, S.; Lai, M.; Yao, W.; Li, Z. Synthesis and electrochemical properties of $LiNi_{0.8}Co_xMn_{0.2-x}O_2$ positive-electrode material for Lithium-ion batteries. *Electrochim. Acta* **2016**, *212*, 343–351. [CrossRef]

18. Zou, Y.; Yang, X.; Lv, C.; Liu, T.; Xia, Y.; Shang, L.; Waterhouse, G.I.; Yang, D.; Zhang, T. Multishelled Ni-Rich $Li(Ni_xCo_yMn_z)O_2$ hollow fibers with low cation mixing as high-performance cathode materials for Li-ion batteries. *Adv. Sci.* **2017**, *4*, 1600262. [CrossRef]

19. Zhang, X.; Jiang, W.J.; Mauger, A.; QILU; Gendron, F.; Julien, C.M. Minimization of the cation mixing in $Li_{1+x}(NMC)_{1-x}O_2$ as cathode material. *J. Power Sources* **2009**, *195*, 1292–1301. [CrossRef]

20. Manthiram, A.; Song, B.; Li, W. A perspective on nickel-rich layered oxide cathodes for Lithium-ion batteries. *Energy Storage Mater.* **2017**, *6*, 125–139. [CrossRef]

21. Zheng, J.; Kan, W.H.; Manthiram, A. Role of Mn content on the electrochemical properties of nickel-rich layered $LiNi_{(0.8-x)}Co_{(0.1)}Mn_{(0.1+x)}O_2$ ($0.0 \leq x \leq 0.08$) cathodes for lithium-ion batteries. *ACS Appl. Mater. Interfaces* **2015**, *7*, 6926–6934. [CrossRef]

22. Oh, P.; Oh, S.M.; Li, W.; Myeong, S.; Cho, J.; Manthiram, A. High-performance heterostructured cathodes for Lithium-ion batteries with a Ni-rich layered oxide core and a Li-rich layered oxide shell. *Adv. Sci.* **2016**, *3*, 1600184. [CrossRef]

23. Song, B. A facile cathode design combining Ni-rich layered oxides with Li-rich layered oxides for Lithium-ion batteries. *J. Power Sources* **2016**, *325*, 620–629. [CrossRef]

24. Schipper, F.; Bouzaglo, H.; Dixit, M.; Erickson, E.M.; Weigel, T.; Talianker, M.; Grinblat, J.; Burstein, L.; Schmidt, M.; Lampert, J. From surface ZrO_2 coating to bulk Zr doping by high temperature annealing of nickel-rich lithiated oxides and their enhanced electrochemical performance in Lithium ion batteries. *Adv. Energy Mater.* **2018**, *8*, 1701682. [CrossRef]

25. Tan, L.; Liu, H. Influence of ZnO coating on the structure, morphology and electrochemical performances for $LiNi_{1/3}Co_{1/3}Mn_{1/3}O_2$ material. *Russ. J. Electrochem.* **2011**, *47*, 156–160. [CrossRef]

26. Li, J.; Fan, M.; He, X.; Zhao, R.; Jiange, C.; Wan, C. TiO_2 coating of $LiNi_{1/3}Co_{1/3}Mn_{1/3}O_2$ cathode materials for Li-ion batteries. *Ionics* **2006**, *12*, 215–218. [CrossRef]

27. Cheng, C.; Yi, H.; Chen, F. Effect of Cr_2O_3 Coating on $LiNi_{1/3}Co_{1/3}Mn_{1/3}O_2$ as cathode for Lithium-ion batteries. *J. Electron. Mater.* **2014**, *43*, 3681–3687. [CrossRef]

28. Hudaya, C.; Ji, H.P.; Lee, J.K.; Choi, W. SnO_2-coated $LiCoO_2$ cathode material for high-voltage applications in Lithium-ion batteries. *Solid State Ionics* **2014**, *256*, 89–92. [CrossRef]

29. Li, X.; Liu, J.; Meng, X.; Tang, Y.; Banis, M.N.; Yang, J.; Hu, Y.; Li, R.; Cai, M.; Sun, X. Significant impact on cathode performance of Lithium-ion batteries by precisely controlled metal oxide nanocoatings via atomic layer deposition. *J. Power Sources* **2014**, *247*, 57–69. [CrossRef]

30. Wang, Z.; Liu, L.; Chen, L.; Huang, X. Structural and electrochemical characterizations of surface-modified $LiCoO_2$ cathode materials for Li-ion batteries. *Solid State Ionics* **2002**, *148*, 335–342. [CrossRef]

31. Kim, J.W.; Kim, D.H.; Oh, D.Y.; Lee, H.; Kim, J.H.; Lee, J.H.; Jung, Y.S. Surface chemistry of $LiNi_{0.5}Mn_{1.5}O_4$ particles coated by Al_2O_3 using atomic layer deposition for Lithium-ion batteries. *J. Power Sources* **2015**, *274*, 1254–1262. [CrossRef]

32. Guan, D.; Wang, Y. Ultrathin surface coatings to enhance cycling stability of $LiMn_2O_4$ cathode in Lithium-ion batteries. *Ionics* **2013**, *19*, 1–8. [CrossRef]

33. Mohanty, D.; Dahlberg, K.; King, D.M.; David, L.A.; Sefat, A.S.; Wood, D.L.; Daniel, C.; Dhar, S.; Mahajan, V.; Lee, M. Modification of Ni-rich FCG NMC and NCA cathodes by atomic layer deposition: preventing surface phase transitions for high-voltage Lithium-ion batteries. *Sci. Rep.* **2016**, *6*, 26532. [CrossRef]

34. Shi, Y.; Zhang, M.; Qian, D.; Meng, Y.S. Ultrathin Al_2O_3 coatings for improved cycling performance and thermal stability of $LiNi_{0.5}Co_{0.2}Mn_{0.3}O_2$ cathode material. *Electrochim. Acta* **2016**, *203*, 154–161. [CrossRef]

35. Zhang, X.; Belharouak, I.; Li, L.; Lei, Y.; Elam, J.W.; Nie, A.; Chen, X.; Yassar, R.S.; Axelbaum, R.L. Structural and electrochemical study of Al_2O_3 and TiO_2 Coated $Li_{1.2}Ni_{0.13}Mn_{0.54}Co_{0.13}O_2$ cathode material using ALD. *Adv. Energy Mater.* **2013**, *3*, 1299–1307. [CrossRef]

36. Zhao, J.; Wang, Y. Atomic layer deposition of epitaxial ZrO_2 coating on $LiMn_2O_4$ nanoparticles for high-rate lithium ion batteries at elevated temperature. *Nano Energy* **2013**, *2*, 882–889. [CrossRef]

37. Aziz, S.; Zhao, J.; Cain, C.; Wang, Y. Nanoarchitectured $LiMn_2O_4$/Graphene/ZnO composites as electrodes for Lithium ion batteries. *J. Mater. Sci. Technol.* **2014**, *30*, 427–433. [CrossRef]

38. Zhao, J.; Wang, Y. Surface modifications of Li-ion battery electrodes with various ultrathin amphoteric oxide coatings for enhanced cycleability. *J. Solid State Electrochem.* **2013**, *17*, 1049–1058. [CrossRef]

39. Zhao, J.; Ying, W. Ultrathin surface coatings for improved electrochemical performance of Lithium ion battery electrodes at elevated temperature. *J. Phys. Chem. C* **2012**, *116*, 11867–11876. [CrossRef]

40. Zhao, J.; Wang, Y. High-capacity full Lithium-ion cells based on nanoarchitectured ternary manganese-nickel-cobalt carbonate and its lithiated derivative. *J. Mater. Chem. A* **2014**, *2*, 14947–14956. [CrossRef]

41. Aykol, M.; Kirklin, S.; Wolverton, C. Thermodynamic aspects of cathode coatings for Lithium-ion batteries. *Adv. Energy Mater.* **2015**, *4*, 1400690. [CrossRef]

42. Zhao, F.; Tang, Y.; Wang, J.; Tian, J.; Ge, H.; Wang, B. Vapor-assisted synthesis of Al_2O_3-coated $LiCoO_2$ for high-voltage lithium ion batteries. *Electrochim. Acta* **2015**, *174*, 384–390. [CrossRef]

43. Araki, K.; Taguchi, N.; Sakaebe, H.; Tatsumi, K.; Ogumi, Z. Electrochemical properties of $LiNi_{1/3}Co_{1/3}Mn_{1/3}O_2$ cathode material modified by coating with Al_2O_3 nanoparticles. *J. Power Sources* **2014**, *269*, 236–243. [CrossRef]

44. Kalluri, S.; Yoon, M.; Jo, M.; Liu, H.K.; Dou, S.X.; Cho, J.; Guo, Z. Feasibility of cathode surface coating technology for high-energy Lithium-ion and beyond-Lithium-ion batteries. *Adv. Mater.* **2017**, *29*, 1605807. [CrossRef]

45. Liu, S.; Xiong, L.; He, C. Long cycle life lithium ion battery with lithium nickel cobalt manganese oxide (NCM) cathode. *J. Power Sources* **2014**, *261*, 285–291. [CrossRef]

46. Jung, S.K.; Gwon, H.; Hong, J.; Park, K.Y.; Seo, D.H.; Kim, H.; Hyun, J.; Yang, W.; Kang, K. Understanding the degradation mechanisms of $LiNi_{0.5}Co_{0.2}Mn_{0.3}O_2$ cathode material in Lithium ion batteries. *Adv. Energy Mater.* **2014**, *4*, 1300787. [CrossRef]

Article

Rational Construction of LaFeO$_3$ Perovskite Nanoparticle-Modified TiO$_2$ Nanotube Arrays for Visible-Light Driven Photocatalytic Activity

Jiangdong Yu [1,2], Siwan Xiang [1], Mingzheng Ge [3], Zeyang Zhang [1], Jianying Huang [4], Yuxin Tang [5,6], Lan Sun [1,2,*], Changjian Lin [1] and Yuekun Lai [4,*]

[1] State Key Laboratory of Physical Chemistry of Solid Surface, Department of Chemistry, College of Chemistry and Chemical Engineering, Xiamen University, Xiamen 361005, China; yujiangdong@stu.xmu.edu.cn (J.Y.); xiangsiwan@stu.xmu.edu.cn (S.X.); 20520171151374@stu.xmu.edu.cn (Z.Z.); cjlin@xmu.edu.cn (C.L.)
[2] Shenzhen Research Institute of Xiamen University, Shenzhen 518000, China
[3] College of Textile and Clothing Engineering, Soochow University, Suzhou 215006, China; mzge1990@ntu.edu.cn
[4] College of Chemical Engineering, Fuzhou University, Fuzhou 350116, China; jyhuang@fzu.edu.cn
[5] School of Materials Science and Engineering, Nanyang Technological University, 50 Nanyang Avenue, Singapore 639798, Singapore; yxtang@ntu.edu.sg
[6] Institute of Applied Physics and Materials Engineering, University of Macau, Macao, China
* Correspondence: sunlan@xmu.edu.cn (L.S.); yklai@fzu.edu.cn (Y.L.)

Received: 27 August 2018; Accepted: 20 October 2018; Published: 23 October 2018

Abstract: LaFeO$_3$ nanoparticle-modified TiO$_2$ nanotube arrays were fabricated through facile hydrothermal growth. The absorption edge of LaFeO$_3$ nanoparticle-modified TiO$_2$ nanotube arrays displaying a red shift to ~540 nm was indicated by the results of diffuse reflectance spectroscopy (DRS) when compared to TiO$_2$ nanotube arrays, which means that the sample of LaFeO$_3$ nanoparticle-modified TiO$_2$ nanotube arrays had enhanced visible light response. Photoluminescence (PL) spectra showed that the LaFeO$_3$ nanoparticle-modified TiO$_2$ nanotube arrays efficiently separated the photoinduced electron–hole pairs and effectively prolonged the endurance of photogenerated carriers. The results of methylene blue (MB) degeneration under simulated visible light illumination showed that the photocatalytic activity of LaFeO$_3$ nanoparticle-modified TiO$_2$ nanotube arrays is obviously increased. LaFeO$_3$ nanoparticle-modified TiO$_2$ nanotube arrays with 12 h hydrothermal reaction time showed the highest degradation rate with a 2-fold enhancement compared with that of pristine TiO$_2$ nanotube arrays.

Keywords: TiO$_2$ nanotube; LaFeO$_3$; perovskite; heterojunction; visible light driven; photocatalysis

1. Introduction

In recent decades, the application of TiO$_2$ nanomaterials in photocatalytic purification of organic contaminants has been widely used owing to its non-toxicity, superior redox potentials, long term thermodynamic stability, chemical inertness, and high photocatalytic activity [1–9]. The main drawbacks of TiO$_2$ are its wide band-gap (3.0–3.2 eV) which limits the light absorption to UV-light (ca. 4%), and fast recombination of photoinduced electron–hole pairs [10]. In order to improve the catalytic efficiency, it is crucial for a catalyst to be sensitive to solar light irradiation. In addition, an efficient separation of photogenerated electrons and holes could better trigger chemical reactions either by electrons or holes [11]. Nowadays, doping is viewed as an advanced strategy of narrowing the bandgap and improving light absorption of a catalyst by forming more delocalized intra-band states, including metal/nonmetal ion doping and co-doping [12–14]. However, this doping method has little effect of inhibiting the photoinduced electron–hole pairs from recombination [15]. Another approach is to develop new heterojunctions using visible light to initiate photocatalytic reactions. As such, one technique to enhance the photocatalytic activity, coupled catalysts, bears the merits of accelerating the partition of electron–hole pairs, and effectively extends the UV light utilization to the visible light absorption region.

Recently, a considerable amount of attention has been paid to perovskite-type ABO_3 transition-metal oxides in the photocatalytic field thanks to its small band gap (2.0–2.5 eV) and high chemical stability [16–18]. Among the well-known ABO_3 perovskites, LaFeO3 (LFO) has been well-known as a visible-light-sensitive photocatalyst due to its suitable bandgap (~2.1 eV). Therefore, the construction of heterojunction composites based on TiO_2 modified with LFO has great potential to improve its photocatalytic performance, as this hybrid design enjoys a favorable alignment between band offset and energy levels [19]. In such an architecture, a narrow bandgap LFO could be easily activated by visible light. In addition, coupling with wide bandgap oxide TiO_2, heterostructures make the division of the desired device's functions (e.g., light absorption and long-range carrier transfer) between coupled materials more feasible [20].

In previous works, highly ordered TiO_2 nanotube array (TNT) film has been explored as an excellent substrate for photoanode due to its fast electron transmission, preferable mechanical properties, and large specific surface area [21–23]. On the other hand, the merit of using smaller band gap LFO nanoparticles is their activation by visible light. However, the synthesis of ABO_3 perovskites have generally been accomplished by solid-state reaction [24], sol-gel [25], and combustion methods [26] which require high temperature (>900 °C) that would destroy the nanotube structure of TNTs [27] and result in severe agglomeration of the micrometer-sized particles. To solve this issue, we have assessed wet chemical methods. By comparing diverse wet chemical routines, a hydrothermal method is one of the best synthetic methods for fabricate nano-materials. The main advantages of this strategy are its simplicity, controllable size, effective cost, and low temperature growth [28]. Therefore, it is a facile and appealing approach for the synthesis of LFO nanoparticle-modified TNTs (LFO/TNTs).

In this work, a LFO/TNTs heterojunction nanocomposite has been fabricated by an electrochemical anodization technique and subsequently by a hydrothermal approach process using a lanthanum citrate and ion citrate coordination complex. The as-prepared materials have been thoroughly characterized by scanning electron microscope (SEM), X-ray diffraction (XRD), X-ray photoelectron spectroscopy (XPS), UV-Vis diffuse reflectance spectroscopy (DRS), and photoluminescence (PL) spectra. The results of the photocatalytic experiment show that LFO/TNTs have outstanding photocatalytic activity of photodegradation of methylene blue (MB) compared to that of pure TNTs in the visible-light range.

2. Materials and Methods

The vertically-aligned and highly ordered TNTs were synthesized using a previously described method. Before anodization, 1×1.5 cm^2 Ti foil (purity > 99%, thickness of 0.1 mm) was soaked in a bath of acetone, alcohol, and distilled water for 15 min under ultrasonic conditions. The processed Ti foil was then dipped in the 0.5 wt % HF aqueous solution with applied voltage. The set-up for anodization was a simple two-electrode cell set at 20 V for 30 min at room temperature, in which Ti foil and Pt foil were applied as the working electrode and counter electrode, respectively. Subsequently, the sample was taken out of the HF solution and rinsed with deionized water immediately to get rid of the residual HF. Afterwards, the achieved amorphous TNTs were annealed in a muffle oven with a heating rate of 5 °C min^{-1} and kept at 450 °C for 2 h, and then cooled to room temperature naturally.

The LaFeO3 nanoparticle-modified TNTs were synthesized by the hydrothermal reaction carried out in a stainless steel autoclave. In a typical experiment, initially, based on the stoichiometric composition of the reactants, both lanthanum nitrate (La(NO3)3·6H2O) and ferric nitrate (Fe(NO3)3·9H2O) were dissolved in deionized water to achieve a mixed solution of 5 mM. Next, citric acid with mole ratio of 1:1:2 was added to the mixed solution as a stabilizer. Subsequently, the mixture was transferred into an autoclave with an as-prepared TNTs sample at the bottom, and kept at 160 °C for 8 h, 10 h, 12 h, and 14 h. After cooling, the sample was washed several times with deionized water, followed by annealing at 550 °C for 2 h to obtain LFO/TNTs. All chemicals and reactants used in this work were analytical reagents (AR).

The structure and morphology of the as-prepared samples were characterized by using a field emission scanning electron microscope (FESEM, Hitachi S4800, Hitachi, Tokyo, Japan). Crystal information was identified by X-ray diffraction (XRD, Phillips, Panalytical X'pert, PANalytical, B.V., the Netherlands, Cu Kα radiation (λ = 1.5417 Å)). The elemental composition was measured by X-ray photoelectron spectroscopy (XPS, VG, Physical Electrons Quantum 2000 Scanning Esca Microprobe, Al Kα radiation, Physical Electronics, Chanhassen, MN, USA). UV-visible diffuse reflectance spectra (DRS) of the as-prepared samples were analyzed using a UV-Vis-NIR spectrophotometer (Varian Cary 5000, Varian Medical Systems, Palo Alto, CA, USA). Photoluminescence (PL) spectra were recorded by a Hitachi F-7000 spectrophotometer (Hitachi F7000, Hitachi, Tokyo, Japan). All degeneration MB dye was performed under aerobic conditions in a home-made quartz glass reactor without bias voltage applied, whose temperature can be controlled through the equipped water jacket. During the degradation process, 3 mL of solution was periodically extracted from the reactor and detected by a UV-Vis spectrophotometer (Unico UV-2102 PC, Unico, Shanghai, China).

3. Results and Discussion

As shown in Figure 1a,b, the top and side view SEM images of pure TNTs demonstrate vertically aligned and highly ordered TiO_2 nanotubes with tube length of about 350 nm and an average tube diameter of around 100 nm. It can be seen from Figure 1b that the inner wall of the TNTs is clean and smooth and some thin, broken TiO_2 films cling to the top surface of the nanotubes. Figure 1c–f shows the SEM images of LFO/TNTs prepared with different hydrothermal reaction times. After loading LFO nanoparticles onto the TNTs, the interstitial space around the surface edge of the walls and between nanotubes is occupied by additional nanoparticles ranging in size from 30 to 50 nm. The LFO nanoparticles manifests as rounded corners with rectangular or rough ellipsoid shape, with a drastically increasing amount almost totally covering all the openings of the nanotubes following hydrothermal reaction of up to 14 h. Interestingly, when the hydrothermal reaction time is 12 h (LFO/TNTs-12h, Figure 1e), clusters of LFO nanoparticles are formed circling around the openings of the nanotubes, with some of them having permeated into the tubes (inset of Figure 1e).

Figure 2 shows the XRD patterns of TNTs, LFO/TNTs, and LFO powder. For TNTs, the diffraction peak at 2θ = 25.3 was attributed to the anatase TiO_2 (101) crystal planes. It is worth noting that the signal intensity of anatase TiO_2 is relatively weak, though the abundance of TiO_2 nanotubes fully covers the surface of Ti foil substrate. Yet, when it comes to LFO, no signal could be detected from LFO/TNTs at all. Nevertheless, the crystal structure and phase composition of LFO were characterized by the powder XRD patterns of the as-prepared sample collected from of the superfluous sediment in the Teflon-liner and annealed at 550 °C However, peaks corresponding to La_2O_3, Fe_2O_3, or other crystalline contaminations were not distinguished, indicating that the product is single phase LFO. We believe the subtlety of the signal of LFO in LFO/TNTs is due to the extremely high signal intensity of Ti foil substrate.

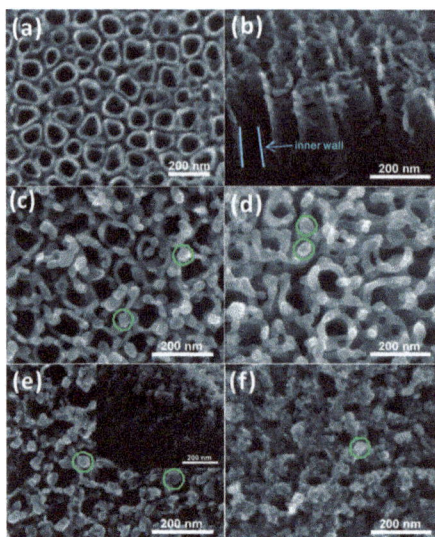

Figure 1. (**a**) Top and (**b**) side view SEM images of TiO₂ nanotube arrays (TNTs). Top view SEM images of LaFeO₃ nanoparticle-modified TNTs (LFO/TNTs) prepared by hydrothermal deposition for (**c**) 8 h, (**d**) 10 h, (**e**) 12 h, and (**f**) 14 h (green circles indicate LFO nanoparticles).

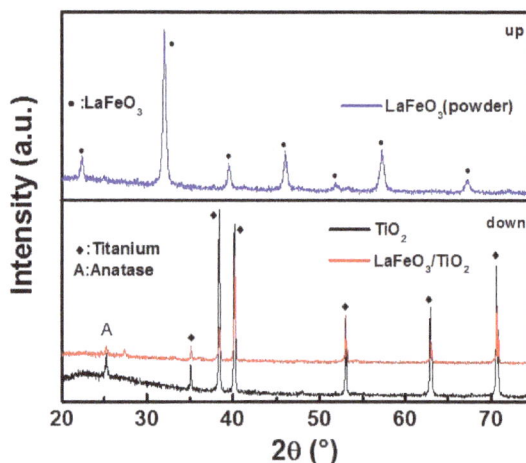

Figure 2. XRD patterns of TNTs, LFO/TNTs, and LaFeO₃ powders, respectively.

Further, the chemical states and elemental composition of LFO nanoparticles loaded on the TNTs were studied using XPS. The binding energy values for these components were adjusted by applying C 1s 284.4 eV peak as the reference. The ordinary XPS survey spectrum collected from the LFO/TNTs-12h shown in Figure 3a confirms the existence of C 1s, Fe 2p, La 3d, and O 1s. The signal of carbon can be attributed to the residual carbon on the surface of the sample. It can be observed from Figure 3c that the peaks located at 724.4 eV and 710.6 eV are attributed to the binding energies of Fe $2p_{1/2}$ and $2p_{3/2}$, respectively [29]. It is notable that shoulder peaks didn't appear in the Fe 2p spectra, which means that Fe primarily shows +3 oxidation state. The high resolution spectrum in Figure 3b shows the representative peaks of La $3d_{3/2}$ and La $3d_{5/2}$ emerging at 851.8 and 855.4 eV and at 834.6 and 838.8 eV, respectively. The difference of spin-orbit splitting between La $3d_{5/2}$ and La $3d_{3/2}$ is about 16.8 eV, which is in accordance with the previous report [30]. The line of the La 3*d* core-level revealed that the La primarily shows a valence state of +3 [29]. The XPS spectra further confirms the prepared nanoparticles to be stoichiometric LaFeO$_3$. The wide and asymmetric O 1*s* XPS spectra in Figure 3d indicates that there are two different chemical states of O, including the anionic oxygen in the TiO$_2$ lattice (O1*s* (1) dominant peak at 529.9 eV) [31] and the chemisorbed oxygen species (O 1*s* (2) peak at 531.54 eV). The O 1*s* binding energy of the lattice oxygen is generally lower than that of the O^{2-} or OH$^-$ species by 2.1–2.5 eV. Lanthanum oxide is known to be hygroscopic, the corresponding higher O 1*s* peak (531.54 eV) is attributed to the absorbed water molecule. The OH-XPS signal observed for the LFO nanoparticles is associated with the adsorbed water species [32].

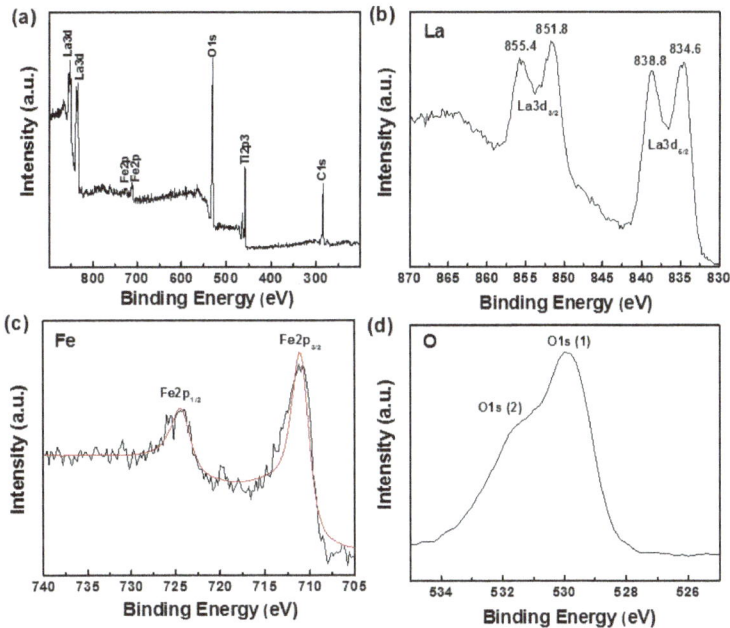

Figure 3. XPS spectra of LaFeO$_3$/TiO$_2$ nanotube arrays: (**a**) a survey XPS spectrum and high resolution spectra of Fe (**b**), La (**c**), and O (**d**).

The DRS spectra of TNTs, LFO/TNTs-8h, LFO/TNTs-10h, LFO/TNTs-12h, and LFO/TNTs-14h are shown in Figure 4a. For the TNTs, the onset position of light absorption is around 380 nm with a band gap of 3.2 eV. Several satellite peaks can be found in the visible region which is supposed to be caused by the sub-bandgap states of the TNTs [33]. Compared to TNTs, the series of LFO/TNTs exhibits an absorption edge that is obviously red-shifted with a clear absorption of visible light. Another way to make this conclusion is simply by visual observation of these samples, as their color regularly turns from violet to yellowish-green. This means that the light absorption of modified samples is expanded to the visible region of the solar spectrum (up to λ = 550 nm), which is a fundamental condition for the fabrication of visible light-driven photocatalysts. By changing the hydrothermal reaction time (from 8 to 14 h), the absorption edge of each LFO/TNTs changes slightly, with LFO/TNTs-12h exhibiting the widest absorption in the visible region. Figure 4b shows the $(\alpha h\nu)^{1/2}$–$h\nu$ curve derived from the equation $(\alpha h\nu)^2 = k(h\nu - E_g)$, where E_g, ν, h, and α represent the band energy, light frequency, Planck constant, and adsorption coefficient, respectively [34]. It is clear that the band gap of the LFO/TNTs composites decreased sharply from 3.2 to 2.1 eV after the deposition of LFO nanoparticles on the TNTs, thanks to the narrow band gap of LFO (2.1 eV) for sensitizing visible light irradiation.

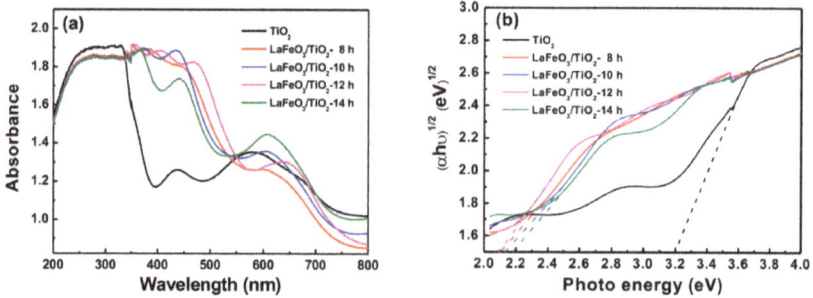

Figure 4. (**a**) DRS spectra and (**b**) $(\alpha h\upsilon)^{1/2}$–$h\upsilon$ of TiO$_2$ nanotube arrays and LaFeO$_3$/TiO$_2$ nanotube arrays with different hydrothermal reaction time.

Photoluminescence (PL) spectroscopy is a powerful tool to study the behavior of charge carrier, such as trapping, immigration, and transfer, as well as to comprehend the processes concerning the recombination of electron–hole pairs [35]. Briefly, upon irradiation, electron–hole pairs experience a recombination process to emit photons and then produce PL [36]. In this work, Figure 5 shows the PL spectra of TNTs and LFO/TNTs composites (excited at 325 nm) in the wavelength range of 400–600 nm. For pristine TNTs, a broad-band emission peak around 450 nm may be related to the recombination of photoexcited electron–hole pairs occupying the singly ionized oxygen vacancies in TiO$_2$. The shoulder peak observed at around 465 nm is attributed to the oxygen vacancies [37]. Besides, compared to pure TNTs, all LFO/TNTs samples exhibit lower PL intensity, suggesting a higher photocatalytic activity due to the slower recombination rate of electron–hole pairs.

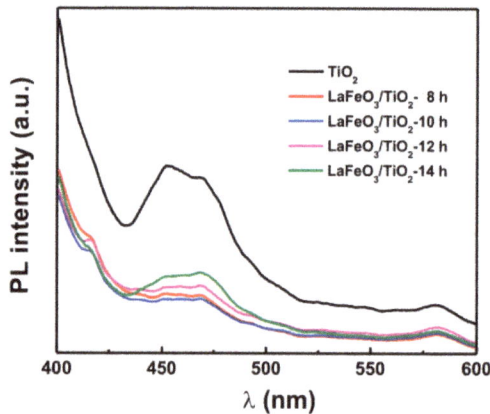

Figure 5. PL spectra of TiO$_2$ nanotube arrays and LaFeO$_3$/TiO$_2$ nanotube arrays.

The photocatalytic activities of pure TNTs and LFO/TNTs composites with different hydrothermal reaction times were estimated by observing a photodegradation of the MB aqueous solution (10 mg L^{-1}) under visible light irradiation and the photodegradation kinetics of the MB dye are shown in Figure 6. All samples show a photocatalytic reaction following a pseudo first-order reaction mechanism, which could be simulated as $\ln(C/C_0) = -kt$, where C_0, C, and t are the initial and reaction concentrations and reaction time of the MB aqueous solution, respectively. k is the apparent first-order reaction constant. It can be seen from Figure 6 that pure TNT (0.00073 min^{-1}) showed no better photocatalytic activity than MB self-degradation (0.00068 min^{-1}) under visible light, indicating that TNTs have little effect on accelerating the decomposition rate of MB dye under visible light illumination. Unlike the tiny amount of MB removal by TNTs, in contrast, the degradation efficiency of the LFO/TNTs samples under the same experiment conditions was much higher, which reveals that LFO/TNTs samples are readily excited under light irradiation, and photogenerated holes accelerate the oxidation of MB dye before recombination with electrons. This result is consistent with our previous work [38]. Obviously, on this occasion, the optimal hydrothermal reaction time is 12 h, and the photocatalytic activity (0.00208 min^{-1}) was triple that of the TNTs. In spite of the LFO/TNTs-14 h sample having a higher amount of LFO than LFO/TNTs-12h, the reason for why LFO/TNTs-12h shows the best photodegradation performance other than LFO/TNTs-14h is that the LFO/TNTs-12h sample can absorb visible light effectively without having the opening of TiO$_2$ TNTs blocked. Whereas, according to the SEM results, a 14 h reaction time causes excessive accumulation of LaFeO$_3$ nanoparticles, thus blocking the openings of TiO$_2$ TNTs and leading to the loss of the high surface ratio of the nanotube arrays.

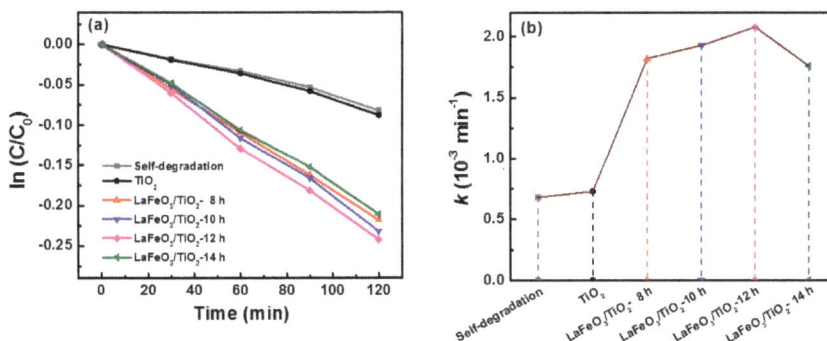

Figure 6. (**a**) Photodegradation rate of MB for TiO$_2$ nanotube arrays and LaFeO$_3$/TiO$_2$ nanotube arrays prepared with different hydrothermal reaction times; (**b**) The reaction constant *k* of each sample.

In this paper, the concepts of electronegativity were introduced to estimate the valence band edge position of FLO/TiO$_2$ heterostructures. Therefore, to calculate the valence band and conduction band edges of FLO/TiO$_2$ composites at the point of zero charge requires two formulas $E_{VB} = X - E^e + 0.5E_g$ and $E_{CB} = E_{VB} - E_g$. [39–41], where E_g, E^e, and X represent the band gap energy of the semiconductor, the energy of free electrons (about 4.5 eV), and the electronegativity of the semiconductor, respectively, which is the geometric mean of the electronegativity of the integral atoms. The X values for TiO$_2$ and LFO are calculated to be ca. 5.8 and 5.7 eV, and the E_{VB} of TiO$_2$ and LFO can be achieved for 2.9 eV and 2.3 eV after substituting X into the equation, respectively. Therefore, the conduction band potentials of TiO$_2$ and LFO were calculated to be −0.3 eV and 0.2 eV, respectively. Figure 7 shows the band gap structure and possible charge carriers transfer between LFO and TiO$_2$ under visible light illumination. In general, LFO shows a feature of p-type semiconductor with the corresponding Fermi position near to the valence band [42], while TiO$_2$ is known as an n-type semiconductor whose Fermi position lies next to the conduction band [43], as shown in Figure 7a. After coupling TiO$_2$ with LFO to form the p-n heterojunction, the Fermi level of LFO is lifted up, while the Fermi level of TiO$_2$ is dragged down, until they are at the same level and reach equilibrium (Figure 7b).

Meanwhile, with the increase/decrease of the Fermi position, the complementary energy bands of LFO and TiO$_2$ move in the opposite directions, i.e., rising for LFO and lowering for TiO$_2$, resulting in the conduction band position of p-type LFO being higher than that of n-type TiO$_2$. Thus, the electron–hole pairs have been effectively separated due to this superior energy band structure, resulting in enhanced photocatalytic efficiency under visible-light irradiation. Upon visible light illumination, the electrons of LFO are excited and flow from the valence band (VB) to the conduction band (CB), creating electron–hole pairs. The holes left on the VB of LFO could react with the surface OH$^-$ to yield ·OH, while the excited electrons could effectively transfer to CB of TiO$_2$ and then convert dissolved O_2 to $\cdot O_2{}^-$. Both $\cdot OH$ and $\cdot O_2{}^-$ are powerful oxidative species that can efficiently decompose MB dye.

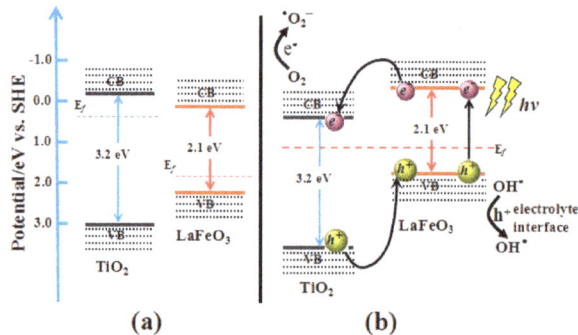

Figure 7. Schematic diagram showing the separation and transfer of charge carriers of p-LaFeO$_3$ and n-TiO$_2$ under visible light irradiation: (**a**) before and (**b**) after Fermi level reach equilibrium.

4. Conclusions

In summary, a p-n heterojunction LFO/TNTs photocatalyst was synthesized by a facile combination of electrochemical anodization and a hydrothermal method. The as-prepared LFO/TNTs could not only harvest UV light, but could also absorb visible light. The LFO/TNTs composites exhibit high photodegradation thanks to the construction of a p–n heterojunction between LFO and TiO$_2$ which can efficiently constrain the recombination of electron–hole pairs and facilitate rapid photoexcited electron transfer from p-type LFO to n-type TNTs. Besides, the advantages of easy separation from the polluted solution without second contamination and re-usability make the prepared LFO/TNTs an excellent photocatalyst for commercial applications. Besides LFO, this general strategy can also provide an effective tool for fabricating other ABO$_3$ perovskite semiconductor nanoparticle-modified TiO$_2$ nanotube array composites which exhibit promising prospects in other areas, such as water splitting and energy storage, etc.

Author Contributions: Conceptualization, Y.T., L.S. and Y.L.; Data Curation, J.Y. and Z.Z.; Formal Analysis, S.X., J.H., Y.T., C.L. and Y.L.; Investigation, J.Y.; Methodology, J.Y. and M.G.; Project Administration, J.Y. and L.S.; Supervision, L.S. and C.L.; Validation, Z.Z.; Writing–Original Draft, J.Y.; Writing–Review & Editing, S.X., M.G., J.H., Y.T., L.S., C.L. and Y.L.

Funding: This research was funded by National Natural Science Foundation of China (Nos. 21621091 and 21501127) and Guangdong Natural Science Foundation (No. 2016A030313845).

Conflicts of Interest: The authors declare no conflict of interest.

References

1. Wang, M.; Iocozzia, J.; Sun, L.; Lin, C.; Lin, Z. Inorganic-modified semiconductor TiO$_2$ nanotube arrays for photocatalysis. *Energy Environ. Sci.* **2014**, *7*, 2182–2202. [CrossRef]
2. Tang, Y.; Jiang, Z.; Xing, G.; Li, A.; Kanhere, P.D.; Zhang, Y.; Sum, T.C.; Li, S.; Chen, X.; Dong, Z.; et al. Efficient Ag@AgCl cubic cages photocatalyst profited from ultrafast plasmon-induced electron transfer process. *Adv. Funct. Mater.* **2013**, *23*, 2932. [CrossRef]
3. Tang, Y.; Zhang, Y.; Malyi, O.I.; Bucher, N.; Xia, H.; Xi, S.; Zhu, Z.; Lv, Z.; Li, W.; Wei, J.; et al. Identifying the origin and contribution of surface storage in TiO$_2$(B) nanotube electrode by in-situ dynamic valence state monitoring. *Adv. Mater.* **2018**, *30*, 1802200. [CrossRef] [PubMed]
4. Riboni, F.; Nguyen, N.T.; So, S.; Schmuki, P. Aligned metal oxide nanotube arrays: Key-aspects of anodic TiO$_2$ nanotube formation and properties. *Nanoscale Horizons* **2016**, *1*, 445–466. [CrossRef]

5. Ge, M.-Z.; Cao, C.-Y.; Li, S.-H.; Tang, Y.-X.; Wang, L.-N.; Qi, N.; Huang, J.-Y.; Zhang, K.-Q.; Al-Deyab, S.S.; Lai, Y.K. In situ plasmonic Ag nanoparticle anchored TiO$_2$ nanotube arrays as visible-light-driven photocatalysts for enhanced water splitting. *Nanoscale* **2016**, *8*, 5226–5234. [CrossRef] [PubMed]

6. Cai, J.; Shen, J.; Zhang, X.; Ng, Y.H.; Huang, J.; Guo, W.; Lin, C.; Lai, Y. Light-driven sustainable hydrogen production utilizing TiO$_2$ nanostructures: A review. *Small Methods* **2018**, *2*, 1800184. [CrossRef]

7. Ye, M.; Gong, J.; Lai, Y.; Lin, C.; Lin, Z. High-efficiency photoelectrocatalytic hydrogen generation enabled by palladium quantum dots-sensitized TiO$_2$ nanotube arrays. *J. Am. Chem. Soc.* **2012**, *134*, 15720–15723. [CrossRef] [PubMed]

8. Macak, J.M.; Zlamal, M.; Krysa, J.; Schmuki, P. Self-organized TiO$_2$ nanotube layers as highly efficient photocatalysts. *Small* **2007**, *3*, 300–304. [CrossRef] [PubMed]

9. Zhang, G.; Huang, H.; Zhang, Y.; Chan, H.L.; Zhou, L. Highly ordered nanoporous TiO$_2$ and its photocatalytic properties. *Electrochem. Commun.* **2007**, *9*, 2854–2858. [CrossRef]

10. Ge, M.; Cao, C.; Huang, J.; Li, S.; Chen, Z.; Zhang, K.-Q.; Al-Deyab, S.S.; Lai, Y. A review of one-dimensional TiO$_2$ nanostructured materials for environmental and energy applications. *J. Mater. Chem. A* **2016**, *4*, 6772–6801. [CrossRef]

11. Ge, M.; Li, Q.; Cao, C.; Huang, J.; Li, S.; Zhang, S.; Chen, Z.; Zhang, K.; Al-Deyab, S.S.; Lai, Y. One-dimensional TiO$_2$ nanotube photocatalysts for solar water splitting. *Adv. Sci.* **2017**, *4*, 1600152. [CrossRef] [PubMed]

12. Kowalski, D.; Kim, D.; Schmuki, P. TiO$_2$ nanotubes, nanochannels and mesosponge: Self-organized formation and applications. *Nano Today* **2013**, *8*, 235–264. [CrossRef]

13. Lai, Y.; Huang, J.; Zhang, H.; Subramaniam, V.-P.; Tang, Y.-X.; Gong, D.-G.; Sundarb, L.; Sun, L.; Chen, Z.; Lin, C.J. Nitrogen-doped TiO$_2$ nanotube array films with enhanced photocatalytic activity under various light sources. *J. Hazard. Mater.* **2010**, *184*, 855–863. [CrossRef] [PubMed]

14. Ge, M.Z.; Cai, J.S.; Iocozzia, J.; Cao, C.; Huang, J.; Zhang, X.; Shen, J.; Wang, S.; Zhang, S.; Zhang, K.-Q.; et al. A review of TiO$_2$ nanostructured catalysts for sustainable H$_2$ generation. *Int. J. Hydrog. Energy* **2017**, *42*, 8418–8449. [CrossRef]

15. Li, H.; Zhou, Y.; Tu, W.; Ye, J.; Zou, Z. State-of-the-art progress in diverse heterostructured photocatalysts toward promoting photocatalytic performance. *Adv. Funct. Mater.* **2015**, *25*, 998–1013. [CrossRef]

16. Jia, L.; Li, J.; Fang, W. Enhanced visible-light active C and Fe co-doped LaCoO$_3$ for reduction of carbon dioxide. *Catal. Commun.* **2009**, *11*, 87–90. [CrossRef]

17. Wheeler, G.P.; Choi, K.-S. photoelectrochemical properties and stability of nanoporous p-Type LaFeO$_3$ photoelectrodes prepared by electrodeposition. *ACS Energy Lett.* **2017**, *2*, 2378–2382. [CrossRef]

18. Dhinesh Kumar, R.; Thangappan, R.; Jayavel, R. Enhanced visible light photocatalytic activity of LaMnO$_3$ nanostructures for water purification. *Res. Chem. Intermed.* **2018**, *44*, 4323–4337. [CrossRef]

19. Truppi, A.; Petronella, F.; Placido, T.; Striccoli, M.; Agostiano, A.; Curri, M.L.; Comparelli, R. Visible-light-active TiO$_2$-based hybrid nanocatalysts for environmental applications. *Catalysts* **2017**, *7*, 100. [CrossRef]

20. Farhadi, S.; Amini, M.M.; Mahmoudi, F. Phosphotungstic acid supported on aminosilica functionalized perovskite-type LaFeO$_3$ nanoparticles: A novel recyclable and excellent visible-light photocatalyst. *RSC Adv.* **2016**, *6*, 102984–102996. [CrossRef]

21. Guo, M.; Xie, K.; Lin, J.; Yong, Z.; Zhou, L.; Wang, Y.; Huang, H. Design and coupling of multifunctional TiO$_2$ nanotube photonic crystal to nanocrystalline titania layer as semi-transparent photoanode for dye-sensitized solar cell. *Energy Environ. Sci.* **2012**, *5*, 9881–9888. [CrossRef]

22. Nguyen, N.T.; Altomare, M.; Yoo, J.E.; Taccardi, N.; Schmuki, P. Noble metals on anodic TiO$_2$ nanotube mouths: Thermal dewetting of minimal Pt Co-catalyst loading leads to significantly enhanced photocatalytic H$_2$ generation. *Adv. Energy Mater.* **2016**, *6*, 1501926. [CrossRef]

23. Xie, K.; Sun, L.; Wang, C.; Lai, Y.; Wang, M.; Chen, H.; Lin, C. Photoelectrocatalytic properties of Ag nanoparticles loaded TiO$_2$ nanotube arrays prepared by pulse current deposition. *Electrochim. Acta* **2010**, *55*, 7211–7218. [CrossRef]

24. Zhou, K.; Wu, X.; Wu, W.; Xie, J.; Tang, S.; Liao, S. Nanocrystalline LaFeO$_3$ preparation and thermal process of precursor. *Adv. Powder Technol.* **2013**, *24*, 359–363.

25. Parida, K.M.; Reddy, K.H.; Martha, S.; Das, D.P.; Biswal, N. Fabrication of nanocrystalline LaFeO$_3$: An efficient sol-gel auto-combustion assisted visible light responsive photocatalyst for water decomposition. *Int. J. Hydrogen Energy* **2010**, *35*, 12161–12168. [CrossRef]

26. Wang, Y.; Zhu, J.; Zhang, L.; Yang, X.; Lu, L.; Wang, X. Preparation and characterization of perovskite LaFeO$_3$ nanocrystals. *Mater. Lett.* **2006**, *60*, 1767–1770. [CrossRef]

27. Yu, J.; Wang, B. Effect of calcination temperature on morphology and photoelectrochemical properties of anodized titanium dioxide nanotube arrays. *Appl. Catal. B* **2010**, *94*, 295–302. [CrossRef]

28. Phoon, B.L.; Lai, C.W.; Pan, G.T.; Yang, T.C.K.; Juan, J.C. One-pot hydrothermal synthesis of strontium titanate nanoparticles photoelectrode using electrophoretic deposition for enhancing photoelectrochemical water splitting. *Ceram. Int.* **2018**, *44*, 9923–9933. [CrossRef]

29. Flynn, B.T.; Zhang, L.; Shutthanandan, V.; Varga, T.; Colby, R.J.; Oleksak, R.P.; Manandhar, S.; Engelhard, M.H.; Chambers, S.A.; Henderson, M.A.; et al. Growth and surface modification of LaFeO$_3$ thin films induced by reductive annealing. *Appl. Surf. Sci.* **2015**, *330*, 309–315. [CrossRef]

30. Zhang, C.; Wang, C.; Zhan, W.; Guo, Y.; Guo, Y.; Lu, G.; Baylet, A.; Giroir-Fendler, A. Catalytic oxidation of vinyl chloride emission over LaMnO$_3$ and LaB$_{0.2}$Mn$_{0.8}$O$_3$ (B = Co, Ni, Fe) catalysts. *Appl. Catal. B* **2013**, *129*, 509–516. [CrossRef]

31. Simmons, G.W.; Beard, B.C. Characterization of acid-base properties of the hydrated oxides on iron and titanium metal surfaces. *J. Phys. Chem.* **1987**, *91*, 1143–1148. [CrossRef]

32. Thirumalairajan, S.; Girija, K.; Hebalkar, N.Y.; Mangalaraj, D.; Viswanathan, C.; Ponpandian, N. Shape evolution of perovskite LaFeO$_3$ nanostructures: A systematic investigation of growth mechanism, properties and morphology dependent photocatalytic activities. *RSC Adv.* **2013**, *3*, 7549–7561. [CrossRef]

33. Lai, Y.; Sun, L.; Chen, Y.; Zhuang, H.; Lin, C.; Chin, J.W. Effects of the structure of TiO$_2$ nanotube array on Ti substrate on its photocatalytic activity. *J. Electrochem. Soc.* **2006**, *153*, D123–D127. [CrossRef]

34. Vijayan, B.K.; Dimitrijevic, N.M.; Wu, J.; Gray, K.A. The effects of Pt doping on the structure and visible light photoactivity of titania nanotubes. *J. Phys. Chem. C* **2010**, *114*, 21262–21269. [CrossRef]

35. Yamashita, H.; Harada, M.; Misaka, J.; Takeuchi, M.; Neppolian, B.; Anpo, M. Photocatalytic degradation of organic compounds diluted in water using visible light-responsive metal ion-implanted TiO$_2$ catalysts: Fe ion-implanted TiO$_2$. *Catal. Today* **2003**, *84*, 191–196. [CrossRef]

36. Zhang, N.; Liu, S.; Fu, X.; Xu, Y.J. Synthesis of M@TiO$_2$ (M = Au, Pd, Pt) core-shell nanocomposites with tunable photoreactivity. *J. Phys. Chem. C* **2011**, *115*, 9136–9145. [CrossRef]

37. Lei, Y.; Zhang, L.D.; Meng, G.W.; Li, G.H.; Zhang, X.Y.; Liang, C.H.; Zhang, X.Y.; Liang, C.H.; Chen, W.; Wang, S.X. Preparation and photoluminescence of highly ordered TiO$_2$ nanowire arrays. *Appl. Phys. Lett.* **2001**, *78*, 1125–1127. [CrossRef]

38. Xiang, S.; Zhang, Z.; Gong, C.; Wu, Z.; Sun, L.; Ye, C.; Lin, C. LaFeO$_3$ nanoparticle-coupled TiO$_2$ nanotube array composite with enhanced visible light photocatalytic activity. *Mater. Lett.* **2018**, *216*, 1–4. [CrossRef]

39. Zhang, X.; Zhang, L.; Xie, T.; Wang, D. Low-temperature synthesis and high visible-light-induced photocatalytic activity of BiOI/TiO$_2$ heterostructures. *J. Phys. Chem. C* **2009**, *113*, 7371–7378. [CrossRef]

40. Lin, X.; Xing, J.; Wang, W.; Shan, Z.; Xu, F.; Huang, F. Photocatalytic activities of heterojunction semiconductors Bi$_2$O$_3$/BaTiO$_3$: A strategy for the design of efficient combined photocatalysts. *J. Phys. Chem. C* **2007**, *111*, 18288–18293. [CrossRef]

41. Butler, M.A.; Ginley, D.S. Prediction of flatband potentials at semiconductor-electrolyte interfaces from atomic electronegativities. *J. Electrochem. Soc.* **1978**, *125*, 228–232. [CrossRef]

42. Fan, H.; Zhang, T.; Xu, X.; Lv, N. Fabrication of N-type Fe$_2$O$_3$ and P-type LaFeO$_3$ nanobelts by electrospinning and determination of gas-sensing properties. *Sens. Actuators B Chem.* **2011**, *153*, 83–88. [CrossRef]

43. Khan, M.; Al-Shahry, M.; Ingler, W.B. Efficient photochemical water splitting by a chemically modified n-TiO$_2$. *Science* **2002**, *297*, 2243–2245. [CrossRef] [PubMed]

coatings

MDPI

Article

Mixed Nickel-Cobalt-Molybdenum Metal Oxide Nanosheet Arrays for Hybrid Supercapacitor Applications

Yin She [1,2], Bin Tang [3], Dongling Li [1,2], Xiaosheng Tang [1,2], Jing Qiu [1], Zhengguo Shang [1,2] and Wei Hu [1,*]

[1] Key Laboratory of Optoelectronic Technology and System of Ministry of Education, College of Optoelectronic Engineering, Chongqing University, Chongqing 400044, China; sheyin@cqu.edu.cn (Y.S.); lidongling@cqu.edu.cn (D.L.); xstang@cqu.edu.cn (X.T.); jingqiu@cqu.edu.cn (J.Q.); zhengry@cqu.edu.cn (Z.S.)
[2] Key Laboratory of Fundamental Science Micro/Nano Device System Technology, Micro System Research Center of Chongqing University, Chongqing 400044, China
[3] Institute of Electronic Engineering, China Academy of Engineering Physics, Mianyang 621999, China; tangbin@caep.cn
* Correspondence: weihu@cqu.edu.cn; Tel.: +86-138-9611-1800

Received: 16 July 2018; Accepted: 19 September 2018; Published: 25 September 2018

Abstract: Mixed metal oxide nanomaterials have been demonstrated to be promising positive electrodes for energy storage applications because of the synergistic enhancement effects. In this work, nickel-cobalt-molybdenum metal oxide (NCMO) nanosheets with hierarchical, porous structures were directly developed on nickel foam (NF) through a hydrothermal method and ensuing annealing treatment. Electrochemical tests in three-electrode configurations revealed that the as-prepared NCMO nanosheets possessed high specific capacitance (1366 F g^{-1} at the current density of 2 A g^{-1}), good rate capability (71.3% at the current density of 40 A g^{-1}), as well as excellent cycling stability (89.75% retention after 5000 cycles). Additionally, a hybrid supercapacitor was assembled and achieved an energy density of 46.2 Wh kg^{-1} at a power density of 713 W kg^{-1}. Based on the systematic analysis of microstructure, morphology, and element compositions, the excellent electrochemical performance of the NCMO nanosheets could be attributed to the mesoporous feature, desirable compositions, excellent mechanical and electrical contacts, and fast ion/electron transportation rates. This study shows that the NCMO nanosheets offer great potentials for application in supercapacitors.

Keywords: mixed metal oxides; nanosheet arrays; nickel-cobalt-molybdenum metal oxide (NCMO); supercapacitor; energy storage

1. Introduction

High performance and environmentally friendly electrochemical energy storage devices and systems have received widespread attention because of the huge demands for clean, efficient, and sustainable energy in recent decades [1–4]. Among the myriad available power sources, supercapacitors, with complementary characteristics of lithium-ion batteries and electrostatic capacitors, have sparked a dramatic expansion of academic research and industrial interests [5–7]. Especially, hybrid supercapacitors, consisting of a pseudocapacitor electrode and an electric double layer capacitor electrode, have aroused more and more attention since they possess the merits of both types of electrodes and display average properties of battery-level energy density and high power density, long cycling stability, and short charging time of conventional capacitors [8–11].

It is well acknowledged that a pseudocapacitor electrode plays a key role in achieving high-performance hybrid supercapacitors, which mainly collect the charges from the Faradic reactions that have occurred on the electrode surface. Thus, considerable efforts have been devoted to developing pseudocapacitor electrode materials that undergo abundant redox reactions [12–14]. Given the properties of richer reaction sites, higher electrochemical activities and electronic conductivities compared with the corresponding binary oxides, sulfides, and hydroxides, ternary metal oxides along with ternary metal sulfides and hydroxides have been widely explored [15–20]. For instance, Guan and co-workers fabricated Co-Mn double hydroxides by an electro-deposition, which showed capacitance of 1062.6 F g^{-1} at 0.7 A g^{-1} and cyclability of 96.3% after 5000 cycles [21]. Luo's group synthesized mesoporous CuCo$_2$O$_4$ nano-grasses on copper foam, showing a specific capacitance of 796 F g^{-1} at 2 A g^{-1} and a cyclic stability of 94.7% after 5000 cycles [22]. Rout et al. developed nickel cobalt sulfide ultrathin nanosheets by electro-deposition method. When used as the pseudo-electrodes, the prepared nickel-cobalt sulfide nanosheets displayed capacitances of 1712 F g^{-1} at 1 A g^{-1} and retained 87% after 3000 cycles [23]. The previous studies indicated that the ternary metal sulfide electrodes displayed outstanding specific capacitance, but suffered from poor cycling stability. While the ternary metal oxides and hydroxides electrodes displayed high cycling stability, the specific capacitances were comparatively low because of their poor intrinsic charge/ionic conductivity and the diffusion-controlled process limitation.

More recently, mixed metal oxides, composed of three or more metal elements with similar radii, are expected to have a stronger synergetic effect in comparison with the corresponding binary and ternary metal oxides [24–26]. Tu's group synthesized spinel Mn-Ni-Co ternary oxide nanowires on nickel foam (NF) for supercapacitor application, which showed high specific capacitance together with good cyclability [27]. Lee et al. fabricated Ni-Co-Mn metal oxide nanoparticles on reduced graphene oxide and intensively studied the high capacitance mechanism of the mixed metal oxide [28]. In our previous work, we have synthesized Ni-Zn-Co oxide nanowire arrays through a hydrothermal reaction, which showed high capacitance and long-term cycling stability [29].

By taking advantage of the synergetic effect to improve electrochemical activity and stability, it is of great significance to explore and achieve high-performance pseudocapacitor electrodes in mixed metal oxide nanostructures. Though ternary CoMoO$_4$ and NiMoO$_4$ nanomaterials and heterostructures have been intensively investigated, there are few reports on mixed nickel-cobalt-molybdenum metal oxide (NCMO) for supercapacitor applications. Herein, we present a facile synthesis of mixed NCMO nanosheets on NF through the combination of a hydrothermal method and ensuing annealing treatment. The NCMO nanosheets manifested exceptional electrochemical performances, including high specific capacitance and good cycling stability, owing to the following features: (1) the ultrathin nanosheets structure and mesoporous morphology provided large specific surface area and in turn enlarge the utilization of the electrode materials and further facilitate the irrigation of electrolyte, prompting Faradic redox reactions; (2) the molybdenum element showed intercalation pseudocapacitive behaviour through reversible intercalation/deintercalation; (3) the poor crystallization resulted in an increase in defect transportations in NCMO nanosheets compared to a material with good crystallinity; (4) moreover, the mesopores on the nanosheets served as "ion-buffering reservoirs" and brought about short effective diffusion paths [30–33]. In a word, the NCMO nanosheets are expected to act as high-performance pseudocapacitor electrode materials for hybrid supercapacitor applications.

2. Materials and Methods

2.1. Materials Preparation

Firstly, 1 mmol Ni(NO$_3$)$_2$·6H$_2$O, 1 mmol Co(NO$_3$)$_2$·6H$_2$O, 1 mmol Na$_2$MoO$_4$·2H$_2$O, 6 mmol NH$_4$F together with 15 mmol CO(NH$_2$)$_2$ were simultaneously dissolved in 70 mL deionized (DI) water and stirred for 1 h under room temperature. After then, we transformed the forming solution into a 100 mL Teflon-lined stainless steel autoclave with a rectangular NF (3 cm × 5 cm) immersed in it. It is worth noting that a sheet of polytetrafluoroethylene tape was adhered onto the top side of the NF in order to avoid the solution invasion. Subsequently, we kept the sealed autoclave in an oven at 120 °C for 5 h. After the autoclave cooled down to room temperature, we obtained the Ni-Co-Mo precursor and washed it by DI water and ethanol for many times, followed by drying at 80 °C for 6 h. Finally, an annealing treatment was conducted under conditions of 350 °C for 2 h in an air atmosphere with a ramp rate of 2 °C min^{-1}. The mass loading of the active materials on NF was about 1.2 mg cm^{-2}.

2.2. Hybrid Supercapacitor Fabrication

A hybrid supercapacitor was composed of NCMO nanosheets as the positive electrode and the activated carbon (AC) mixture as the negative electrode. The negative electrode was fabricated by the following two steps: (1) Initially, a mixture consisting of activated carbon, acetylene black, and polyvinylidene fluoride with a mass ratio of 8:1:1 was poured into the *N*-methyl-2-pyrrolidone solvent to form a slurry; (2) then we coated the slurry on a rectangular shaped NF and dried at 80 °C for 12 h. Based on the charge balance theory, the mass balancing of both negative and positive electrodes can be optimized as the relationship: $\frac{m_+}{m_-} = \frac{C_- \times \Delta E_-}{C_+ \times \Delta E_+}$, where C_+ and C_- stand for the specific capacitances of the positive and negative electrodes, respectively. ΔE is the potential range [34–37].

2.3. Materials Characterization

An X-ray diffractometer (XRD; RIGAKU D-MAX 2200; Cu Kα radiation, Tokyo, Japan) was utilized to characterize the crystal phases. Scanning electron microscopy (SEM; QUANTA 400F, Thermo Fisher Scientific, Waltham, MA, USA) and transmission electron microscopy (TEM; JEOL JEM-2010HR, JEOL Ltd., Tokyo, Japan) were performed to check the structures and morphologies. X-ray photoelectron spectroscopy (XPS; ESCALAB 250, Al Kα radiation, Thermo Fisher Scientific, Waltham, MA, USA) was applied to investigate the chemical states. A surface area analyzer (Micromeritics ASAP 2020, Norcross, GA, USA) was employed to investigate the surface characteristics via nitrogen adsorption/desorption measurements.

2.4. Measurements of Electrochemical Performance

The electrochemical measurements were implemented by a three-electrode electrochemical setup. The as-fabricated Ni-Co-Mo oxide nanosheet arrays, a platinum foil, a saturated calomel electrode (SCE), and 2.0 M KOH solution were applied as the working electrode, the counter electrode, the reference electrode, and the electrolyte, respectively. Cyclic voltammetry (CV) and galvanostatic charge/discharge (GCD) tests, electrochemical impedance spectroscopy (EIS), and cycling characteristics were performed on an electrochemical workstation (CHI 660E, Chenhua Instrument, Shanghai, China). The specific capacitance (F g^{-1}), areal capacitance (F cm^{-2}), energy density (Wh kg^{-1}), and power density (W kg^{-1}) could be calculated from the galvanostatic discharge curves in light of the following equation: $C_s = \frac{I \times \Delta t}{V \times m}$; $C_a = \frac{I \times \Delta t}{V \times S}$; $E = \frac{1}{2} \times C_s \times V^2$; $P = E/\Delta t$, where I, Δt, m, V, and S are designated as the discharge current and time, the mass, the potential window, and the area of active electrode, respectively [35].

3. Results and Discussion

3.1. Structure and Morphology Characterizations

The crystal structure of the obtained product scratched from NF was examined by XRD, and the corresponding pattern is shown in Figure 1. The observed diffraction peaks at the degrees of 36.6 and 62.9 could be indexed to the (202) and (260) planes of the cubic phase of NiMoO$_4$ (JCPDF card No. 45-0142), and the other diffraction peaks can be indexed to CoMoO$_4$ (JCPDF card No. 21-0868) [36,37]. Moreover, the peak intensities are relatively weak and broad, which indicates the poor crystallization of the NCMO nanosheets. The poor crystallization is beneficial for increased electrochemical performance because more defect transportations are present in materials with weak crystallinity [38]. It is also worth noting that, differing from the previous reported Ni-Co-Mn oxides, Ni-Co-Zn oxides, and other mixed metal oxide nanostructures prepared by a similar hydrothermal method, the hydrothermal processed NCMO nanosheets do not show the spinel structure. It could be caused by the substitution of Mo, which affected the spinel crystal structure significantly [26,27,39].

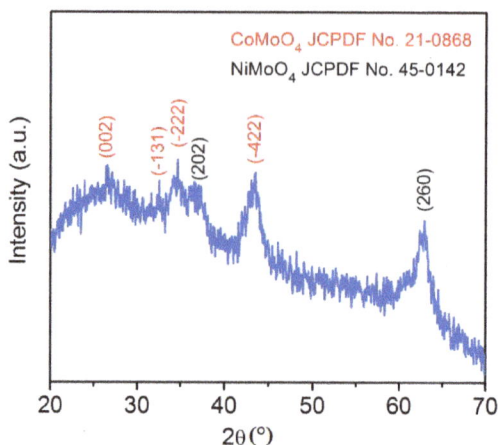

Figure 1. X-ray diffraction (XRD) pattern of the nickel-cobalt-molybdenum metal oxide (NCMO) nanosheets sample.

From the SEM images in Figure 2a–c, it can be seen that the crinkly NCMO nanosheets are uniformly and continuously deposited on NF. The NCMO nanosheets interlace with each other to arrange vertically on the substrate, forming an open-up network structure. The average thickness of the nanosheets is approximately 40 nm. The corresponding elemental mapping images, including Ni, Co, Mo, and O elements, have been separately listed in Figure 2d and indicate the uniform dispersion of all the elements. Further insight into the detailed microstructure was elucidated by TEM. Figure 2e,f shows the low- and high-resolution TEM images of NCMO nanosheets. The surface of the NCMO nanosheets is continuous and relatively thin with a porous architecture, which could be ascribed to the successive release and loss of the produced gas during thermal reactions [40–42]. The thin and mesoporous characteristics of the nanosheets may provide a large surface area and more electroactive sites for Faradaic reactions, facilitating charge transfer between the electrode interface and electrolyte. The high-resolution TEM image of Figure 2f demonstrated the polycrystalline characteristics.

XPS measurements were carried out to further confirm the element compositions and the chemical states of the NCMO nanosheets. The corresponding experimental and fitted spectra are presented in Figure 3. As shown in Figure 3a, peaks of Ni, Co, Mo, and O elements are visualized. The deconvolutions of Ni 2p, Mo 2p, Co 2p, and O 1s are presented in Figure 3b–e, respectively. As indicated in Figure 3b, the Ni 2p XPS spectrum presents two spin-orbit doublets and two satellites. Ni $2p_{1/2}$ and Ni $2p_{3/2}$ spin-orbit peaks are located at 873.4 and 856.1 eV, respectively, with a splitting energy of 17.3 eV, implying the existence of the Ni^{2+} oxidation state in the nanosheet arrays [42]. The Co 2p region can be best fitted into two prominent peaks in Figure 3c with the binding energies at 786.3 and 804.2 eV, corresponding to the Co $2p_{1/2}$ and Co $2p_{3/2}$ levels, which is a signature of Co^{2+} and Co^{3+} oxidation states [43]. In Figure 3d, the peaks at 235.3 and 232.3 eV are ascribed to Mo $3d_{3/2}$ and Mo $3d_{5/2}$ levels, respectively [44]. The spin-orbit splitting value of Mo $3d_{3/2}$ and Mo $3d_{5/2}$ is 2.0 eV, which signifies the Mo^{6+} oxidation state in NCMO nanosheets. The fitted peaks of the O 1s XPS spectrum in Figure 3e centered at 531.5 and 532.8 eV. The main peak at 531.5 eV confirms the formation M–O (M = Co, Ni, Mo) bonds and the dwarf peak at 532.8 eV may be associated with the deficient oxygen or the surface-absorbed oxygen species in NCMO nanosheets [45].

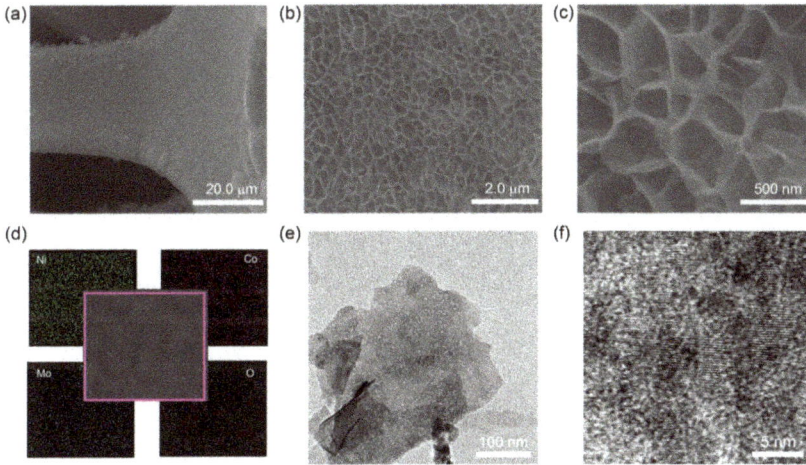

Figure 2. Morphology and microstructure characterizations. (**a**–**c**) Scanning electron microscopy (SEM) images of the NCMO nanosheets on nickel foam (NF) with different magnifications. (**d**) Elemental mappings of Ni, Co, Mo, and O atoms of the NCMO nanosheets. (**e**,**f**) Transmission electron microscopy (TEM) images of the NCMO nanosheets sample.

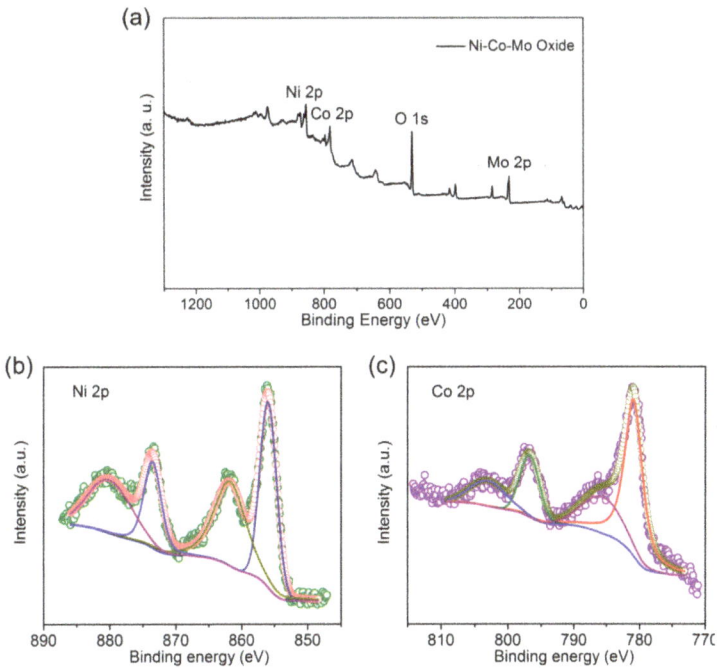

Figure 3. *Cont.*

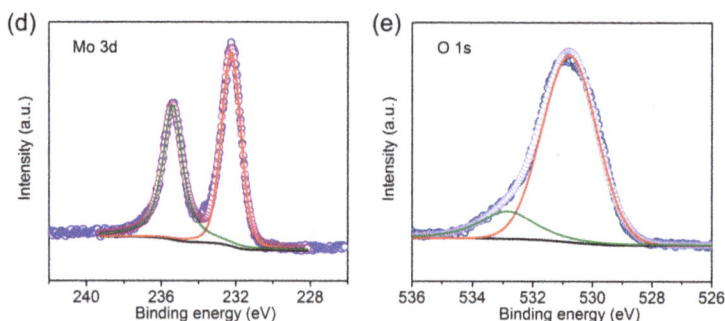

Figure 3. (**a**) X-ray photoelectron spectroscopy (XPS) survey spectrum of NCMO nanosheets. High-resolution XPS spectra of (**b**) Ni 2p, (**c**) Co 2p, (**d**) Mo 3d, and (**e**) O 1s of the NCMO nanosheets sample.

Figure 4 shows the nitrogen adsorption/desorption isotherms and the corresponding pore size distribution plot of the Ni-Co-Mo oxide nanosheet arrays. Notably, Figure 4a displays a distinct hysteresis loop, indicating the mesoporous characteristics. The Ni-Co-Mo oxide nanosheet arrays exhibited a high Brunauer-Emmett-Teller (BET) specific surface area of 94.4 m^2 g^{-1} and the corresponding pore size distribution curve scattered within the range from 2 to 10 nm, as shown in Figure 4b. The high surface area and mesoporous structures will facilitate the redox reaction between the electrolyte ions and Ni-Co-Mo oxide electrode not only by affording sufficient reaction sites and large surface for the electrolyte to permeate but also by accelerating ion diffusion and electron transportation.

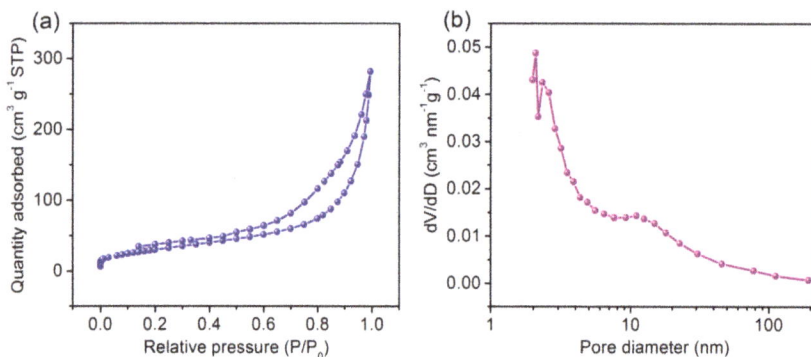

Figure 4. (**a**) Nitrogen (77 K) adsorption/desorption isotherms and (**b**) the corresponding pore size distribution profile of the NCMO nanosheets.

3.2. Electrochemical Properties

Enlightened by the advantageous structural features and compositional merits, the electrochemical properties of the as-fabricated NCMO nanosheets on NF were appraised by CV and GCD measurements. Figure 5a illustrates the typical CV curves with a 0 to 0.6 V potential range from 5 to 30 mV s^{-1}. Clearly, the CV curves manifest a pair of distinct redox peaks during the cathodic and anodic sweepings, which are governed by Faradaic reversible reactions mainly related to M-O/M-O-OH reactions (M represents Ni or Co). In other words, the electrochemical capacitance of NCMO is attributed to the quasi-reversible electron transfer process that mainly involves the Ni^{2+}/Ni^{3+} and Co^{2+}/Co^{3+} redox couples, and is probably mediated by the OH$^-$ ions in the alkaline electrolyte. In fact, the Mo atoms do not participate in any redox reaction; the redox behavior of Mo has no contribution to the measured capacitance. Even though the Mo element did not participate in the Faraday reaction, the introduction of Mo affected the crystal structure. Additionally, the molybdenum oxide showed intercalation pseudocapacitive behavior through reversible intercalation/deintercalation of K^+ ions into/out of the nanostructures [46,47]. Moreover, when the scan rate increases from 5 to 30 mV s^{-1}, the profiles of the CV curves are well maintained and the oxidation peak shifts moderately from 0.39 to 0.54 V. This observation implies good electronic conductivity, outstanding electrochemical reversibility, and fast ion/electron transfer kinetics during Faradaic reactions. Figure 5b depicts the GCD curves at the current densities from 2 to 40 A g^{-1}. The GCD curves are approximately symmetrical and display a distinct plateau region at charge/discharge processes with no obvious internal ohmic (iR) drop, which suggests Faradaic reactions with low internal resistance [48].

The specific and areal capacitances calculated from the GCD plots are pictured in Figure 5c,d, respectively. The specific capacitances are as high as 1366, 1340, 1323, 1298, 1277, 1156, 1066, and 973 F g^{-1} and the areal capacitances are 1.64, 1.61, 1.59, 1.56, 1.53, 1.39, 1.28, and 1.17 F cm^{-2} at current densities from 2 to 40 A g^{-1}, respectively [35]. Moreover, the NCMO nanosheets hold 71.3% retention of the capacitance, revealing the good rate capability and diffusion limitation effect [16]. The cycling stability measured by the repetitive charge and discharge processes at the current density of 30 A g^{-1} was depicted in Figure 5e. Impressively, the specific capacitance had a retention of 89.75% even after 5000 cycles.

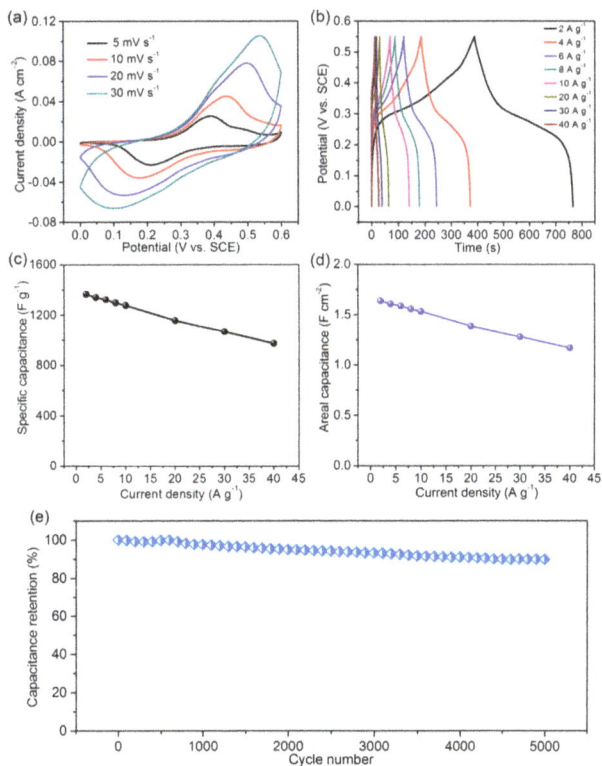

Figure 5. (**a**) Cyclic voltammetry (CV) properties of NCMO sample at various scan rates. (**b**) Galvanostatic charge/discharge (GCD) measurements at different current densities. (**c,d**) The specific capacitances and areal capacitances calculated from the GCD plots. (**e**) Long-term cycling manifestation at a current density of 30 A g^{-1}.

With the aim to assess the potential energy storage applications of the NCMO nanosheets, a hybrid supercapacitor was assembled. The CV plots shown in Figure 6a were performed at various sweep rates from 5 to 50 mV s^{-1} in a potential range from 0 to 1.8 V. It was observed that all the curves behave similarly in shape with a conspicuous pair of redox peaks. The wide potential window is beneficial for promoting the energy density. With the increase of the scan rates, the CV curves displayed very slight shifts of the redox peaks, indicating the device possesses outstanding rate capability. Figure 6b illustrates GCD results of the hybrid supercapacitor cell. The nearly symmetrical shape of the curves announces the superior reversibility of Faradaic reactions and good electrochemical capacitive characteristics [49]. Figure 6c records the Nyquist impedance spectrum of the hybrid supercapacitor. The semicircle region of the spectrum demonstrated that the charge transfer resistance was about 1.0 Ω, implying good charge-transfer kinetics and fast electron transport [50]. The Ragone plots correlated containing/covering the energy density with the power density of the hybrid supercapacitor cell were depicted in Figure 6d. These plots revealed that the device manifested a high energy density of 46.2 Wh kg^{-1} at 713.0 W kg^{-1}, which remained a high energy density of 38.0 Wh kg^{-1} at 7130 W kg^{-1}.

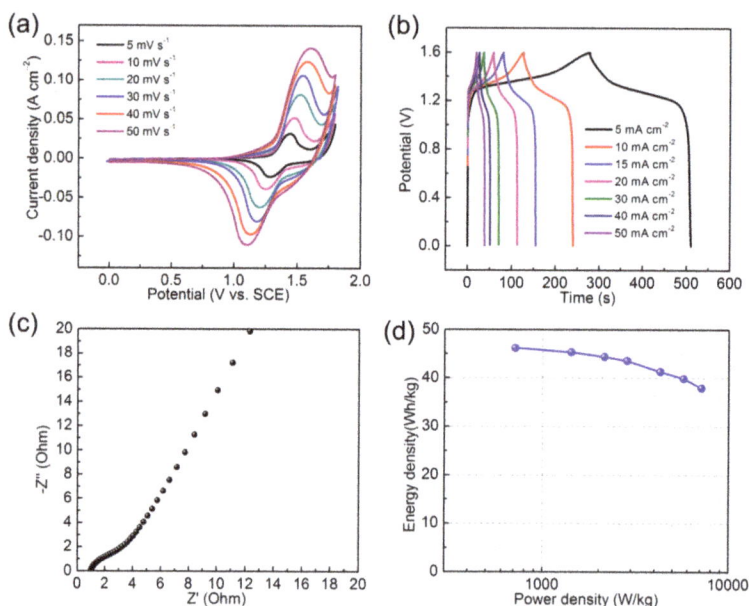

Figure 6. Electrochemical manifestation of the hybrid supercapacitor cell. (**a,b**) display the CV and GCD curves at different scan rates and current densities, respectively; (**c**) Nyquist impedance spectrum; (**d**) Ragone plot.

In addition, we make an electrochemical performance comparison of recent synthesized mixed metal oxide, sulfide, and hydroxide for supercapacitors applications, listed in Table 1. From the comparison, we could conclude that the NCMO nanosheets in this work possess high specific capacitance along with long cycling stability.

Table 1. Comparison of the electrochemical performance of various mixed metal pseudocapacitor electrodes.

Electrode Materials	Specific Capacitance	Rate Capability	Cycling Performance	Ref.
Ni-Zn-Co oxide nanowire arrays	776 F g^{-1} at 2 A g^{-1}	73.8% at 32 A g^{-1}	88.9% of the maximum value after 10,000 cycles	[29]
Ni-Co-Mo oxy-hydroxide nanoflakes	2562 F g^{-1} at 1 A g^{-1}	88.4% at 10 A g^{-1}	about 91% of its original capacitance after 1000 cycles	[24]
Ni-Co-Mo sulfide nanosheets	2717 F g^{-1} at 1 A g^{-1}	83.6% at 10 A g^{-1}	about 80% capacitance retention after 1000 cycles	[25]
Zn-Ni-Co oxides nanowire arrays	2482 F g^{-1} at 1 A g^{-1}	91.9% at 5 A g^{-1}	94% capacitance retention over 3000 cycles	[26]
Mn-Ni-Co oxide nanowire array	638 F g^{-1} at 1 A g^{-1}	63.3% at 20 A g^{-1}	93.6% of the maximum value after 6000 cycles	[27]
Ni$_{0.8}$-Co$_{0.2}$-Se nanowires	86 F g^{-1} at 1 A g^{-1}	Not reported	exceeding 95% over the 2000 cycles test	[48]
Ni$_{1/3}$Co$_{2/3}$MoO$_4$ nanosheets	1103 F g^{-1} at 1 A g^{-1}	84.3% at 10 A g^{-1}	remaining 85.18% at after 1000 cycles	[49]
Hollow Ni-Al-Mn layered hydroxide nanospheres	1756 F g^{-1} at 4 A g^{-1}	Not reported	89.5% of initial values after 4000 cycles	[50]
Ni-Co-Mo oxide nanosheet arrays	1366 F g^{-1} at 2 A g^{-1}	71.3% at 40 A g^{-1}	89.75% of the maximum value after 5000 cycles	This work

4. Conclusions

We successfully manufactured hierarchical porous NCMO nanosheets on NF by combining a hydrothermal method and an annealing procedure. Due to its nanosheet structure, mesoporous features, and the synergistic enhancement effect, the NCMO nanosheets delivered a high specific capacitance of 1366 F g^{-1} and long cyclability of 89.75% after 5000 cycles. Additionally, the NCMO nanosheets-based hybrid supercapacitor cell exhibited an energy density of 46.2 Wh kg^{-1} at 713.0 W kg^{-1}. In conclusion, the outstanding electrochemical performance suggests that the as-fabricated NCMO nanosheets could serve as potential electrode materials in practical energy storage applications.

Author Contributions: Conceptualization, W.H.; Formal analysis, Y.S. and W.H.; Investigation, Y.S., D.L. and Z.S.; Methodology, W.H.; Supervision, W.H.; Writing–original draft, Y.S.; Writing–review & editing, B.T., X.T. and J.Q.

Funding: This research was funded by the Natural Science Foundation of China (No. 51602033), Chongqing Research Program of Basic Research and Frontier Technology (No. cstc2015jcyjBX0038), Fundamental Research Funds for the Central Universities (Nos. 2018CDQYGD0008 and 10611CDJXZ238826), and National Key Research and Development Program of China (Nos. 2016YFE0125200 and 2016YFC0101100).

Conflicts of Interest: The authors declare no conflict of interest.

References

1. Dunn, B.; Kamath, H.; Tarascon, J.M. Electrical energy storage for the grid: A battery of choices. *Science* **2011**, *334*, 928–935. [CrossRef] [PubMed]
2. Simon, P.; Gogotsi, Y.; Dunn, B. Where do batteries end and supercapacitors begin? *Science* **2014**, *343*, 1210–1211. [CrossRef] [PubMed]
3. Wang, Y.; Song, Y.; Xia, Y. Electrochemical capacitors: Mechanism, materials, systems, characterization and applications. *Chem. Soc. Rev.* **2016**, *45*, 5925–5950. [CrossRef] [PubMed]
4. Qi, D.; Liu, Y.; Liu, Z.; Zhang, L.; Chen, X. Design of architectures and materials in in-plane micro-supercapacitors: Current status and future challenges. *Adv. Mater.* **2017**, *29*, 1602802. [CrossRef] [PubMed]
5. Wang, G.; Zhang, L.; Zhang, J. A review of electrode materials for electrochemical supercapacitor. *Chem. Soc. Rev.* **2012**, *41*, 797–828. [CrossRef] [PubMed]
6. Yu, Z.; Tetard, L.; Zhai, L.; Thomas, J. Supercapacitor electrode materials: Nanostructures from 0 to 3 dimensions. *Energy Environ. Sci.* **2014**, *8*, 702–730. [CrossRef]
7. González, A.; Goikolea, E.; Barrena, J.A.; Mysyk, R. Review on supercapacitors: Technologies and materials. *Renew. Sustain. Energy Rev.* **2016**, *58*, 1189–1206. [CrossRef]
8. Yu, M.; Wang, Z.; Han, Y.; Tong, Y.; Lu, X.; Yang, S. Recent progress in the development of anodes for asymmetric supercapacitors. *J. Mater. Chem. A* **2016**, *4*, 4634–4658. [CrossRef]

9. Dong, X.W.; Zhang, Y.Y.; Wang, W.J.; Zhao, R. Rational construction of 3D NiCo$_2$O$_4$@CoMoO$_4$ core/shell nanoarrays as a positive electrode for asymmetric supercapacitor. *J. Alloys Compd.* **2017**, *729*, 716–723. [CrossRef]

10. Zhang, L.; Hui, K.N.; Hui, K.S.; Lee, H. High-performance hybrid supercapacitor with 3D hierarchical porous flower-like layered double hydroxide grown on nickel foam as binder-free electrode. *J. Power Sources* **2016**, *318*, 76–85. [CrossRef]

11. Yang, Y.; Cheng, D.; Chen, S.; Guan, Y.; Xiong, J. Construction of hierarchical NiCo$_2$S$_4$@Ni(OH)$_2$ core-shell hybrid nanosheet arrays on Ni foam for high-performance aqueous hybrid supercapacitors. *Electrochim. Acta* **2016**, *193*, 116–127. [CrossRef]

12. Yin, C.; Yang, C.; Jiang, M.; Deng, C.; Yang, L.; Li, J.; Qian, D. A novel and facile one-pot solvothermal synthesis of PEDOT-PSS/Ni-Mn-Co-O hybrid as an advanced supercapacitor electrode material. *ACS Appl. Mater. Interfaces* **2016**, *8*, 2741–2752. [CrossRef] [PubMed]

13. Sk, M.M.; Yue, C.Y.; Ghosh, K.; Jena, R.K. Review on advances in porous nanostructured nickel oxides and their composite electrodes for high-performance supercapacitors. *J. Power Sources* **2016**, *308*, 121–140. [CrossRef]

14. Guan, C.; Liu, X.; Ren, W.; Li, X.; Cheng, C.; Wang, J. Rational design of metal-organic framework derived hollow NiCo$_2$O$_4$ arrays for flexible supercapacitor and electrocatalysis. *Adv. Energy Mater.* **2017**, *7*, 1602391. [CrossRef]

15. Meher, S.K.; Rao, G.R. Effect of microwave on the nanowire morphology, optical, magnetic, and pseudocapacitance behavior of Co$_3$O$_4$. *J. Phys. Chem. C* **2011**, *115*, 25543–25556. [CrossRef]

16. Raj, S.; Srivastava, S.K.; Kar, P.; Roy, P. Three-dimensional NiCo$_2$O$_4$/NiCo$_2$S$_4$ hybrid nanostructures on Ni-foam as high-performance supercapacitor electrode. *RSC Adv.* **2016**, *6*, 95760–95767. [CrossRef]

17. Hu, W.; Chen, R.; Xie, W.; Zou, L.; Qin, N.; Bao, D. CoNi$_2$S$_4$ nanosheet arrays supported on nickel foams with ultrahigh capacitance for aqueous asymmetric supercapacitor applications. *ACS Appl. Mater. Interfaces* **2014**, *6*, 19318–19326. [CrossRef] [PubMed]

18. Guan, B.Y.; Yu, L.; Wang, X.; Song, S.; Lou, X.W. Formation of onion-like NiCo$_2$S$_4$ particles via sequential ion-exchange for hybrid supercapacitors. *Adv. Mater.* **2017**, *29*, 1605051. [CrossRef] [PubMed]

19. Wang, H.; Casalongue, H.S.; Liang, Y.; Dai, H. Ni(OH)$_2$ nanoplates grown on graphene as advanced electrochemical pseudocapacitor materials. *J. Am. Chem. Soc.* **2010**, *132*, 7472–7477. [CrossRef] [PubMed]

20. Liu, Y.; Fu, N.; Zhang, G.; Xu, M.; Lu, W.; Zhou, L.; Huang, H. Design of hierarchical Ni-Co@Ni-Co layered double hydroxide core-shell structured nanotube array for high-performance flexible all-solid-sate battery-type supercapacitors. *Adv. Funct. Mater.* **2017**, *27*, 1605307. [CrossRef]

21. Jagadale, A.D.; Guan, G.; Li, X.; Du, X.; Ma, X.; Hao, X.; Abudula, A. Ultrathin nanoflakes of cobalt-manganese layered double hydroxide with high reversibility for asymmetric supercapacitor. *J. Power Sources* **2016**, *306*, 526–534. [CrossRef]

22. Cheng, J.; Yan, H.; Lu, Y.; Qiu, K.; Hou, X.; Xu, J.; Han, L.; Liu, X.; Kim, J.; Luo, Y. Mesoporous CuCo$_2$O$_4$ nanograss as multi-functional electrodes for supercapacitors and electro-catalysts. *J. Mater. Chem. A* **2015**, *3*, 9769–9776. [CrossRef]

23. Sahoo, S.; Naik, K.K.; Late, D.J.; Rout, C.S. Electrochemical synthesis of a ternary transition metal sulfide nanosheets on nickel foam and energy storage application. *J. Alloys Compd.* **2017**, *695*, 154–161. [CrossRef]

24. Duan, C.; Zhao, J.; Qin, L.; Yang, L.; Zhou, Y. Ternary Ni-Co-Mo oxy-hydroxide nanoflakes grown on carbon cloth for excellent supercapacitor electrodes. *Mater. Lett.* **2017**, *208*, 65–68. [CrossRef]

25. Sahoo, S.; Mondal, R.; Late, D.J.; Rout, C.S. Electrodeposited nickel cobalt manganese based mixed sulfide nanosheets for high performance supercapacitor application. *Microporous Mesoporous Mater.* **2017**, *244*, 101–108. [CrossRef]

26. Wu, C.; Cai, J.; Zhang, Q.; Zhou, X.; Zhu, Y.; Shen, P.; Zhang, K. Hierarchical mesoporous zinc-nickel-cobalt ternary oxide nanowire arrays on nickel foam as high-performance electrodes for supercapacitors. *ACS Appl. Mater. Interfaces* **2015**, *7*, 26512–26521. [CrossRef] [PubMed]

27. Li, L.; Zhang, Y.; Shi, F.; Zhang, Y.; Zhang, J.; Gu, C.; Wang, X.; Tu, J. Spinel manganese-nickel-cobalt ternary oxide nanowire array for high-performance electrochemical capacitor applications. *ACS Appl. Mater. Interface* **2014**, *6*, 18040–18047. [CrossRef] [PubMed]

28. Lee, H.J.; Lee, J.H.; Chung, S.Y.; Choi, J.W. Enhanced pseudocapacitance in multicomponent transition-metal oxides by local distortion of oxygen octahedral. Angew. *Chem. Int. Ed.* **2016**, *55*, 3958–3962. [CrossRef] [PubMed]

29. Hu, W.; Wei, H.; She, Y.; Tang, X.; Zhou, M.; Zang, Z.; Du, J.; Gao, C.; Guo, Y.; Bao, D. Flower-like nickel-zinc-cobalt mixed metal oxide nanowire arrays for electrochemical capacitor applications. *J. Alloys Compd.* **2017**, *708*, 146–153. [CrossRef]

30. Cai, D.; Liu, B.; Wang, D.; Liu, Y.; Wang, L.; Li, H.; Wang, Y.; Wang, C.; Li, Q.; Wang, T. Facile hydrothermal synthesis of hierarchical ultrathin mesoporous $NiMoO_4$ nanosheets for high performance Supercapacitors. *Electrochim. Acta* **2014**, *115*, 358–363. [CrossRef]

31. Singh, A.K.; Sarkar, D.; Khan, G.G.; Mandal, K. Unique hydrogenated Ni-NiO core-shell 1D nano-heterostructures with superior electrochemical performance as supercapacitor. *J. Mater. Chem. A* **2013**, *1*, 12759–12767. [CrossRef]

32. Sevilla, M.; Fuertes, A.B. Direct synthesis of highly porous interconnected carbon nanosheets and their application as high-performance supercapacitors. *ACS Nano* **2014**, *8*, 5059–5078. [CrossRef] [PubMed]

33. Li, H.; Gao, Y.; Wang, C.; Yang, G. A simple electrochemical route to access amorphous mixed-metal hydroxides for supercapacitor electrode materials. *Adv. Energy Mater.* **2015**, *5*, 1401767. [CrossRef]

34. Li, L.; Hui, K.S.; Hui, K.N.; Xia, Q.; Fu, J. Facile synthesis of NiAl layered double hydroxide nanoplates for high-performance asymmetric supercapacitor. *J. Alloys Compd.* **2017**, *721*, 803–812. [CrossRef]

35. Hsu, A.R.; Chien, H.H.; Liao, C.Y.; Lee, C.C.; Tsai, J.H.; Hsu, C.C.; Cheng, I.C.; Chen, J.Z. Scan-mode atmospheric-pressure plasma jet processed reduced graphene oxides for quasi-solid-state gel-electrolyte supercapacitors. *Coatings* **2018**, *8*, 52. [CrossRef]

36. Zhang, Z.; Liu, Y.; Huang, Z.; Ren, L.; Qi, X.; Wei, X.; Zhong, J. Facile hydrothermal synthesis of $NiMoO_4$@$CoMoO_4$ hierarchical nanospheres for supercapacitor applications. *Phys. Chem. Chem. Phys.* **2015**, *17*, 20795–20804. [CrossRef] [PubMed]

37. Senthilkumar, B.; Meyrick, D.; Lee, Y.S.; Selvan, R.K. Synthesis and improved electrochemical performances of nano b-$NiMoO_4$-$CoMoO_4$·xH_2O composites for asymmetric supercapacitors. *RSC Adv.* **2013**, *3*, 16542–16548. [CrossRef]

38. Yin, Z.; Chen, Y.J.; Zhao, Y.; Li, C.; Zhu, C.; Zhang, X. Hierarchical nanosheet-based $CoMoO_4$-$NiMoO_4$ nanotubes for applications in asymmetric supercapacitor and oxygen evolution reaction. *J. Mater. Chem. A* **2015**, *3*, 22750–22758. [CrossRef]

39. Godillot, G.; Taberna, P.L.; Daffos, B.; Simon, P.; Delmas, C.; Guerlou-Demourgues, L. High power density aqueous hybrid supercapacitor combining activated carbon and highly conductive spinel cobalt oxide. *J. Power Sources* **2016**, *331*, 277–284. [CrossRef]

40. Li, L.; Cheah, Y.; Ko, Y.; Teh, P.; Wee, G.; Wong, C.; Peng, S.; Srinivasan, M. The facile synthesis of hierarchical porous flower-like $NiCo_2O_4$ with superior lithium storage properties. *J. Mater. Chem. A* **2013**, *1*, 10935–10941. [CrossRef]

41. Peng, S.; Li, L.; Wu, H.B.; Srinivasan, M.; Lou, X.W. Controlled growth of $NiMoO_4$ nanosheet and nanorod arrays on various conductive substrates as advanced electrodes for asymmetric supercapacitors. *Adv. Energy Mater.* **2015**, *5*, 1401172. [CrossRef]

42. Cai, D.; Wang, D.; Liu, B.; Wang, Y.; Liu, Y.; Wang, L.; Li, H.; Huang, H.; Li, Q.; Wang, T. Comparison of the electrochemical performance of $NiMoO_4$ nanorods and hierarchical nanospheres for supercapacitor applications. *ACS Appl. Mater. Interfaces* **2013**, *5*, 12905–12910. [CrossRef] [PubMed]

43. Shen, L.; Che, Q.; Li, H.; Zhang, X. Metal oxides: Mesoporous $NiCo_2O_4$ nanowire arrays grown on carbon textiles as binder-free flexible electrodes for energy storage. *Adv. Funct. Mater.* **2014**, *24*, 2630–2637. [CrossRef]

44. Veerasubramani, G.K.; Krishnamoorthy, K.; Sang, J.K. Improved electrochemical performances of binder-free $CoMoO_4$ nanoplate arrays@Ni foam electrode using redox additive electrolyte. *J. Power Sources* **2016**, *306*, 378–386. [CrossRef]

45. Ding, R.; Qi, L.; Jia, M.; Wang, H. Facile synthesis of mesoporous spinel $NiCo_2O_4$ nanostructures as highly efficient electrocatalysts for urea electro-oxidation. *Nanoscale* **2013**, *6*, 1369–1376. [CrossRef] [PubMed]

46. Ghosh, K.; Yue, C.Y. Development of 3D MoO_3/graphene aerogel and sandwich-type polyaniline decorated porous MnO_2-graphene hybrid film based high performance all-solid-state asymmetric supercapacitors. *Electrochim. Acta* **2018**, *276*, 47–63. [CrossRef]

47. Riley, L.A.; Lee, S.H.; Gedvilias, L.; Dillon, A.C. Optimization of MoO_3 nanoparticles as negative-electrode material in high-energy lithium ion batteries. *J. Power Sources* **2010**, *195*, 588–592. [CrossRef]
48. Guo, K.; Cui, S.; Hou, H.; Chen, W.; Mi, L. Hierarchical ternary Ni-Co-Se nanowires for high-performance supercapacitor device design. *Dalton Trans.* **2016**, *45*, 19358–19465. [CrossRef] [PubMed]
49. Zhang, J.; Zhang, R.; Song, P.; Zhao, J.; Guo, X.; Zhang, D.; Yuan, B. $CoMoO_4$ and $Ni_{1/3}Co_{2/3}MoO_4$ nanosheets with high performance supercapacitor and nonenzymatic glucose detection properties. *RSC Adv.* **2015**, *5*, 84451–84456. [CrossRef]
50. Chandrasekaran, N.I.; Muthukumar, H.; Sekar, A.D.; Manickam, M. Hollow nickel-aluminium-manganese layered triple hydroxide nanospheres with tunable architecture for supercapacitor application. *Mater. Chem. Phys.* **2017**, *195*, 247–258. [CrossRef]

coatings

Article

Influence of Ge Incorporation from GeSe$_2$ Vapor on the Properties of Cu$_2$ZnSn(S,Se)$_4$ Material and Solar Cells

Chao Gao [1],*, Yali Sun [2] and Wei Yu [2]

[1]   Institute of Photovoltaics, Nanchang University, Nanchang 330031, China
[2]   College of Physics Science and Technology, Hebei University, Baoding 071002, China;
      yali_sun@outlook.com (Y.S); weiyu_hbu@126.com (W.Y.)
*   Correspondence: cgao@ncu.edu.cn

Received: 18 July 2018; Accepted: 24 August 2018; Published: 28 August 2018

Abstract: Cu$_2$ZnSn(S,Se)$_4$ (CZTSSe) and Cu$_2$Zn(Sn,Ge)(S,Se)$_4$ (CZTGSSe) thin films were prepared based on a non-vacuum solution method. The CZTSSe films were obtained by annealing the solution-deposited precursor films with Se, while the CZTGSSe films were obtained by annealing the similar precursor films with Se and GeSe$_2$. We found that Ge could be incorporated into the annealed films when GeSe$_2$ was present during the annealing process. The Ge incorporation obviously enlarged the sizes of the crystalline grains in the annealed films. However, the energy dispersive spectrometry (EDS) measurements revealed that the element distribution was not uniform in the CZTGSSe films. We fabricated solar cells based on the CZTSSe and CZTGSSe films. It was found the Ge incorporation decreases the E_u energy of the absorber material. The solar cell efficiency was increased from 5.61% (CZTSSe solar cell) to 7.14% (CZTGSSe solar cell) by the Ge incorporation. Compared to CZTSSe solar cells, the CZTGSSe solar cells exhibited a lower diode ideality factor and lower reverse saturation current density.

Keywords: Cu$_2$ZnSn(S,Se)$_4$; Ge incorporation; annealing; solar cells

1. Introduction

The Cu$_2$ZnSn(S,Se)$_4$ (CZTSSe) material has gained a great deal of attention as a promising photovoltaic material due to its properties, including suitable optical bandgap (1.0–1.5 eV), high absorption coefficient (>10^{-4} cm^{-1} in the visible light region), and Earth-abundant nature [1]. To-date, the efficiencies of CZTSSe solar cells have increased to 12.6% [2]. However, this efficiency is much lower than the efficiencies of mature Cu(In,Ga)Se$_2$ (CIGS) or CdTe solar cells. The efficiencies of CZTSSe solar cells are mainly limited by the low open-circuit voltage, which could be related with the poor properties (e.g., unfavorable electrical defects, secondary phases, large band-tail energy) of the CZTSSe material [3].

Recently, Giraldo et al. found that the properties of CZTSSe material can be greatly improved by Ge incorporation [4–6]. The Ge incorporation in CZTSSe material can increase the sizes of the crystalline grains, increase the carrier lifetime, decrease the band-tail energy, prevent the formation of deep-level defects, and more [4]. Moreover, the GeO formed in the grain-boundaries was believed to passivate the grain boundaries [5]. Several groups have incorporated Ge into CZTSSe, and they found the Ge incorporation can indeed increase the open-circuit voltage and efficiency of the solar cell [6,7]. For most of the Ge-incorporated CZTSSe materials produced by the two-step method, the Ge element was introduced during the precursor preparation. The Ge may also be introduced into the material during the annealing process. For example, Umehara et al. reported on the preparation of Cu$_2$Sn$_x$Ge$_{1-x}$S$_3$ material by annealing co-sputtered Cu-Sn films with S and GeS$_2$ vapor [8].

In this paper, we prepared CZTSSe and Ge-incorporated $Cu_2Zn(Sn,Ge)(S,Se)_4$ (CZTGSSe) thin films based on a non-vacuum solution method. The Ge incorporation was realized by annealing the solution-deposited precursor films with Se and $GeSe_2$, while the CZTSSe films were prepared by annealing precursor films with only Se. By comparing the morphologies, compositions, and crystallinities of the CZTSSe and CZTGSSe films, we investigated the influence of the Ge incorporation by this method on the properties of CZTSSe material. Based on the prepared CZTSSe and CZTGSSe films, solar cells were fabricated and their performances were compared.

2. Materials and Methods

Mo-coated soda-lime glass (Mo/SLG) was used as substrate for the preparation of the films. After the cleaning of the substrates, precursor films were prepared on the substrates by spin-coating the precursor solution, which was prepared by successively dissolving copper(II)-acetate monohydrate (99.99%, 2.3 mmol), tin(II)-chloride dehydrate (99.99%, 1.6 mmol), zinc(II)-chloride (99.99%, 1.6 mmol), and thiourea (99.5%, 5.5 mmol) into 2 mL of dimethyl sulfoxide (99.9%). For the spin-coating process, 200 μL of the precursor solution was first dipped on the substrate, then the substrate was spun at 500 r/min for 30 s and 2000 r/min for 120 s successively. After the spin-coating process, the films were dried at 300 °C for 2 min. These processes were repeated several times until the film thickness reached approximately 1 μm. Both the spin-coating process and the drying process were carried out in a glove box filled with nitrogen.

After the preparation of precursor films, they were annealed in a furnace to prepare the CZTSSe or CZTGSSe films. For the preparation of CZTGSSe films, the precursor films with Se pellets and $GeSe_2$ powder were first placed in a graphite box. The graphite box was then transferred into a tube furnace. After that, the tube was evacuated and refilled with nitrogen to a pressure of 100 Torr. Finally, the furnace was heated to 600 °C for 15 min. The preparation of CZTSSe films was the same as with CZTGSSe films, except no $GeSe_2$ was used during the annealing process.

Solar cells based on the CZTSSe or CZTGSSe films were fabricated using the following structure: Mo/CZT(G)SSe/CdS/i-ZnO/AZO (Al-doped ZnO)/Ag (i-ZnO is short for intrinsic ZnO). A mixed solution of 3 mM cadmium acetate, 0.25 M thiourea, and 1 M ammonia hydroxide was used for the buffer deposition. The CdS buffer layer was deposited by placing the substrate into the deposition solution with a temperature of 75 °C. After 10 min deposition, a CdS buffer layer with a thickness of ~50 nm was deposited on the absorber layer. Then, an i-ZnO layer with a thickness of ~50 nm and an AZO (Al-doped ZnO) layer with a thickness of ~200 nm was successively deposited on the buffer layer by sputtering. The pressure of the Ar gas during the sputtering was 0.3 Pa, and the sputtering powers were 50 W and 100 W for i-ZnO and AZO, respectively. Ag grids were deposited on the AZO layer using a thermal evaporation setup, and the thickness of the Ag grid was approximately 200 nm. Finally, the prepared solar cell was divided into small individual solar cells with area ~1 cm^2.

The constitutions of the films were analyzed by a Bruker D8 Advance X-ray diffractometer (Billerica, MA, USA) and a JY LabRAM HR Raman spectrometer (Horiba, Kyoto, Japan) equipped with a 532 nm laser. The surface morphologies of the films were characterized by a FEI JSM-7500F scanning electron microscope (SEM) (JEOL, Ltd., Tokyo, Japan). The compositions of the thin films were analyzed by an energy dispersive spectrometer (EDS) that was attached in the SEM setup. The solar cell performances were characterized by an Agilent B1500A semiconductor parametric analyzer (Agilent Technologies, Santa Clara, CA, USA) under AM 1.5 global solar irradiations. The quantum efficiency (QE) curves of the solar cells were measured by a QTesT 1000ADX setup (Crowntech, Macungie, PA, USA).

3. Results and Discussion

Figure 1 shows the typical SEM images of the precursor film and the films after different annealing processes. The precursor film looked smooth on the surface, and no obvious crystalline grains were found in the film (Figure 1a,b). After annealing with Se, crystalline grains with sizes 1–4 μm could be seen in the film (Figure 1c,d). The film showed dense morphology, but the surface became coarse.

When GeSe$_2$ was added during the annealing process, the sizes of the crystalline grains in the annealed film became greater than those in the film annealed with only Se.

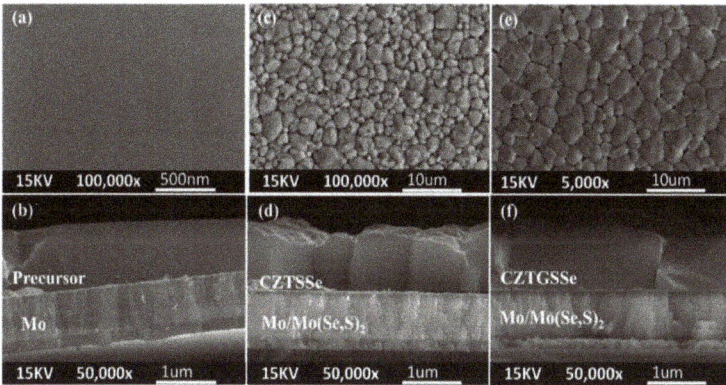

Figure 1. Surface and cross section SEM images of the (**a,b**) precursor film; (**c,d**) Cu$_2$ZnSn(S,Se)$_4$ (CZTSSe) film; and (**e,f**) Cu$_2$Zn(Sn,Ge)(S,Se)$_4$ (CZTGSSe) film.

We counted the crystalline grains in the annealed films and measured their sizes. Based on our counting, the average sizes of the crystalline grains in CZTSSe and CZTGSSe films were 1.86 μm and 2.84 μm. Therefore, the use of GeSe$_2$ during the annealing process enlarged the grain sizes by 53%. We also counted the distribution of the grain sizes in CZTSSe and CZTGSSe films. The statistics were fitted using Gauss curves (Figure 2b,c). The peak positions of the fitting curves for CZTSSe and CZTGSSe films were at 1.67 μm and 2.62 μm, respectively. These values were close to the average grain sizes in CZTSSe and CZTGSSe films.

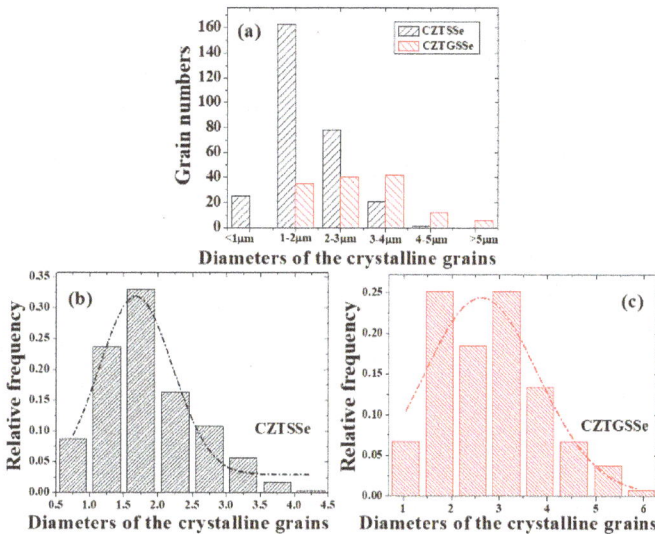

Figure 2. (**a**) Grain size counting for CZTSSe and CZTGSSe films; (**b**) Distribution of the grain sizes in CZTSSe film; (**c**) Distribution of the grain sizes in CZTGSSe film.

Table 1 shows the compositions of the precursor film and the films after different annealing processes. It should be noted that our XRD measurement (shown below) conformed with the formation of $Mo(S,Se)_2$ between the substrate and the annealed films, as $Mo(S,Se)_2$ contains S and Se, and the Mo and S peaks coincided in the EDS spectrum. The presence of $Mo(S,Se)_2$ may have influenced the EDS measurement in the CZTSSe or CZTGSSe films, and the atomic percent of S and Se may slightly deviate from the real values in the films. However, the ratios between the metal elements were not influenced by the formation of $Mo(S,Se)_2$. The Cu/(Zn + Sn) and Zn/Sn ratios in the precursor film were similar to that in the precursor solution. After annealing with Se, both the Cu/(Zn + Sn) and Zn/Sn ratios in the film increased, which could be caused by the Sn loss during the annealing process. The composition of the film that annealed with Se was close to the optimum composition (Cu/(Zn + Sn) = 0.8, Zn/Sn = 1.2) for Cu_2ZnSnS_4-based photovoltaic material [9]. When $GeSe_2$ was used for the annealing, Ge was found to be present in the annealed film (Ge/(Sn + Ge) = 0.37). Due to the incorporation of Ge, both the ratio of Cu/(Zn + Sn + Ge) and Zn/(Sn + Ge) in the film decreased, and thus the composition of the films slightly deviated from the optimum composition for CZTSSe-based material.

Table 1. Composition of the precursor film, CZTSSe film, and CZTGSSe film.

Samples	Cu (at.%)	Zn (at.%)	Sn (at.%)	Ge (at.%)	S (at.%)	Se (at.%)	Cu/(Zn + Sn + Ge)	Zn/(Sn + Ge)	Ge/(Sn + Ge)
Precursor	19.6	14.0	13.3	–	53.1	–	0.72	1.05	–
CZTSSe	21.9	15.1	12.3	–	3.8	46.9	0.80	1.23	–
CZTGSSe	20.3	14.9	9.4	5.5	3.6	46.3	0.68	1.00	0.37

The EDS results proved that Ge was incorporated into the films that were selenized with Se and $GeSe_2$. As the precursor films contained no Ge, the Ge element in the annealed film must have come from the $GeSe_2$ powder which was placed into the graphite box before the selenization. However, the $GeSe_2$ powder did not directly contact the films during the annealing. Therefore, the mechanism by which the Ge element transferred from the $GeSe_2$ powder to the annealed film could be interesting. To determine this mechanism, we checked the properties of $GeSe_2$ and found that $GeSe_2$ can vaporize at around 600 °C [10]. Therefore, we believe that $GeSe_2$ vapor could be present in the graphite box during the annealing process, which would result in the diffusion of Ge from the surface to the bulk of the films. As discussed in [6], the incorporated Ge may react with Se to form a specific liquid phase. The liquid phase can assist in the diffusion of the elements and the crystallization of the material, so the films selenized with Se and $GeSe_2$ showed larger crystalline grains.

Figure 3 shows the XRD patterns and Raman spectra of the precursor and annealed films. The results proved that the main phase in the annealed film was CZTSSe or CZTGSSe (XRD peaks at around 27°, 45°, 53° and Raman peaks at around 175, 198, and 235 cm^{-1} could be attributed to CZTSSe or CZTGSSe) [9,11]. However, a small amount of secondary phases such as Zn(S,Se) and Cu_xSe (Raman peak at around 250 cm^{-1}, and 265 cm^{-1}, respectively) may exist in the films [9]. Besides, $Mo(S,Se)_2$ (XRD peak at around 32° and 56°) was formed between CZTSSe/CZTGSSe and Mo contact. Compared to the CZTSSe film, the XRD and Raman peak of CZTGSSe film shifted to a high angle or high wavenumber (shift from 26.9° to 27.0° for (112) diffraction peak, and from 198.1 cm^{-1} to 200.6 cm^{-1} for the main Raman peak), which agrees with previous reports [11–13]. The reason for this phenomenon could be that the Ge incorporation changes the lattice parameters and modifies the phonon mode of the material. Based on the XRD data, we calculated the lattice parameters of CZTSSe and CZTGSSe materials (a = 5.71 Å, c = 11.63 Å for CZTSSe; a = 5.69 Å, c = 11.48 Å for CZTGSSe). These values are in coincide with the published results [11–13].

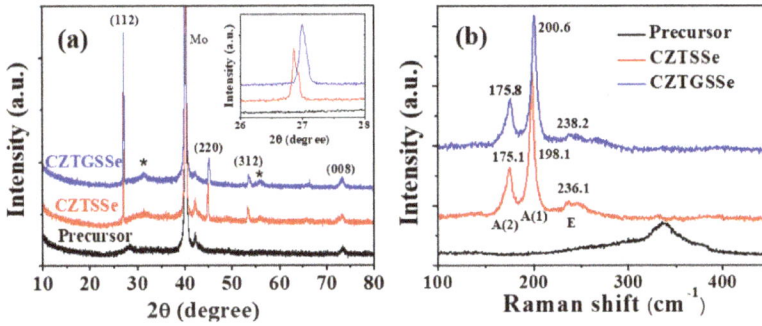

Figure 3. (**a**) XRD patterns and (**b**) Raman spectra of the precursor, CZTSSe, and CZTGSSe films.

Surprisingly, we found that the XRD and Raman peaks of the film were broadened when GeSe$_2$ was used for the annealing. For the CZTSSe film, the full width at half maxima (FWHM) of the (112) diffraction peak and the main Raman peak were 0.10° and 5.4 cm^{-1}, respectively. For the CZTGSSe film, the FWHM of the (112) peak and the main Raman peak were 0.13° and 6.1 cm^{-1}, respectively. This result seems to contradict the fact that the usage of GeSe$_2$ greatly improved the size of the crystalline grains in the film. We checked the literature and found that similar behavior of Ge incorporation has been reported. In Reference [14], the authors found the Ge incorporation broadened the XRD peaks and attributed this to the non-uniform distribution of the Ge element in the material (i.e., some of the region may be Ge-rich but other regions may be Ge-poor). So, we carried out EDS mapping measurement on the Ge-incorporated film, and the results are shown in Figure 4. From the figure, we see the that the distributions of Cu, Zn, and Sn were relatively uniform, but S and Se were not uniformly distributed, which is similar to the report in Reference [15]. Besides, the distribution of Ge was also not uniform. In some of the regions, Ge was rich at the grain boundaries. In other regions, however, Ge was rich inside the crystalline grains. Therefore, the lattice parameters of the material in different regions would be different due to different compositions. Since the XRD pattern of the whole film is the combination of the XRD patterns from different regions of the film, the inhomogeneous distribution of the elements may result in the enlarged FWHM in XRD and Raman results.

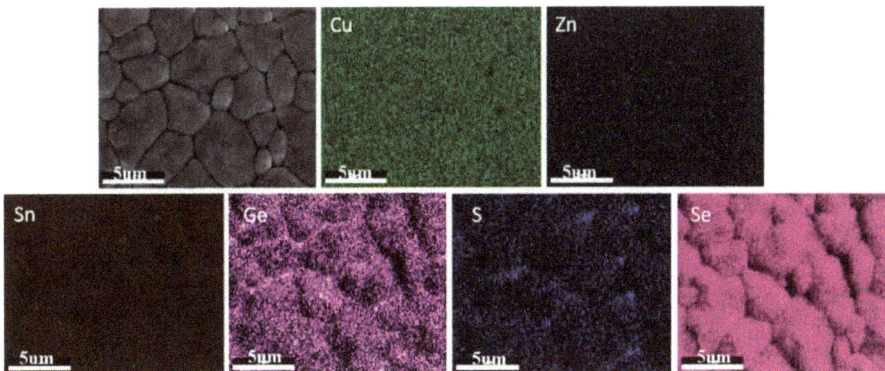

Figure 4. Element mapping analysis of the CZTGSSe film.

We fabricated solar cells based on the CZTSSe and CZTGSSe films. Figure 5 shows the *J*–*V* and QE results of the fabricated solar cells. Compared to the CZTSSe solar cell, the CZTGSSe solar cell showed better performances, even though the distribution of Ge in the CZTGSSe material was inhomogeneous and the composition of CZTGSSe was not optimal. The open-circuit voltage, short-circuit current density, and fill factor for the CZTGSSe solar cell were all larger than the parameters for the CZTSSe solar cell. Therefore, the efficiency of the solar cell was improved from 5.61% to 7.14% by Ge incorporation. QE measurements revealed that the CZTGSSe solar cell could respond better than the CZTSSe solar cell in the wavelength region of 500–1000 nm. This means less recombination of the photo-carriers was achieved in the bulk of the CZTGSSe absorber layer, so more photo-carriers could be collected for the CZTGSSe solar cell [16]. We estimated the band-gap energies of the absorber materials based on the EQE (external quantum efficiency) data (Figure 5c), which were 1.10 eV for CZTSSe and 1.13 eV for CZTGSSe. A larger band-gap energy means less absorption of the light with long wavelength, and therefore the EQE of CZTGSSe solar cell was lower than the QE of the CZTSSe solar cell in the wavelength region above 1000 nm. Moreover, we calculated the E_u energies of the absorber materials based on the EQE data in the long wavelength region, which were 29 meV for the CZTSSe absorber and 22 meV for the CZTGSSe absorber. For materials with low crystallinity or disordered structure, there were large amounts of localized states which could be extended in the band gap. This could cause the exponential part (Urbach tail) near the optical band gap in the absorption coefficient curve. Urbach energy (E_u) is often interpreted as the width of the Urbach tail. A large E_u energy normally means the material has poor crystallinity or disordered micro-structures. A [Cu_{Zn} + Zn_{Cu}] disorder often exists in Cu_2ZnSnS_4-based materials, which is believed to be an important factor limiting the performance of the solar cells [17]. The calculated E_u energies of CZTSSe and CZTGSSe indicated that the Ge incorporation could alleviate the formation of localized states in the material, which is beneficial to the solar cells. The results above prove that the Ge incorporation by our method is a promising way to improve the quality of solar cell absorber material. By further optimizing the preparation process of this method, we believe that high-quality CZTGSSe material with optimized composition and uniform Ge distribution could be obtained.

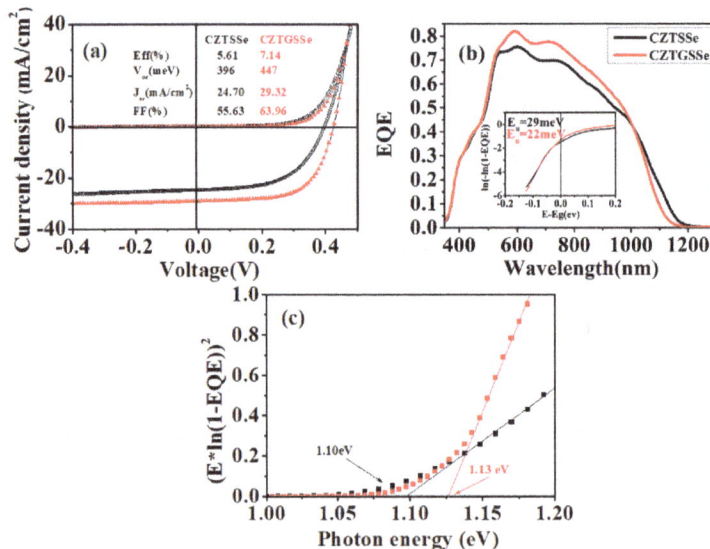

Figure 5. (**a**) *J*–*V* curve, (**b**) external quantum efficiency (EQE), and (**c**) band-gap estimation for CZTSSe and CZTGSSe solar cells.

To further analyze the electrical properties of the solar cells, we estimated the shunt resistance (R_{sh}), series resistance (R_s), diode ideality factor (A), and reverse saturation current density (J_0) of the solar cells based on the $J-V$ data [16]. Figure 6a shows the estimation of G_{sh} ($G_{sh} = 1/R_{sh}$). The G_{sh} of the CZTGSSe solar cell was higher than that of the CZTSSe cell, which may be related to the inhomogeneous distribution of the elements—especially in the grain boundaries (therefore, a small shunt path may exist in the grain boundaries). The estimation in Figure 6b,c gives similar values of ideality factor (2.63 and 2.56 for the CZTSSe solar cell, 1.92 and 1.88 for the CZTGSSe solar cell). The value of the ideality factor can give information about the dominant recombination mechanism in the solar cell. According to [18], an ideality factor lower than but close to 2 means that the main recombination mechanism in the solar cell is the recombination in the space charge region via deep defect levels, while an ideality factor larger than 2 indicates that the main recombination mechanism in the solar cell is the tunnel-enhanced recombination in the space charge region (the tunnel effect can happen for materials with a high concentration of localized states). Therefore, the incorporation of Ge in the absorber layer modifies the main recombination mechanism of the solar cells. This indicates that the Ge incorporation may modify the defect formations in the absorber material, as recombination is closed related with the electrical defects in the material. Figure 6c shows that the J_0 values were 6.3×10^{-2} mA/cm^2 and 6.7×10^{-3} mA/cm^2 for CZTSSe and CZTGSSe solar cells, respectively. The J_0 was dominated by the recombination of the carriers in the solar cells. If the absorber material contains a large amount of deep-level defects, the solar cells can exhibit a severe recombination, which result in a high J_0 [19]. The comparison between the J_0 values for CZTSSe and CZTGSSe solar cells reveals that the incorporation of Ge may decrease the concentration of deep-level defects in the absorber layer. Based on the analysis of E_u energy, diode ideality factor (A), and reverse saturation current density (J_0), we can conclude that the incorporation of Ge in CZTSSe material could modify the formation of defects in the material. The formation of some unfavorable defects for solar cells can be prevented by the Ge incorporation, resulting in the improvement of the solar cell performance.

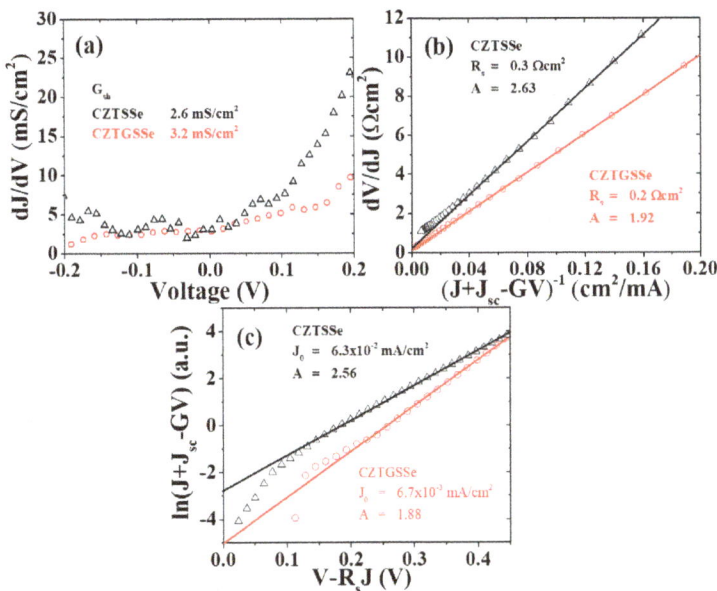

Figure 6. $J-V$ analysis of the CZTSSe and CZTGSSe solar cells: (**a**) G_{sh}; (**b**) R_s and A; (**c**) J_0 and A.

4. Conclusions

CZTSSe films were prepared by annealing the solution-deposited precursor films with Se. When the precursor films were annealed with Se and GeSe$_2$, Ge-incorporated CZTGSSe films were obtained (Ge/(Sn + Ge) as much as 37% in the material). The Ge incorporation greatly increased the sizes of the crystalline grains in the annealed films. However, the EDS analysis revealed that the distribution of elements (especially Ge) in the CZTGSSe films was not uniform. The Ge incorporation also modified the composition of the CZTGSSe film, making the composition deviate from the optimal one. Even given this, the solar cell based on CZTGSSe films performed better than the solar cells based on CZTSSe films. The EQE analysis results proved the Ge incorporation could decrease the E_u energy of the absorber material. Compare to CZTSSe solar cells, the CZTGSSe solar cells exhibited lower diode ideality factor and lower reverse saturation current density, implying that Ge incorporation can prevent the recombination of the photo-carriers in the absorber material. These results indicate that the Ge incorporation by using GeSe$_2$ during the annealing process is a promising way to improve the quality of CZTSSe based material.

Author Contributions: Conceptualization, C.G. and W.Y.; Methodology, Y.S.; Writing, Review, and Editing, C.G.; Supervision, W.Y.; Project Administration, C.G.

Funding: This research was funded by the National Natural Science Foundation of China (No. 61504054) and Natural Science Foundation of Jiangxi Province (No. 20171BAB212018).

Acknowledgments: The authors would like to thank Linyuan Du and Linlin Liu from Hebei University for their help in sample characterization.

Conflicts of Interest: The authors declare no conflict of interest. The funders had no role in the design of the study; in the collection, analyses, or interpretation of data; in the writing of the manuscript, and in the decision to publish the results.

References

1. Liu, X.; Feng, Y.; Cui, H.; Liu, F.; Hao, X.; Conibeer, G.; Mitzi, D.B.; Green, M. The current status and future prospects of kesterite solar cells: A brief review. *Prog. Photovolt. Res. Appl.* **2016**, *24*, 879–898. [CrossRef]
2. Wang, W.; Winkler, M.T.; Gunawan, O.; Gokmen, T.; Todorov, T.K.; Zhu, Y.; Mitzi, D.B. Device characteristics of CZTSSe thin-film solar cells with 12.6% efficiency. *Adv. Energy Mater.* **2014**, *4*, 1301465. [CrossRef]
3. Shin, D.; Saparov, B.; Mitzi, D.B. Defect engineering in multinary earth-abundant chalcogenide photovoltaic materials. *Adv. Energy Mater.* **2017**, *7*, 1602366. [CrossRef]
4. Giraldo, S.; Saucedo, E.; Neuschitzer, M.; Oliva, F.; Placidi, M.; Alcobé, X.; Izquierdo-Roca, V.; Kim, S.; Tampo, H.; Shibata, H.; et al. How small amounts of Ge modify the formation pathways and crystallization of kesterites. *Energy Environ. Sci.* **2018**, *11*, 582–593. [CrossRef]
5. Giraldo, S.; Thersleff, T.; Larramona, G.; Neuschitzer, M.; Pistor, P.; Leifer, K.; Perez-Rodriguez, A.; Moisan, C.; Dennler, G.; Saucedo, E. Cu$_2$ZnSnSe$_4$ solar cells with 10.6% efficiency through innovative absorber engineering with Ge superficial nanolayer. *Prog. Photovolt. Res. Appl.* **2016**, *24*, 1359–1367. [CrossRef]
6. Giraldo, S.; Neuschitzer, M.; Thersleff, T.; Lopez-Marino, S.; Sanchez, Y.; Xie, H.; Colina, M.; Placidi, M.; Pistor, P.; Izquierdo-Roca, V.; et al. Large Efficiency Improvement in Cu$_2$ZnSnSe$_4$ Solar Cells by Introducing a Superficial Ge Nanolayer. *Adv. Energy Mater.* **2015**, *5*, 1501070. [CrossRef]
7. Kim, S.; Kim, K.M.; Tampo, H.; Shibata, H.; Niki, S. Improvement of voltage deficit of Ge-incorporated kesterite solar cell with 12.3% conversion efficiency. *Appl. Phys. Express* **2016**, *9*, 102301. [CrossRef]
8. Umehara, M.; Takeda, Y.; Motohiro, T.; Sakai, T.; Awano, H.; Maekawa, R. Cu$_2$Sn$_{1-x}$Ge$_x$S$_3$ (x = 0.17) Thin-Film Solar Cells with High Conversion Efficiency of 6.0%. *Appl. Phys. Express* **2013**, *6*, 045501. [CrossRef]
9. Mitzi, D.B.; Gunawan, O.; Todorov, T.K.; Wang, K.; Guha, S. The path towards a high-performance solution-processed kesterite solar cell. *Sol. Energy Mater. Sol. Cells* **2011**, *95*, 1421–1436. [CrossRef]
10. Zhang, L.; Yu, H.; Yang, Y.; Yang, K.; Dong, Y.; Huang, S.; Dai, N.; Zhu, D.-M. Synthesis of GeSe$_2$ nanobelts using thermal evaporation and their photoelectrical properties. *J. Nanomater.* **2014**, *2014*, 310716. [CrossRef]
11. Khadka, D.; Kim, S.; Kim, J. Ge-alloyed CZTSe thin film solar cell using molecular precursor adopting spray pyrolysis approach. *RSC Adv.* **2016**, *6*, 37621–37627. [CrossRef]

12. Khadka, D.; Kim, J. Band gap engineering of alloyed $Cu_2ZnGe_xSn_{1-x}Q_4$ (Q = S,Se) films for solar cells. *J. Phys. Chem. C* **2015**, *119*, 1706–1713. [CrossRef]

13. Khadka, D.; Kim, J. Study of structural and optical properties of kesterite $Cu_2ZnGeX4$ (X = S,Se) thin films synthesized by chemical spray pyrolysis. *CrystEngComm* **2013**, *15*, 10500–10509. [CrossRef]

14. Márquez, J.; Stange, H.; Hages, C.J.; Schaefer, N.; Levcenko, S.; Giraldo, S.; Saucedo, E.; Schwarzburg, K.; Abou-Ras, D.; Redinger, A.; et al. Chemistry and dynamics of Ge in kesterite: Toward band-gap-graded absorbers. *Chem. Mater.* **2017**, *29*, 9399–9406. [CrossRef]

15. Hsu, W.; Zhou, H.; Luo, S.; Song, T.; Hsieh, Y.; Duan, H.; Ye, S.; Yang, W.; Hsu, C.-J.; Jiang, C.; et al. Spatial element distribution control in a fully solution-processed nanocrystals-based 8.6% $Cu_2ZnSn(S,Se)_4$ device. *ACS Nano* **2014**, *8*, 9164–9172. [CrossRef] [PubMed]

16. Hegedus, S.; Shafarman, W. Thin-film solar cells: Device measurements and analysis. *Prog. Photovolt. Res. Appl.* **2004**, *12*, 155–176. [CrossRef]

17. Gokmen, T.; Gunawan, O.; Todorov, T.K.; Mitzi, D.B. Band tailing and efficiency limitation in kesterite solar cells. *Appl. Phys. Lett.* **2013**, *103*, 103506. [CrossRef]

18. Rau, U.; Jasenek, A.; Schock, H.; Engelhardt, F.; Meyer, T. Electronic loss mechanisms in chalcopyrite based heterojunction solar cells. *Thin Solid Films* **2000**, *361–362*, 298–302. [CrossRef]

19. Polizzotti, A.; Repins, I.; Noufi, R.; Wei, S.; Mitzi, D. The state and future prospects of kesterite photovoltaics. *Energy Environ. Sci.* **2013**, *6*, 3171–3182. [CrossRef]

coatings

MDPI

Article

Electrochemical Performance of Few-Layer Graphene Nano-Flake Supercapacitors Prepared by the Vacuum Kinetic Spray Method

Mohaned Mohammed Mahmoud Mohammed and Doo-Man Chun *

School of Mechanical Engineering, University of Ulsan, Ulsan 44610, Korea; hano125@gmail.com
* Correspondence: dmchun@ulsan.ac.kr; Tel.: +82-52-259-2706; Fax: +82-52-259-1680

Received: 1 July 2018; Accepted: 24 August 2018; Published: 27 August 2018

Abstract: A few-layer graphene nano-flake thin film was prepared by an affordable vacuum kinetic spray method at room temperature and modest low vacuum conditions. In this economical approach, graphite microparticles, a few layers thick, are deposited on a stainless-steel substrate to form few-layer graphene nano-flakes using a nanoparticle deposition system (NPDS). The NPDS allows for a large area deposition at a low cost and can deposit various metal oxides at room temperature and low vacuum conditions. The morphology and structure of the deposited thin films are alterable by changing the scan speed of the deposition. These changes were verified by field emission scanning electron microscopy (FE-SEM), transmission electron microscopy (TEM), X-ray diffraction (XRD), and Raman spectroscopy. The electrochemical performances of the supercapacitors, fabricated using the deposited films and H_3PO_4–PVA gel electrolytes with different concentrations, were measured using a 2-electrode cell. The electrochemical performance was evaluated by cyclic voltammetry, galvanostatic Charge–discharge, and electrochemical impedance spectroscopy. The proposed affordable fabricated supercapacitors show a high areal capacitance and a small equivalent series resistance.

Keywords: nanoparticle deposition system; few-layer graphene nano-flakes; supercapacitor

---

1. Introduction

Supercapacitors, also called ultracapacitors, are promising electrochemical storage devices due to their high-power density, fast charge/discharge rates, and long Charge–discharge cycles [1–4]. Supercapacitors have the potential to supplement or replace the use of batteries for energy storage applications, namely those for wearable and portable electronics, energy storage systems, and electrical and hybrid vehicles [5]. There are two representative types of supercapacitor: (1) the electric double layer capacitor (EDLC); and (2) the pseudocapacitor. EDLCs store energy via ion adsorption/desorption on the electrode surface, exhibit an excellent cycle life and power density, but are restrained by limited adsorption capacity, which adversely impacts their energy density [6]. Carbon materials with a large specific surface area and excellent electrical conductivity, such as active carbon (AC), carbon nanotubes (CNTs), and graphene, have been used for EDLCs. In contrast, pseudocapacitors store energy via fast and reversible surface redox reactions. Typical pseudo-capacitive materials include transition metal oxides/hydroxides and conducting polymers. Pseudocapacitors hold a much higher energy density but have unsatisfactory cycle stability and rate capability, so their power density is generally low [6]. Graphene is a 2D plane of sp^2 bonded carbon atoms, organized in a honeycomb lattice. There are two reasons why graphene is a particularly suitable material for storage devices. The first reason is that graphene has good electrical conductivity. The electrical conductivity is a result of graphene's unique electronic properties, which include a massless Dirac fermion, an ambipolar electric field effect, and an extremely high carrier mobility [7]. These properties arise from the unique electronic band structure of graphene, which is considered a zero-gap conductor [8]. Moreover, due to the high quality of its 2D crystal structure, graphene exhibits fast transport properties, which also results in a low defect density, allowing graphene to behave like a metal with a high constant mobility [9]. The second reason for graphene's suitability for electronic storage is its high surface area. The theoretical calculation of the specific surface area of graphene yields a value of 2630 m^2/g. This is a high specific surface area even compared with the graphite ($\sim$10 m^2/g) [1] or CNTs (1315 m^2/g) [10]. These advantages make the graphene and their composites suitable for many applications, such as all-solid-state laser scribed flexible graphene supercapacitor [11], flexible and durable graphene oxide fabricated on cotton textile [12], electrochemically doped graphene [13], graphene/polyaniline nanofiber composite [14], and activation of microwave exfoliated GO (MEGO) and thermally exfoliated GO (TEGO) by KOH [15] for supercapacitor applications, in addition to cuprous oxide/reduced graphene oxide (CuO_2/rGO) nanocomposites for light-controlled conductive switching [16]. Also CuO_2/rGO exhibits excellent photocatalytic activity [17], and AgInZnS-graphene oxide (GO) nanocomposites can be used as active biomarkers for noninvasive biomedical imaging [18]. Besides that, there are other properties about graphene and their composites such as the excellent mechanical properties and chemical tunability of graphene oxide paper reported by Dikin et al. [19]. However, there is drawback which limits the use of graphene in some applications. The measured values of the graphene specific surface area are much lower than the theoretical value, in the range of 1000–1800 m^2/g [20]. The lower specific surface area is a result of graphene agglomeration during fabrication of the electrode material, resulting in a loss of surface area [21]. In 2004, Geim and co-workers synthesized the first graphene single layers by mechanical exfoliation using the Scotch tape method [22]. Because of the manual operation, this method has a very small yield, but the graphene obtained by this method was useful for fundamental studies. This method is neither scalable nor can it be adapted for mass production [22,23]. There are two major approaches for producing single layer graphene: the top-down and bottom-up approaches. The purpose of the top-down approach is to reduce the graphite to single or few-layer graphene platelets. Graphite conversion into graphite oxide is a common method in the top-down strategies [19,24]. This familiar top-down approach often requires strong reaction conditions, such as high reduction temperatures, while using concentrated sulfuric acid with highly hazardous procedures [25–27]. Moreover, the obtained graphene always suffers from many defects in the structure, rendering its actual electric conductivity poor, thus restricting its applications [28]. The electrochemical exfoliation method has been reported to be promising for the large-scale production of high-quality graphene, however, this approach consumes a large amount of electric energy and the use of high voltages also facilitates the generation of oxygen groups, as well as structural damage of the graphene [25,29]. Also, liquid-phase exfoliation [30] and graphite intercalation compounds (GICs) [31] are all included in the top-down approaches. The advantages of these approaches are that they produce a large area, with a low cost of fabrication. However, many defects are formed through the exfoliation, so it is impossible to get high-quality single-layer graphene [32]. The bottom-up methods require atom by atom growth, which includes growth on SiC, molecular beam epitaxy, chemical vapor deposition, and chemical synthesis [33]. CVD is a method in which graphene is grown directly on a transition metal substrate through saturation of carbon upon exposure to a hydrocarbon gas at a high temperature [34–40]. One of the major advantages of the CVD technique is its high compatibility with the current complementary metal–oxide–semiconductor (CMOS) technology. However, controlling the film thickness is difficult, and the formation of secondary crystals cannot be avoided [41]. Another disadvantage of the CVD method is that it requires expensive substrate materials for graphene growth, which limits the applications for large-scale production. Further, the transfer to another substrate is a complex process. To avoid the disadvantages of the previous strategies, a promising vacuum kinetic spray method was used to deposit few-layer graphene nanoflakes from graphite microparticles, while under room temperature and low vacuum pressure conditions. This vacuum kinetic spray method is known as a nanoparticle deposition system (NPDS). Previously, a NPDS was developed to deposit metals and ceramics, such as tin, nickel, Al_2O_3, TiO_2, and WO_3, at a low temperature. Additionally, few-layer graphene nanoflake layers have been successfully deposited [42–47]. The NPDS uses relatively low pressure compressed air as a carrier gas and a low vacuum pressure (around 0.04 MPa), for low equipment and processing costs. The impact velocity of the micro-sized graphite particles deposited by NPDS is one of the critical process parameters, and its importance was studied with regard the formation of few-layer graphene flakes on copper foil without binders by Nasim and Chun [46]. Further, these researchers carried out computational fluid dynamics (CFD) simulations to study the relationship between the stand-off distance and the impact velocity. The CFD analysis predicts the critical velocity for fragmentation and interlayer separation of graphite particles for the deposition of a few-layer graphene flake structure on copper, without the presence of unfragmented graphite particles [46]. Another important factor that can affect the deposition of micro-sized graphite particles is the substrate hardness. The substrate-dependent behavior of deposition showed that the degree of fragmentation and interlayer separation increase with an increase in the substrate hardness. By using the NPDS, thick and fragmented graphite particles were deposited on soft substrates, while the deposition on the hard substrates result in a few-layer graphene flake structure [47].

This research introduced a new deposition method for few layered graphene nano-flakes thin film from micro-sized graphite powders without any additives or chemicals, so that this deposition method can be low-cost, and eco-friendly manufacturing technique. The few layered graphene nano-flakes thin film prepared by the NPDS was applied for supercapacitor applications. Herein, the effect of the scan speed on the formation of a few-layer graphene flake structure was studied, and graphite microparticles were deposited on stainless steel using the NPDS for application in relatively low-cost supercapacitors. The surface morphology of the formed films was analyzed by a field-emission scanning electron microscope (FE-SEM), X-ray diffraction (XRD), and Raman spectroscopy. Then, the few-layer graphene nanoflake thin film was used as an electrode of a supercapacitor, with a polyvinyl alcohol with phosphoric acid (PVA–H_3PO_4) gel electrolyte serving as a separator and binding material [48]. A polymer gel electrolyte is often used to provide anions and cations, which participate in the surface process, contributing to the electrochemical capacitance [49]. The electrochemical performance was evaluated in a 2-electrode system using cyclic voltammetry (CV), galvanostatic Charge–discharge, and electrochemical impedance spectroscopy (EIS).

2. Materials and Methods

2.1. Materials

Commercially available graphite powder (MGF 10 995A, Samjung C&G, Gyeongsan, Korea) with a particle size of 10 μm or larger was used in this research for deposition on a stainless steel substrate (SUS304, Nilaco Corporation, Tokyo, Japan), with a thickness of 0.5 mm. Additionally, polyvinyl alcohol (PVA) with an average molecular weight of 88,000 g/mol (#6716-1405, Daejung Chemicals & Metals Co., Ltd., Siheung, Korea), and phosphoric acid (H_3PO_4) with 85% purity (#7664-38-2, Daejung Chemicals & Metals Co., Ltd., Siheung, Korea) were used in the preparation of the supercapacitors.

2.2. Graphene Electrode Preparation

The NPDS consists of a compressor, cylindrical powder feeder, nozzle, vacuum chamber, vacuum pump, vibrator, and controller, as shown in Figure 1. The compressor supplies a compressed air of pressure 0.2 MPa to carry the micro-sized graphite powder from the cylindrical powder feeder to the nozzle which is inside the vacuum chamber. In this research, a slit converging-diverging nozzle of dimensions 50 × 0.2 mm^2 was used to accelerate the powder for a large area deposition. A vibrator was used to shake the powder feeder to carry the powder easily to the nozzle. The distance between the nozzle and the stainless-steel substrate was set to be 5 mm. The y-stage is connected to a permanent magnet AC (PMAC) motor to control the scan speed of the substrate. In this study the deposition was carried out directly on a stainless-steel substrate by three different scan speeds without any additional chemical or binders, as shown in Table 1.

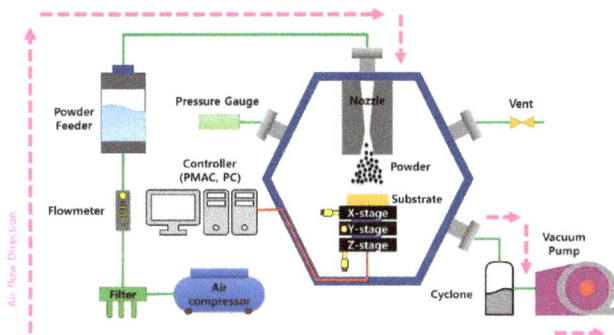

Figure 1. Schematic diagram of the nanoparticle deposition system.

Table 1. Process parameters for graphite deposition on the stainless-steel substrate.

Parameter	Value
Compressor pressure (MPa)	0.2
Chamber pressure (MPa)	0.036–0.04
Scan speed (mm/min)	0.4, 0.8, and 1.2
Distance between substrate and nozzle (mm)	5
Deposition shape	2×5 cm^2
Substrate material	Stainless-steel

2.3. Characterization

The morphology and structure of the deposited film are characterized by field emission scanning electron microscopy (FE-SEM) (JSM-6500F, JEOL, Tokyo, Japan). The crystalline structures of the as-purchased graphite powder and the deposited film on a stainless-steel substrate were examined by XRD (Ultra 4, Rigaku, Tokyo, Japan). Raman Spectra were recorded using Raman spectroscopy (Alpha 300R, WITec GmbH, Ulm, Germany) with a 532 nm wavelength laser operating at 1 mW as an excitation source. Focused ion beam (FIB) (Helios NanoLab 450, FEI, Eindhoven, The Netherlands) milling was carried out to prepare the high-resolution transmission electron microscopy (HR-TEM) sample. A FIB lift-out TEM grid (FIB lift-out Cu TEM grid with 3 posts, Omniprobe, Inc., Dallas, TX, USA) was used to hold the milled HR-TEM sample. Finally, the sample prepared by the FIB was used for HR-TEM (JEM-2100F, JEOL, Tokyo, Japan).

2.4. Preparation of the Polymer Gel Electrolyte

The polymer gel electrolyte was prepared as follows. 1 g of PVA was stirred in 10 mL of deionized water at 80 °C for 2 hours until completely dissolved. After cooling, 0.03 mol (2.94 gm) of H_3PO_4 were added to the PVA and stirring was continued until the solution became a viscous gel. The sane procedure was followed to prepare 0.06 and 0.09 mol of H_3PO_4–PVA gel electrolyte by adding 5.88 and 8.82 gm of H_3PO_4 to the 1 gm of PVA respectively.

2.5. Assembly of Supercapacitor Device

The graphene thin film was deposited on the stainless-steel with area 10×10 mm^2. To fabricate a supercapacitor, the deposited area was dipped on the H_3PO_4–PVA gel electrolyte for a minute, the electrolyte on the electrode was left in air overnight to dry. After this, the two identical electrodes with the dried electrolytes were sandwiched together by applying a 1 kg weight, resting on the top, to ensure a full contact. The fabrication process is shown in Figure 2.

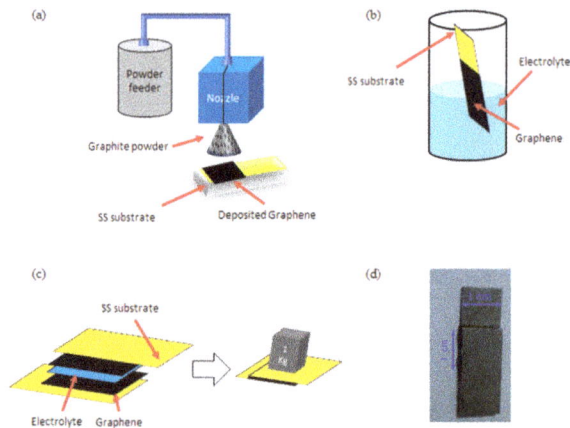

Figure 2. (**a**) Thin film deposition by nanoparticle deposition system (NPDS); (**b**) dipping deposited electrode into the gel electrolyte; (**c**) sandwiching the two identical electrodes and applying mechanical force; and (**d**) actual image of the fabricated supercapacitor.

2.6. Electrochemical Measurements

The electrochemical properties were measured using a 2-electrode cell. The electrochemical performance was measured using a CorrTest electrochemical workstation (C350, Wuhan Corrtest Instruments Corp. Ltd., Wuhan, China), using cyclic voltammetry, galvanostatic Charge–discharge, and electrochemical impedance spectroscopy. The cyclic voltammetry measurement was scanned between 0 and 1 V with different scan rates between 2 and 100 mV/s. The galvanostatic Charge–discharge tests were performed at different current densities. Impedance data was collected, ranging from 1 MHz to 1 Hz, with an open circuit potential (OCP), with an AC signal 10 mV in amplitude.

3. Results and Discussion

3.1. Morphology Properties

After deposition, the masses per unit area and thicknesses of the deposited films with different scan speeds were calculated. The mass per unit area of the deposited films decreased as the scan speed increased from 0.4 to 0.8 and 1.2 mm/min. The thickness of each deposited film was measured by confocal microscopy, showing that as the scan speed changed from 0.4 to 1.2 mm/min the thickness decreased. The values of mass per unit area and the corresponding thicknesses are shown in Table 2 and Figure 3.

Table 2. Thickness and mass values of the deposited films with different scan speeds.

Scan Speed (mm/min)	Thickness (μm)	Mass (mg/cm^2)
0.4	2.53	0.3
0.8	1.54	0.17
1.2	0.85	0.101

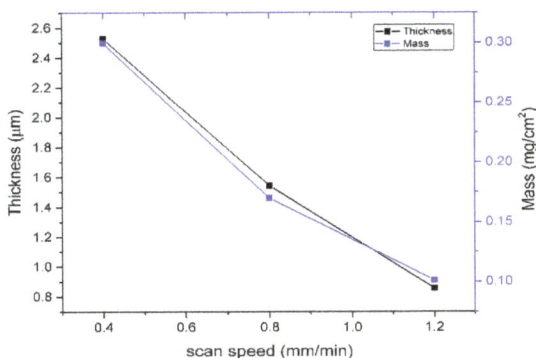

Figure 3. Mass per unit area verses the thickness of the deposited films with scan speeds 0.4, 0.8 and 1.2 mm/min.

The surface morphology of the original graphite powder and the deposited thin film, on a stainless-steel substrate, was observed with an FE-SEM. Figure 4a is the FE-SEM image of the original graphite powder, showing that the sizes of the original powder particles are around 6 μm and the shapes are irregular. Figure 4b–g shows the surface morphology of the deposited films on stainless steel, by comparing Figure 4a with Figure 4b,c, we can see that most of the graphite powder fragmented to small particles; some of these particles are still in microscale graphite form, but most of the other particles are converted to very small flakes. As seen in Figure 4d–g, as the scan speed increased, the relative proportion of the microscale graphite particles on the deposited films become smaller relative to the slow scan speed case (0.4 mm/min). For 1.2 mm/min scan speed of deposition, the sizes of most of the fragmented particles ranged from about 100–180 nm, however, there were some fragmented particles with much smaller size. In Figure 4c,e,g, red arrows indicate some of the graphene nano-flakes. The thin layers of graphite were separated during the NPDS deposition due to their high impact velocity. The orientation of the particle may be random during deposition, and particles with different orientations collided with the substrate. Due to the high-velocity impact, interlayer sliding and interlayer separation occurred and very thin structures were deposited on the stainless-steel substrate. In addition, large particles broke into small pieces to make nano-sized, thin structures.

Figure 5 shows the TEM image of the deposited film with a scan speed of 1.2 mm/min. The HR-TEM image shown in Figure 5a indicates the crystalline structure of randomly oriented graphene nano-flakes with some areas of amorphous structure. The Fast Fourier transformer (FFT) of the HR-TEM image in Figure 5b shows a polycrystalline structure which has arisen from the short-range randomly oriented graphene flakes. The obtained histogram plot is shown in Figure 5c at the surrounded position by the yellow rectangle in Figure 5a indicates an in-plane lattice spacing of 0.26 nm corresponding to the graphene lattice spacing [50,51]. Other crystalline areas showed the same lattice spacing of 0.26 nm, as all positions were measured using the histogram plot. This suggests that all crystalline areas are graphene structured. In Figure 5d, the highly magnified image of the surrounded position by the yellow rectangle in Figure 5a shows a triangular sublattice pattern of carbon atoms instead of a hexagonal pattern. This triangular pattern may come from few-layer graphene flake structures [52,53]. The corresponding FFT clearly shows a hexagonal pattern of the graphene structure corresponding to the (100) plane.

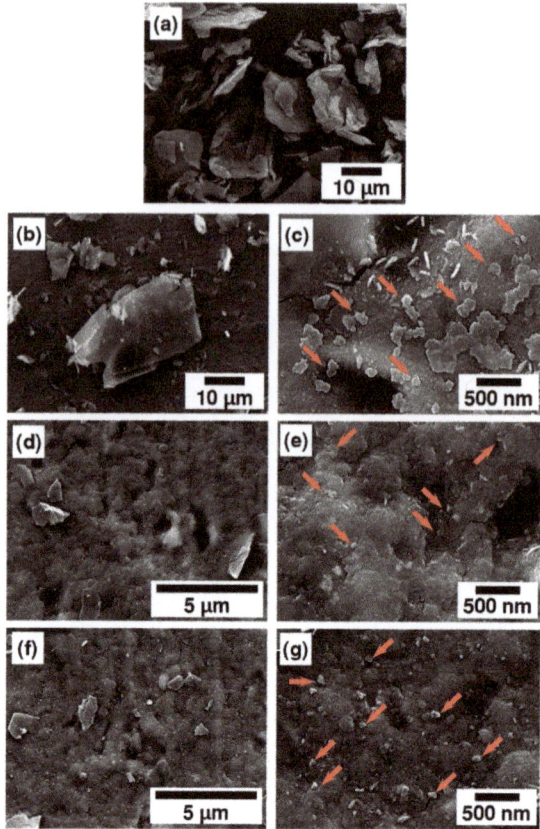

Figure 4. Field emission scanning electron microscopy (FE-SEM) of (**a**) graphite powder, deposited film at scan speeds of (**b**,**c**) 0.4, (**d**,**e**) 0.8, and (**f**,**g**) 1.2 mm/min, respectively.

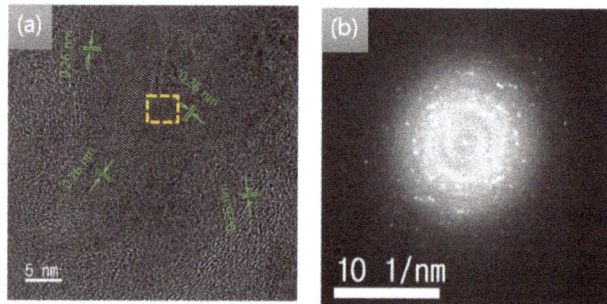

Figure 5. *Cont.*

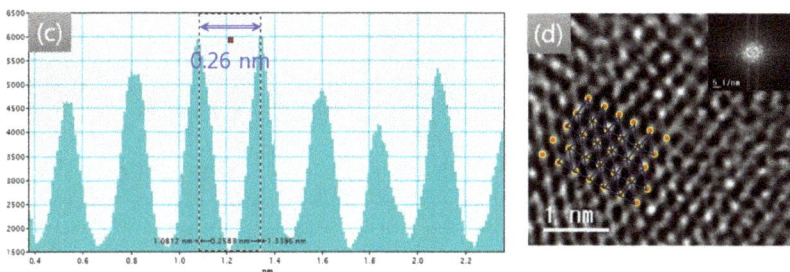

Figure 5. (**a**) High-resolution transmission electron micropscopy (HR-TEM) image of the deposited film with scan speed 1.2 mm/min; (**b**) Fast Fourier transform (FFT) of image (**a**); (**c**) Histogram plot of the area surrounded by the yellow rectangular in (**a**); (**d**) Magnified HR-TEM of the area surrounded by the yellow rectangular in (**a**) (in the top right, FFT of image (**d**) was inserted).

3.2. X-ray Diffraction Analysis

The structural properties of original graphite powder, and the deposited films with different scan speeds (0.4, 0.8, and 1.2 mm/min) and stainless-steel substrate were characterized using XRD analysis. Figure 6 shows the XRD spectrum of the original graphite powder, the sharp and high amplitude peak (002) at 2θ = 26.5°, which is the strongest peak obtained in the diffraction pattern in the direction perpendicular (c-axis) to the graphite hexagonal plates, indicates a highly organized layered structure with an interplanar spacing of 0.365 nm. The XRD spectrum of the deposited films shows peaks γ(111), γ(200), and α(211) at 43.6°, 50.5°, and 82° corresponding to the stainless steel substrate. In the of 0.4 mm/min case, there is (002) graphite peak at 26.3°. This peak disappeared when the scan speed was increased to 0.8 and 1.2 mm/min, which means that the graphene/graphite structure in case of 0.4 mm/min scan speed of deposition transformed to graphene structure as the scan speed increased. Also, the deposited films show a broad carbon (002) peak between 15° and 25°, the broadness of the (002) peak and the shift to a smaller angle indicates a decrease in the particle size due to defragmentation of graphite microparticles to graphene nano-flakes and an increase in the interplanar spacing.

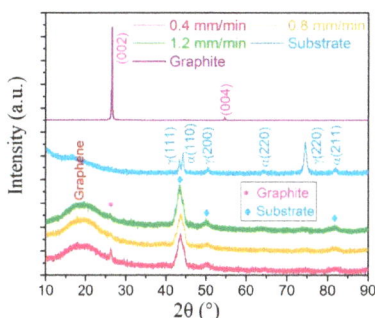

Figure 6. X-ray diffraction (XRD) of the graphite powder, SUS304 substrate, and deposited films at different scan speeds.

3.3. Raman Spectroscopy

The Raman spectroscopy results of the original graphite powder and deposited films are shown in Figure 7. It is obvious that the graphite powder has mainly three bands, D, G, and 2D, at 1352.4, 1582.1, and 2720.77 cm^{-1}, respectively. The G peak is due to the doubly degenerate zone center E_{2g} mode [54].

The G band has the most intense peak and the D band has negligible intensity. The negligible intensity of the D band suggests a crystalline structure with almost no disorder. The 2D band for graphite consists of two bands $2D_1$ and $2D_2$, as shown in the enlarged view of the 2D peak, the double band structure of the 2D peak was explained by other researchers [55–60]. For all of the deposited films under all scan speeds (0.4, 0.8 and 1.2 mm/min), the deposited film on stainless steel substrate showed two different behaviors. During measurement, the laser beam was focused in two different positions for a more accurate measurement. When the laser beam was focused on a larger size, fewer fragmented particles were produced, as shown in Figure 7a. In this case, the D peak was located at 1352.4 cm^{-1} for all deposited films, the G peak positions, for the slower to faster scan speeds, were at 1586, 1592 and 1596 cm^{-1}. We can observe that the ratio of I_D/I_G, as shown in Table 3, becomes higher when the scan speed of deposition increases, which confirms that the amount of graphite structure on the deposited films decreases as the scan speed of the deposition is increases. 2D peaks appear around 2718, 2708 and 2699 cm^{-1}, respectively for the different scan speeds of deposition. The shift of the G band to a higher energy compared to the graphite powder, the sharpness of 2D peak, and the shift in the 2D band to a lower energy, all indicate that while increasing the scan speed of deposition, more of micro-sized graphite particles gradually transform from graphite to a few-layer graphene structure.

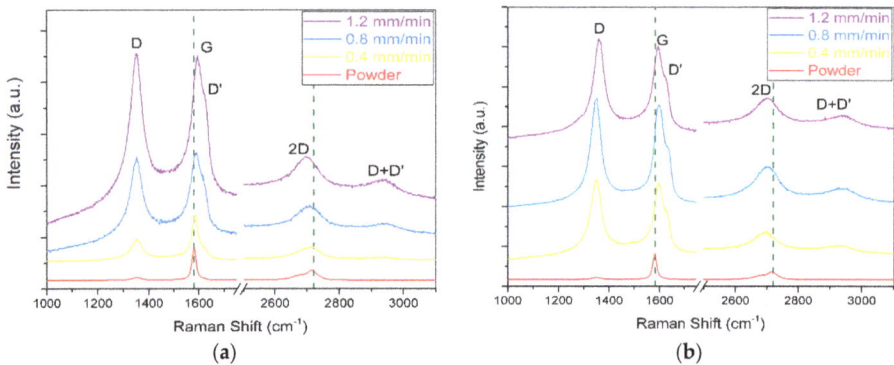

Figure 7. Raman spectroscopy of (**a**) the graphite powder and deposited films at different scan speeds on larger size particles with less fragmentation and (**b**) graphite powder and deposited films at different scan speeds on the more fragmented particles.

When the laser beam was focused on the fragmented particles, the Raman spectrum of the deposited thin films under all scan speeds (0.4, 0.8 and 1.2 mm/min) has a different interesting behavior; as seen in Figure 7b, the intensity of D band peaks at 1352.4, 1352.4, and 1350.8 cm^{-1} respectively, which represents the disorder was higher than the G band appears around 1598.5, 1597, and 1595 cm^{-1}, respectively, for the different scan speeds of deposition, which means that the I_D/I_G for the deposited film is much higher than that of the graphite powder. This increase in I_D/I_G can be described because of the fragmentation of the micro-sized graphite particles to graphene nano-flakes, which increase the edge defects and disorder of the graphene. The G band is shifted to a higher energy compared to the graphite powder, as shown in Figure 7b. The G band has a shoulder at 1654 cm^{-1} (D′ band). Also, the 2D band, which appears around 2698 cm^{-1}, for all deposited films, is shifted to lower energy compared to the graphite powder. The 2D band of the deposited film is sharper than that of the original powder. Moreover, the shift in the 2D band indicates that the thickness of the deposited film is decreases after deposition of the original graphite powder on stainless steel using the NPDS. Based on the shifts in the G and 2D peaks, we concluded that the layers of micron-sized graphite powder were separated due to the high impact velocity during deposition, and that few-layer graphene structures have formed on the stainless steel [55,61]. Furthermore, the strong and sharp D band peak and D′ band peak suggest a nanocrystalline structure, as well as the presence of graphene edges and defects [62–64]. These graphene edges may appear due to fragmented graphene flake structures. Furthermore, defects may have occurred during the high impact velocity deposition, or by interactions between the stainless steel and graphene, which may have produced vacancies, dislocation, and/or dangling bonds [65]. The positions of the D, G, and 2D peaks for the graphite powder and deposited films are shown in Table 3.

Table 3. D, G, and 2D peaks and I_D/I_G for the graphite powder and deposited films.

Sample	Large Size & Less Fragmented Particles				Fragmented Particles			
	D Peak (cm^{-1})	G Peak (cm^{-1})	2D Peak (cm^{-1})	I_D/I_G	D Peak (cm^{-1})	G Peak (cm^{-1})	2D Peak (cm^{-1})	I_D/I_G
Powder	1352.4	1582.1	2720.8	0.269	1352.4	1582.1	2720.8	0.269
0.4 mm/min	1352.4	1586	2718	0.647	1352.4	1598.5	2698	1.054
0.8 mm/min	1352.4	1592	2708	0.970	1352.4	1597	2698	1.064
1.2 mm/min	1352.4	1596	2699	1.009	1350.8	1595	2698	1.069

3.4. Electrochemical Properties

The electrochemical performance of graphene supercapacitors, fabricated from the deposited films (0.4, 0.8 and 1.2 mm/min scan speeds), with the addition of 0.03 mol of H_3PO_4–PVA gel electrolyte, were investigated as part of a 2-electrode system. Figure 8a–c shows the cyclic voltammetry curves measured at the scan rates 2–100 mV/s and potential window 0–1 V. The areal capacitance calculated from the CV curves using the formula

$$C_a = \frac{\int IdV}{sA\Delta V} \tag{1}$$

where I is the response current, s is the scan rate, ΔV is the potential window and A is the area of graphene electrode in contact with the electrolyte. The CV curves of graphene supercapacitor are symmetrical and have a near-rectangular shape over the 1 V potential window, which is typical of an ideal capacitor. The value of areal capacitance was increased from 0.68 mF/cm^2 at 2 mV/s for 0.4 mm/min scan speed to 1.28 mF/cm^2 at 2 mV/s for 1.2 mm/min scan speed, as shown in Figure 8d, two factors explain this improvement of the capacitance: In case of a fast scan speed, after the graphite particles were accelerated to supersonic speeds through the converging-diverging nozzle, the graphite particles impact with the stainless steel substrate, which has relatively high hardness, so a thin layer of graphene will be formed on the substrate, the deposition quickly will cover another location until it covers the required area of substrate. On the other hand, for the slow scan speed, after the acceleration of graphite powder, the impact takes place at first between the powder and stainless-steel substrate. As in the case of a fast scan speed, a thin graphene layer will form over the substrate, but due to the slow scan speed the impact will happen between the graphite powder and the deposited film, which has a lower hardness than the stainless steel. Therefore, the amount of fragmentation of graphite powders becomes lower than that which occurs with a fast deposition scan speed. This is clear from the (002) graphite peak intensity found in the XRD analysis, which decreased as the scan speed increased. Also, this is clear from the Raman spectroscopy, as previously mentioned. In summary, a slow scan speed shows graphene/graphite structure that has a smaller surface area and lower conductivity than graphene structure in the fast scan speed. As such, the improvement of the capacitance can be explained as a result of two reasons. The first reason is that the graphite structure of the deposited film forms more readily with a slow scan speed deposition, such that the specific surface area becomes higher than in the case of a fast scan speed deposition and allow more charge transfer on the surface of the electrode [66]. The second reason is that the number of graphene layers is decreases with an increase in the scan speed, since the value of I_D/I_G for a fast scan speed is larger than that of low scan speed and the decreased number of graphene layers can improve the charge transfer inside the electrode material [67,68].

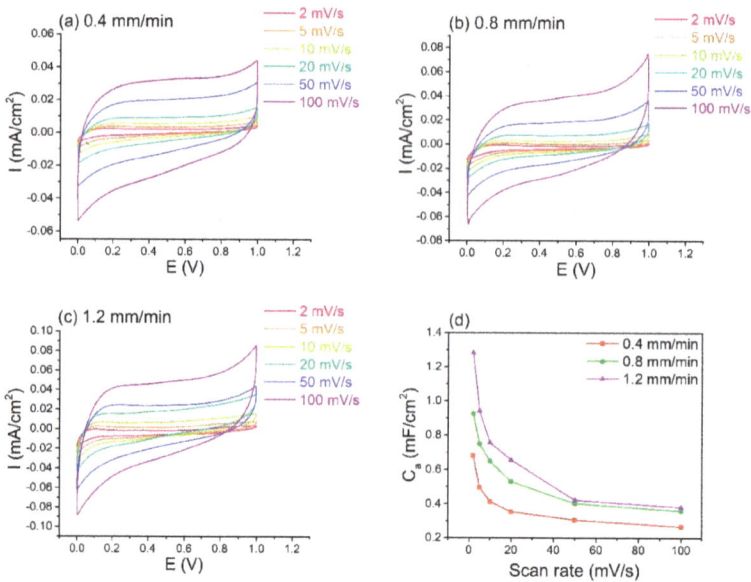

Figure 8. (**a**–**c**) Cyclic voltammetry (CV) curves of 0.03 mol of H_3PO_4 gel electrolyte with different scan speeds; (**d**) Calculated specific areal capacitance from CV curves.

In Figure 9, the cyclic voltammetry was measured again for the deposited films using 0.06 and 0.09 mol H_3PO_4–PVA gel electrolytes. Both electrolytes have quasi-rectangular CV curves and the areal capacitance was 1.28 mF/cm^2 at 2 mV/s for the case of a 1.2 mm/min scan speed with 0.03 mol of H_3PO_4–PVA gel electrolyte. This value became 1.47 mF/cm^2 at 2 mV/s when the concentration of H_3PO_4 was increased to 0.06 mol and increased to 1.67 mF/cm^2 at 2 mV/s for the case of 0.09 mol of H_3PO_4. It appears that the areal capacitance increases gradually while increasing the concentration in the range of 0.03–0.09 mol. These results confirm the assumption that the high ion concentration cloud improves the areal capacitance value, which also agrees with previous research [69,70]. The capacitance of graphene electrode is mainly due to the diffusion of ions from the electrolyte to the surface of the electrode, and the rate of the whole electrode reaction will be enhanced if the diffusion resistance is decreased. In this sense, a high concentration of H_3PO_4–PVA gel electrolyte with a high ionic conductivity will provide a low diffusion resistance. The diffusion resistance (R_{ct}: charge transfer resistance) can be seen in the Nyquist plot. Previous studies have also found that there is a relationship between the specific capacitance of the activated carbons and the electrolyte conductivity, i.e., the specific capacitance increased with increasing electrolyte conductivity [71]. For 0.06 mol of H_3PO_4, the oxidation and reduction peak at ~0.4 and 0.25 V in Figure 9a are corresponding to the pseudocapacitance after phosphorus-doping in graphene layers [72–74]. The doping of phosphorus may be happened during the electrochemical measurement due to the porous structure of the carbon atoms. The existence of phosphorus atoms introduces a positive charge on the neighboring carbon atoms and creates centers for oxygen reduction reaction [73]. Also, in case of 0.09 mol of H_3PO_4, there is a small oxidation and reduction humps at the same potentials due to the same reason. The samples with relatively low concentration of H_3PO_4 (0.03 mol) did not show the oxidation and reduction peak in the CV curves. Therefore, a relatively high concentration of H_3PO_4 may easily create the doping of phosphorus during the electrochemical measurement. The reactions related to pseudocapacitance can be considered as follows [75,76]:

$$\rangle C - OH \rightleftharpoons C = O + H^+ + e^- \qquad (2)$$

$$\rangle C = O + e^- \rightleftharpoons C - O^- \tag{3}$$

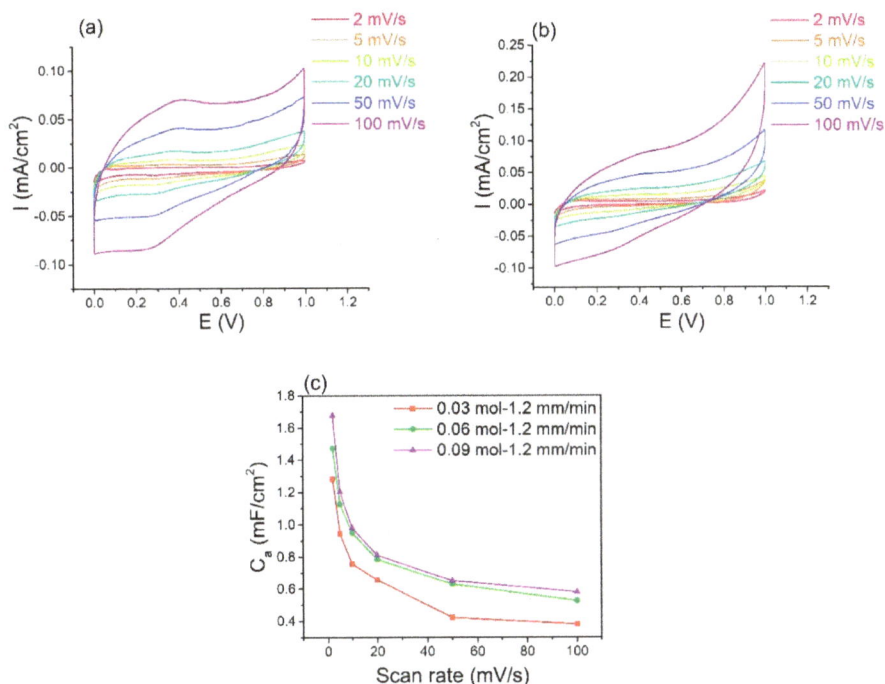

Figure 9. CV curves of (**a**) 0.06 and (**b**) 0.09 mol of H_3PO_4 gel electrolyte with a 1.2 mm/min scan speed electrode; (**c**) The calculated areal capacitance for 0.06 and 0.09 mol of H_3PO_4 gel electrolyte with a 1.2 mm/min scan speed electrode.

The redox reaction attributed to the reversible quinone/hydroquinone pair.

The galvanostatic Charge–discharge curves for the supercapacitors assembled from the deposited films, with different scan speeds (0.4, 0.8 and 1.2 mm/min), in Figure 10a–c, and 0.03, 0.06 and 0.09 mol of H_3PO_4–PVA gel electrolyte, in Figure 11, were investigated at current densities 0.01–0.1 mA. The areal capacitance was calculated from the CV curves using the formula

$$C_a = \frac{I}{\frac{dV}{dt}} \tag{4}$$

where I is the current density and dV/dt is the slope of discharge curve after the IR drop, where R is the internal resistance of the supercapacitor. Figure 10a–c shows the galvanostatic Charge–discharge curves for all supercapacitors, linear charge and discharge curves with a neglectable IR drop, indicating that the electrodes have a lower internal resistance, which leads to better EDL performance. The areal capacitance calculated from discharge curves is shown in Figure 10d. The results are in agreement with cyclic voltammetry, where the areal capacitance is increased by controlling the thickness of the deposited films by the scan speed of the deposition, for 0.03 mol of H_3PO_4–PVA gel electrolyte the areal capacitance was 0.39 mF/cm² for the case of a 0.4 mm/min scan speed. This value increased to 1.07 mF/cm² when the thickness of the deposited film decreased by increasing the scan speed to 1.2 mm/min. Also increasing the concentration of H_3PO_4 in the gel electrolyte from 0.03 to 0.09 mol improves the areal capacitance, which increased to 1.2 mF/cm² for the case of a 1.2 mm/min scan speed with 0.09 mol of H_3PO_4–PVA gel electrolyte. All areal capacitances are calculated from galvanostatic Charge–discharge at 0.01 mA of current.

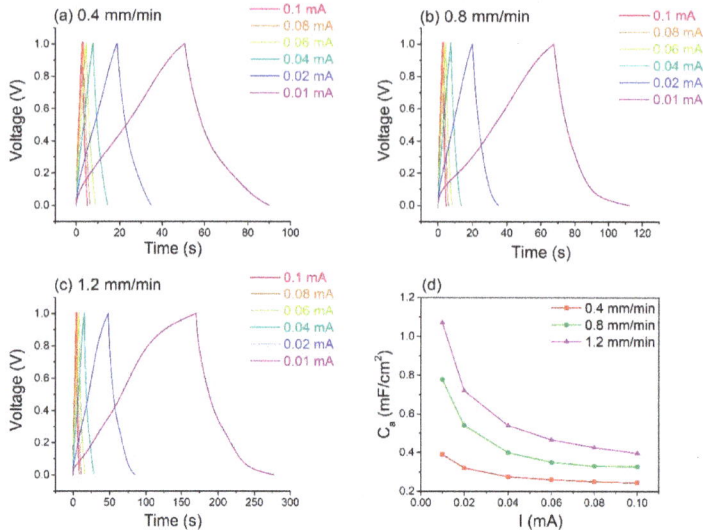

Figure 10. (**a**–**c**) Galvanostatic Charge–discharge curves of 0.03 mol of H_3PO_4 gel electrolyte with different scan speeds; (**d**) Calculated specific areal capacitance from Charge–discharge curves.

Other researchers previously have reported areal capacitances of 3.7 and 2.13 mF/cm^2 by CVD [48] and CVD with chemical treatment [77]. The supercapacitor prepared by NPDS shows an areal capacitance 1.67 mF/cm^2; even though the value is smaller than those found by other researchers, but the fabrication process is much easier, cheaper, and it can be easily applied in industry because it allows a large area deposition in a small amount of time. The supercapacitor prepared by NPDS is superior to those produced by many other, techniques such as LBL (394 µF/cm^2) [78], electrochemical reduction of GO (487 µF/cm^2) [45], vertically oriented graphene (0.087 mF/cm^2) [79], graphene–CNT carpets (0.23 mF/cm^2) [45,80], and graphene–PEDOT: PSS hybrid films (0.179 mF/cm^2) [45,81].

Electrochemical impedance spectroscopy (EIS) measurements were carried out to understand the electrochemical details in the frequency range from 1 Hz to 1 MHz with an open-circuit voltage and an AC amplitude of 10 mV.

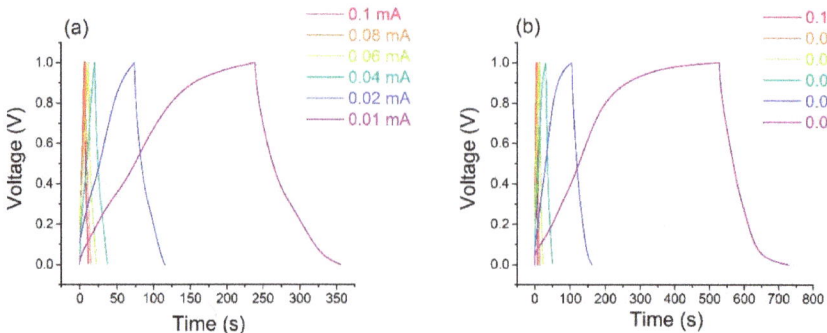

Figure 11. *Cont.*

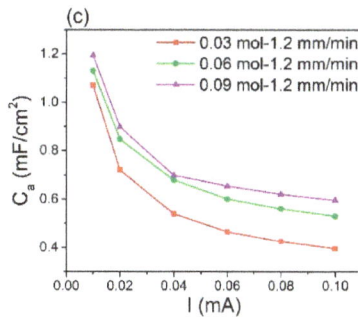

Figure 11. Galvanostatic Charge–discharge curves of (**a**) 0.06 and (**b**) 0.09 mol of H_3PO_4 gel electrolyte with a 1.2 mm/min scan speed electrode; (**c**) The calculated areal capacitance for 0.06 and 0.09 mol of H_3PO_4 gel electrolyte with a 1.2 mm/min scan speed electrode.

As shown in Figure 12a, the fabricated supercapacitors from the three electrodes deposited with different scan speed (0.4, 0.8 and 1.2 mm/min) and 0.03 mol H_3PO_4–PVA all exhibited typical AC impedance characteristics of supercapacitors [82]. In the high-frequency region (Figure 12a enlarge view), the intercept with a real impedance (Z') represents an equivalent series resistance (R_s), which includes the resistance of electrode materials, ionic resistance of electrolyte, and contact resistance between the electrode and current collector; the radius of semicircle is indicative of the electrode conductivity and the charge-transfer resistance (R_{ct}) of electrode materials. R_s is estimated to be around 1.1 Ω for a 0.4 mm/min scan speed electrode, which was larger than that given by a 0.8 mm/min electrode scan speed (0.97 Ω) and 1.2 mm/min scan speed electrode (0.85 Ω). The semicircle diameter represents the value of R_{ct} obtained for the three electrodes and has a small diameter. This may be due to the formation of graphene-graphite structures on the surface of the 0.4 mm/min scan speed electrode; the amount of this formation may have decreased when the scan speed is increased to 0.8 and 1.2 mm/min, thus improving the impedance of the electrodes. In the intermediate frequency region, the slope of the 45° portion of the curves is the Warburg resistance, which represents the ion diffusion/transport in the electrolyte. In the low-frequency region, the capacitive behavior of the 1.2 mm/min scan speed electrode is evident in the curve. The curve is slightly bent but overall exhibits a good capacitive performance.

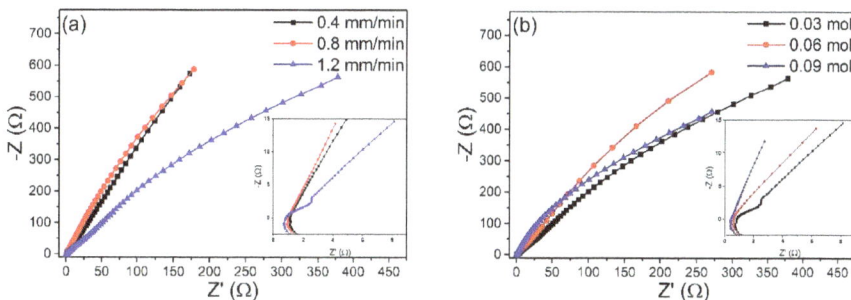

Figure 12. Nyquist plots of (**a**) 0.03 mol of H_3PO_4 gel electrolyte with different scan speeds and (**b**) 0.03, 0.06 and 0.09 mol of H_3PO_4 gel electrolyte with a 1.2 mm/min scan speed electrode.

Nyquist plots of the three typical 1.2 mm/min scan speed electrodes with 0.03, 0.06 and 0.09 mol H_3PO_4–PVA electrolytes were measured, as shown in Figure 12b. R_s values were 0.85, 0.62, and 0.42 Ω respectively, the supercapacitor fabricated with 0.09 mol H_3PO_4–PVA electrolyte exhibits a lower internal resistance and a smaller semicircular diameter at high frequencies, indicating a better interface between the electrolyte and electrode. The conductivity is dependent on the concentration of the H_3PO_4 in the gel electrolyte; as the H_3PO_4 concentration increases, the conductivity also increases, meaning the value of R_s decreases as the concentration increases. As a result of increasing the conductivity, mobility of the H_3PO_4 is increased; this can consequently decrease the value of R_{ct}, which is represented by the radius of semicircle. This result agrees with previous reports [48,83].

The cyclic stability of the deposited film with a 1.2 mm/min scan speed and 0.09 mol of H_3PO_4–PVA was investigated at current 0.06 mA, repeating the galvanostatic charge/discharge measurement ranging from 0 to 1 V. As shown in Figure 13, the supercapacitor has a coulombic efficiency of 98% at the beginning, which decreased to 95% after 5000 cycles. The good cycling performance of the electrode material indicated a good stability and strong adherence to the stainless-steel substrate. The good cycling stability can be attributed to two reasons. The first reason is that the few-layer graphene nanoflake structure is randomly oriented on the substrate, which allows the electrolyte ions to diffuse through the graphene layers, leading to a good improvement in the electrochemical performance of the supercapacitor. The second reason is the mechanical stability caused by the strong adherence of the few-layer graphene nanoflake structures on the stainless-steel substrate. The degradation in the coulombic efficiency after 5000 cycles is likely due to the consumption of the electrolyte as a result of irreversible reactions between the graphene electrode and the H_3PO_4–PVA gel electrolyte.

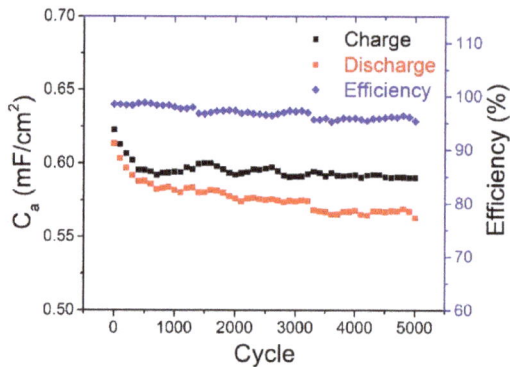

Figure 13. Charge–discharge stability over 5000 cycles for a 1.2 mm/min scan speed and 0.09 mol of H_3PO_4–PVA at a current of 0.06 mA.

4. Conclusions

In this research, a vacuum kinetic spray was used to deposit micro-sized graphite powder on stainless steel substrate at room temperature and low-pressure conditions. The NPDS results in the formation of randomly oriented few-layer graphene nano-flakes; as confirmed by FE-SEM, HR-TEM, XRD, and Raman spectroscopy. Some of the graphite particles fragmented to graphene nano-flakes and some particles did not fragment and were converted to graphite particles with a smaller size than the particles in the original graphite powder. The amount of graphite structure on the deposited films decreases with an increase in the deposition scan speed.

The electrochemical capacitive behavior of the supercapacitors assembled from the deposited films using NPDS, with 0.03 mol PVA-H_3PO_4, showed an areal capacitance comparable to other research, and it was improved by increasing the scan speed of the deposition. Also, increasing the amount of H_3PO_4 in the PVA gel electrolyte can improve the areal capacitance. The very affordably fabricated supercapacitor with 0.09 mol H_3PO_4–PVA gel electrolyte, produced with a 1.2 mm/min deposition scan speed, showed a good stability over 5000 cycles.

Author Contributions: Conceptualization, D.-M.C. and M.M.M.M.; Methodology, M.M.M.M.; Formal Analysis, M.M.M.M.; Investigation, M.M.M.M.; Resources, D.-M.C.; Data Curation, M.M.M.M.; Writing–Original Draft Preparation, M.M.M.M.; Writing–Review & Editing, D.-M.C. and M.M.M.M.; Visualization, M.M.M.M.; Supervision, D.-M.C.; Project Administration, D.-M.C.; Funding Acquisition, D.-M.C.

Funding: This research was supported by a National Research Foundation of Korea (NRF) (NRF-2018R1A2B6004012), Nano-Convergence Foundation (www.nanotech2020.org), and the Ministry of Science and ICT (MSIT, Korea) & Ministry of Trade, Industry and Energy (MOTIE, Korea) (Project name: Establishment of Battery/Ess-Based Energy Industry Innovation Ecosystem).

Conflicts of Interest: The authors declare no conflicts of interest.

References

1. Miller, J.R.; Simon, P. Electrochemical capacitors for energy management. *Science* **2008**, *321*, 651–652. [CrossRef] [PubMed]

2. Simon, P.; Gogotsi, Y. Materials for electrochemical capacitors. *Nat. Mater.* **2008**, *7*, 845–854. [CrossRef] [PubMed]

3. Liu, C.; Li, F.; Ma, L.-P.; Cheng, H.-M. Advanced materials for energy storage. *Adv. Mater.* **2010**, *22*, E28–E62. [CrossRef] [PubMed]

4. Kötz, R.; Carlen, M. Principles and applications of electrochemical capacitors. *Electrochim. Acta* **2000**, *45*, 2483–2498. [CrossRef]

5. Kaempgen, M.; Chan, C.K.; Ma, J.; Cui, Y.; Gruner, G. Printable thin film supercapacitors using single-walled carbon nanotubes. *Nano Lett.* **2009**, *9*, 1872–1876. [CrossRef] [PubMed]

6. Shakir, I. High energy density based flexible electrochemical supercapacitors from layer-by-layer assembled multiwall carbon nanotubes and graphene. *Electrochim. Acta* **2014**, *129*, 396–400. [CrossRef]

7. Geim, A.K.; Novoselov, K.S. The rise of graphene. *Nat. Mater.* **2007**, *6*, 183–191. [CrossRef] [PubMed]

8. Geim, A.K. Graphene: Status and prospects. *Science* **2009**, *324*, 1530–1534. [CrossRef] [PubMed]

9. Novoselov, K.S.; Geim, A.K.; Morozov, S.V.; Jiang, D.; Katsnelson, M.I.; Grigorieva, I.V.; Dubonos, S.V.; Firsov, A.A. Two-dimensional gas of massless Dirac fermions in graphene. *Nature* **2005**, *438*, 197–200. [CrossRef] [PubMed]

10. Peigney, A.; Laurent, C.; Flahaut, E.; Bacsa, R.R.; Rousset, A. Specific surface area of carbon nanotubes and bundles of carbon nanotubes. *Carbon* **2001**, *39*, 507–514. [CrossRef]

11. El-Kady, M.F.; Strong, V.; Dubin, S.; Kaner, R.B. Laser scribing of high-performance and flexible graphene-based electrochemical capacitors. *Science* **2012**, *335*, 1326–1330. [CrossRef] [PubMed]

12. Abdelkader, A.M.; Karim, N.; Vallés, C.; Afroj, S.; Novoselov, K.S.; Yeates, S.G. Ultraflexible and robust graphene supercapacitors printed on textiles for wearable electronics applications. *2D Mater.* **2017**, *4*, 035016. [CrossRef]

13. Abdelkader, A.M.; Fray, D.J. Controlled electrochemical doping of graphene-based 3D nanoarchitecture electrodes for supercapacitors and capacitive deionisation. *Nanoscale* **2017**, *9*, 14548–14557. [CrossRef] [PubMed]

14. Wu, Q.; Xu, Y.; Yao, Z.; Liu, A.; Shi, G. Supercapacitors based on flexible graphene/polyaniline nanofiber composite films. *ACS Nano* **2010**, *4*, 1963–1970. [CrossRef] [PubMed]

15. Zhu, Y.; Murali, S.; Stoller, M.D.; Ganesh, K.; Cai, W.; Ferreira, P.J.; Pirkle, A.; Wallace, R.; Cychosz, K.A.; Thommes, M.; et al. Carbon-based supercapacitors produced by activation of graphene. *Science* **2011**, *332*, 1537–1541. [CrossRef] [PubMed]

16. Wei, J.; Zang, Z.; Zhang, Y.; Wang, M.; Du, J.; Tang, X. Enhanced performance of light-controlled conductive switching in hybrid cuprous oxide/reduced graphene oxide (Cu_2O/rGO) nanocomposites. *Opt. Lett.* **2017**, *42*, 911–914. [CrossRef] [PubMed]

17. Huang, H.; Zhang, J.; Jiang, L.; Zang, Z. Preparation of cubic Cu_2O nanoparticles wrapped by reduced graphene oxide for the efficient removal of rhodamine B. *J. Alloys Compd.* **2017**, *718*, 112–115. [CrossRef]

18. Zang, Z.; Zeng, X.; Wang, M.; Hu, W.; Liu, C.; Tang, X. Tunable photoluminescence of water-soluble AgInZnS–graphene oxide (GO) nanocomposites and their application in-vivo bioimaging. *Sens. Actuators B Chem.* **2017**, *252*, 1179–1186. [CrossRef]

19. Dikin, D.A.; Stankovich, S.; Zimney, E.J.; Piner, R.D.; Dommett, G.H.B.; Evmenenko, G.; Nguyen, S.T.; Ruoff, R.S. Preparation and characterization of graphene oxide paper. *Nature* **2007**, *448*, 457–460. [CrossRef] [PubMed]

20. Segal, M. Selling graphene by the ton. *Nat. Nanotechnol.* **2009**, *4*, 612–614. [CrossRef] [PubMed]

21. Wu, Z.-S.; Wang, D.-W.; Ren, W.; Zhao, J.; Zhou, G.; Li, F.; Cheng, H.-M. Anchoring hydrous RuO₂ on Graphene sheets for high-performance electrochemical capacitors. *Adv. Funct. Mater.* **2010**, *20*, 3595–3602. [CrossRef]

22. Novoselov, K.S.; Geim, A.K.; Morozov, S.V.; Jiang, D.; Zhang, Y.; Dubonos, S.V.; Grigorieva, I.V.; Firsov, A.A. Electric Field effect in atomically thin carbon films. *Science* **2004**, *306*, 666–669. [CrossRef] [PubMed]

23. Singh, V.; Joung, D.; Zhai, L.; Das, S.; Khondaker, S.I.; Seal, S. Graphene based materials: Past, present and future. *Prog. Mater. Sci.* **2011**, *56*, 1178–1271. [CrossRef]

24. Park, S.; Ruoff, R.S. Chemical methods for the production of graphenes. *Nat. Nanotechnol.* **2009**, *4*, 217–224. [CrossRef] [PubMed]

25. Ambrosi, A.; Pumera, M. Electrochemically exfoliated graphene and graphene oxide for energy storage and electrochemistry applications. *Chem. A Eur. J.* **2016**, *22*, 153–159. [CrossRef] [PubMed]

26. Chen, C.-H.; Yang, S.-W.; Chuang, M.-C.; Woon, W.-Y.; Su, C.-Y. Towards the continuous production of high crystallinity graphene via electrochemical exfoliation with molecular in situ encapsulation. *Nanoscale* **2015**, *7*, 15362–15373. [CrossRef] [PubMed]

27. Abdelkader, A.M.; Cooper, A.J.; Dryfe, R.A.W.; Kinloch, I.A. How to get between the sheets: A review of recent works on the electrochemical exfoliation of graphene materials from bulk graphite. *Nanoscale* **2015**, *7*, 6944–6956. [CrossRef] [PubMed]

28. Ossonon, B.D.; Bélanger, D. Functionalization of graphene sheets by the diazonium chemistry during electrochemical exfoliation of graphite. *Carbon* **2017**, *111*, 83–93. [CrossRef]

29. Yang, S.; Brüller, S.; Wu, Z.-S.; Liu, Z.; Parvez, K.; Dong, R.; Richard, F.; Samorì, P.; Feng, X.; Müllen, K. Organic Radical-assisted electrochemical exfoliation for the scalable production of high-quality graphene. *J. Am. Chem. Soc.* **2015**, *137*, 13927–13932. [CrossRef] [PubMed]

30. Hernandez, Y.; Nicolosi, V.; Lotya, M.; Blighe, F.M.; Sun, Z.; De, S.; McGovern, I.T.; Holland, B.; Byrne, M.; Gun'Ko, Y.K.; et al. High-yield production of graphene by liquid-phase exfoliation of graphite. *Nat. Nanotechnol.* **2008**, *3*, 563–568. [CrossRef] [PubMed]

31. Behabtu, N.; Lomeda, J.R.; Green, M.J.; Higginbotham, A.L.; Sinitskii, A.; Kosynkin, D.V.; Tsentalovich, D.; Parra-Vasquez, A.N.G.; Schmidt, J.; Kesselman, E.; et al. Spontaneous high-concentration dispersions and liquid crystals of graphene. *Nat. Nanotechnol.* **2010**, *5*, 406–411. [CrossRef] [PubMed]

32. Sun, Y.; Wu, Q.; Shi, G. Graphene based new energy materials. *Energy Environ. Sci.* **2011**, *4*, 1113–1132. [CrossRef]

33. Ferrari, A.C.; Bonaccorso, F.; Fal'ko, V.; Novoselov, K.S.; Roche, S.; Boggild, P.; Borini, S.; Koppens, F.H.L.; Palermo, V.; Pugno, N.; et al. Science and technology roadmap for graphene, related two-dimensional crystals, and hybrid systems. *Nanoscale* **2015**, *7*, 4598–4810. [CrossRef] [PubMed]

34. Reina, A.; Jia, X.; Ho, J.; Nezich, D.; Son, H.; Bulovic, V.; Dresselhaus, M.S.; Kong, J. Large area, few-layer graphene films on arbitrary substrates by chemical vapor deposition. *Nano Lett.* **2009**, *9*, 30–35. [CrossRef] [PubMed]

35. Bae, S.; Kim, H.; Lee, Y.; Xu, X.; Park, J.-S.; Zheng, Y.; Balakrishnan, J.; Lei, T.; Kim, H.R.; Song, Y.I.; et al. Roll-to-roll production of 30-inch graphene films for transparent electrodes. *Nat. Nanotechnol.* **2010**, *5*, 574–578. [CrossRef] [PubMed]

36. Cai, W.; Moore, A.L.; Zhu, Y.; Li, X.; Chen, S.; Shi, L.; Ruoff, R.S. Thermal transport in suspended and supported monolayer graphene grown by chemical vapor deposition. *Nano Lett.* **2010**, *10*, 1645–1651. [CrossRef] [PubMed]

37. Cai, W.; Piner, R.D.; Zhu, Y.; Li, X.; Tan, Z.; Floresca, H.C.; Yang, C.; Lu, L.; Kim, M.J.; Ruoff, R.S. Synthesis of isotopically-labeled graphite films by cold-wall chemical vapor deposition and electronic properties of graphene obtained from such films. *Nano Res.* **2009**, *2*, 851. [CrossRef]

38. Fallahazad, B.; Hao, Y.; Lee, K.; Kim, S.; Ruoff, R.S.; Tutuc, E. Quantum hall effect in Bernal stacked and twisted bilayer graphene grown on Cu by chemical vapor deposition. *Phys. Rev. B* **2012**, *85*, 201408. [CrossRef]

39. Li, X.; Magnuson, C.W.; Venugopal, A.; Tromp, R.M.; Hannon, J.B.; Vogel, E.M.; Colombo, L.; Ruoff, R.S. Large-area graphene single crystals grown by low-pressure chemical vapor deposition of methane on copper. *J. Am. Chem. Soc.* **2011**, *133*, 2816–2819. [CrossRef] [PubMed]

40. Suk, J.W.; Kitt, A.; Magnuson, C.W.; Hao, Y.; Ahmed, S.; An, J.; Swan, A.K.; Goldberg, B.B.; Ruoff, R.S. Transfer of CVD-grown monolayer graphene onto arbitrary substrates. *ACS Nano* **2011**, *5*, 6916–6924. [CrossRef] [PubMed]

41. Allen, M.J.; Tung, V.C.; Kaner, R.B. Honeycomb carbon: A review of graphene. *Chem. Rev.* **2010**, *110*, 132–145. [CrossRef] [PubMed]

42. Chun, D.M.; Kim, M.H.; Lee, J.C.; Ahn, S.H. TiO$_2$ coating on metal and polymer substrates by nano-particle deposition system (NPDS). *CIRP Ann. Manuf. Technol.* **2008**, *57*, 551–554. [CrossRef]

43. Chun, D.-M.; Kim, M.-H.; Lee, J.-C.; Ahn, S.-H. A nano-particle deposition system for ceramic and metal coating at room temperature and low vacuum conditions. *Int. J. Precis. Eng. Manuf.* **2008**, *9*, 51–53.

44. Jung, K.; Song, W.; Chun, D.-M.; Kim, Y.-H.; Yeo, J.-C.; Kim, M.-S.; Ahn, S.-H.; Lee, C.-S. Nickel line patterning using silicon supersonic micronozzle integrated with a nanoparticle deposition system. *Jpn. J. Appl. Phys.* **2010**, *49*, 05EC09. [CrossRef]

45. Sheng, K.; Sun, Y.; Li, C.; Yuan, W.; Shi, G. Ultrahigh-rate supercapacitors based on eletrochemically reduced graphene oxide for ac line-filtering. *Sci. Rep.* **2012**, *2*, 247. [CrossRef] [PubMed]

46. Nasim, M.N.E.A.A.; Chun, D.-M. Formation of few-layer graphene flake structures from graphite particles during thin film coating using dry spray deposition method. *Thin Solid Films* **2017**, *622*, 34–40. [CrossRef]

47. Nasim, M.N.E.A.A.; Chun, D.-M. Substrate-dependent deposition behavior of graphite particles dry-sprayed at room temperature using a nano-particle deposition system. *Surf. Coat. Technol.* **2017**, *309*, 172–178. [CrossRef]

48. Chen, Q.; Li, X.; Zang, X.; Cao, Y.; He, Y.; Li, P.; Wang, K.; Wei, J.; Wu, D.; Zhu, H. Effect of different gel electrolytes on graphene-based solid-state supercapacitors. *RSC Adv.* **2014**, *4*, 36253–36256. [CrossRef]

49. Li, S.; Wang, X.; Xing, H.; Shen, C. Micro supercapacitors based on a 3D structure with symmetric graphene or activated carbon electrodes. *J. Micromech. Microeng.* **2013**, *23*, 114013. [CrossRef]

50. Yu, K.; Zhao, W.; Wu, X.; Zhuang, J.; Hu, X.; Zhang, Q.; Sun, J.; Xu, T.; Chai, Y.; Ding, F.; et al. In situ atomic-scale observation of monolayer graphene growth from SiC. *Nano Res.* **2018**, *11*, 2809–2820. [CrossRef]

51. Muthurasu, A.; Dhandapani, P.; Ganesh, V. Facile and simultaneous synthesis of graphene quantum dots and reduced graphene oxide for bio-imaging and supercapacitor applications. *New J. Chem.* **2016**, *40*, 9111–9124. [CrossRef]

52. Kumar, G.S.; Thupakula, U.; Sarkar, P.K.; Acharya, S. Easy extraction of water-soluble graphene quantum dots for light emitting diodes. *RSC Adv.* **2015**, *5*, 27711–27716. [CrossRef]

53. Robertson, A.W.; Warner, J.H. Atomic resolution imaging of graphene by transmission electron microscopy. *Nanoscale* **2013**, *5*, 4079–4093. [CrossRef] [PubMed]

54. Tuinstra, F.; Koenig, J.L. Raman spectrum of graphite. *J. Chem. Phys.* **1970**, *53*, 1126–1130. [CrossRef]

55. Ferrari, A.C. Raman spectroscopy of graphene and graphite: Disorder, electron–phonon coupling, doping and nonadiabatic effects. *Solid State Commun.* **2007**, *143*, 47–57. [CrossRef]

56. Cançado, L.G.; Pimenta, M.A.; Neves, B.R.A.; Medeiros-Ribeiro, G.; Enoki, T.; Kobayashi, Y.; Takai, K.; Fukui, K.; Dresselhaus, M.S.; Saito, R.; et al. Anisotropy of the Raman spectra of nanographite ribbons. *Phys. Rev. Lett.* **2004**, *93*, 047403. [CrossRef] [PubMed]

57. Nemanich, R.J.; Solin, S.A. First- and second-order Raman scattering from finite-size crystals of graphite. *Phys. Rev. B* **1979**, *20*, 392–401. [CrossRef]

58. Vidano, R.P.; Fischbach, D.B.; Willis, L.J.; Loehr, T.M. Observation of Raman band shifting with excitation wavelength for carbons and graphites. *Solid State Commun.* **1981**, *39*, 341–344. [CrossRef]

59. Pócsik, I.; Hundhausen, M.; Koós, M.; Ley, L. Origin of the D peak in the Raman spectrum of microcrystalline graphite. *J. Non Cryst. Solids* **1998**, *227*, 1083–1086. [CrossRef]

60. Maultzsch, J.; Reich, S.; Thomsen, C.; Requardt, H.; Ordejón, P. Phonon dispersion in graphite. *Phys. Rev. Lett.* **2004**, *92*, 075501. [CrossRef] [PubMed]

61. Wang, H.; Wang, Y.; Cao, X.; Feng, M.; Lan, G. Vibrational properties of graphene and graphene layers. *J. Raman Spectrosc.* **2009**, *40*, 1791–1796. [CrossRef]

62. Hiramatsu, M.; Shiji, K.; Amano, H.; Hori, M. Fabrication of vertically aligned carbon nanowalls using capacitively coupled plasma-enhanced chemical vapor deposition assisted by hydrogen radical injection. *Appl. Phys. Lett.* **2004**, *84*, 4708–4710. [CrossRef]

63. Hiramatsu, M.; Hori, M. Fabrication of carbon nanowalls using novel plasma processing. *Jpn. J. Appl. Phys.* **2006**, *45*, 5522. [CrossRef]

64. Mori, T.; Hiramatsu, M.; Yamakawa, K.; Takeda, K.; Hori, M. Fabrication of carbon nanowalls using electron beam excited plasma-enhanced chemical vapor deposition. *Diam. Relat. Mater.* **2008**, *17*, 1513–1517. [CrossRef]

65. Ni, Z.; Wang, Y.; Yu, T.; Shen, Z. Raman spectroscopy and imaging of graphene. *Nano Res.* **2008**, *1*, 273–291. [CrossRef]

66. Wang, M.; Liu, Q.; Sun, H.; Stach, E.A.; Zhang, H.; Stanciu, L.; Xie, J. Preparation of high-surface-area carbon nanoparticle/graphene composites. *Carbon* **2012**, *50*, 3845–3853. [CrossRef]

67. Liu, J.; Notarianni, M.; Will, G.; Tiong, V.T.; Wang, H.; Motta, N. Electrochemically exfoliated graphene for electrode films: effect of graphene flake thickness on the sheet resistance and capacitive properties. *Langmuir* **2013**, *29*, 13307–13314. [CrossRef] [PubMed]

68. Galindo, B.; Alcolea, S.G.; Gómez, J.; Navas, A.; Murguialday, A.O.; Fernandez, M.P.; Puelles, R.C. Effect of the number of layers of graphene on the electrical properties of TPU polymers. *IOP Conf. Ser. Mater. Sci. Eng.* **2014**, *64*, 012008. [CrossRef]

69. Srinivasan, V.; Weidner, J.W. Studies on the capacitance of nickel oxide films: effect of heating temperature and electrolyte concentration. *J. Electrochem. Soc.* **2000**, *147*, 880–885. [CrossRef]

70. Zheng, J.P.; Jow, T.R. The effect of salt concentration in electrolytes on the maximum energy storage for double layer capacitors. *J. Electrochem. Soc.* **1997**, *144*, 2417–2420. [CrossRef]

71. Torchała, K.; Kierzek, K.; Machnikowski, J. Capacitance behavior of KOH activated mesocarbon microbeads in different aqueous electrolytes. *Electrochim. Acta* **2012**, *86*, 260–267. [CrossRef]

72. Yu, X.; Kim, H.J.; Hong, J.-Y.; Jung, Y.M.; Kwon, K.D.; Kong, J.; Park, H.S. Elucidating surface redox charge storage of phosphorus-incorporated graphenes with hierarchical architectures. *Nano Energy* **2015**, *15*, 576–586. [CrossRef]

73. Wen, Y.; Wang, B.; Huang, C.; Wang, L.; Hulicova-Jurcakova, D. Synthesis of phosphorus-doped graphene and its wide potential window in aqueous supercapacitors. *Chem. A Eur. J.* **2014**, *21*, 80–85. [CrossRef] [PubMed]

74. Thirumal, V.; Pandurangan, A.; Jayavel, R.; Venkatesh, K.S.; Palani, N.S.; Ragavan, R.; Ilangovan, R. Single pot electrochemical synthesis of functionalized and phosphorus doped graphene nanosheets for supercapacitor applications. *J. Mater. Sci. Mater. Electron.* **2015**, *26*, 6319–6328. [CrossRef]

75. Pan, H.; Poh, C.K.; Feng, Y.P.; Lin, J. Supercapacitor electrodes from tubes-in-tube carbon nanostructures. *Chem. Mater.* **2007**, *19*, 6120–6125. [CrossRef]

76. Xie, B.; Chen, Y.; Yu, M.; Zhang, S.; Lu, L.; Shu, Z.; Zhang, Y. Phosphoric acid-assisted synthesis of layered MoS$_2$/graphene hybrids with electrolyte-dependent supercapacitive behaviors. *RSC Adv.* **2016**, *6*, 89397–89406. [CrossRef]

77. Li, X.; Zhao, T.; Chen, Q.; Li, P.; Wang, K.; Zhong, M.; Wei, J.; Wu, D.; Wei, B.; Zhu, H. Flexible all solid-state supercapacitors based on chemical vapor deposition derived graphene fibers. *Phys. Chem. Chem. Phys.* **2013**, *15*, 17752–17757. [CrossRef] [PubMed]

78. Yoo, J.J.; Balakrishnan, K.; Huang, J.; Meunier, V.; Sumpter, B.G.; Srivastava, A.; Conway, M.; Mohana Reddy, A.L.; Yu, J.; Vajtai, R.; et al. Ultrathin planar graphene supercapacitors. *Nano Lett.* **2011**, *11*, 1423–1427. [CrossRef] [PubMed]

79. Miller, J.R.; Outlaw, R.A.; Holloway, B.C. Graphene double-layer capacitor with ac line-filtering performance. *Science* **2010**, *329*, 1637–1639. [CrossRef] [PubMed]

80. Lin, J.; Zhang, C.; Yan, Z.; Zhu, Y.; Peng, Z.; Hauge, R.H.; Natelson, D.; Tour, J.M. 3-Dimensional graphene carbon nanotube carpet-based microsupercapacitors with high electrochemical performance. *Nano Lett.* **2013**, *13*, 72–78. [CrossRef] [PubMed]

81. Wu, Z.S.; Liu, Z.; Parvez, K.; Feng, X.; Müllen, K. Ultrathin printable graphene supercapacitors with AC line-filtering performance. *Adv. Mater.* **2015**, *27*, 3669–3675. [CrossRef] [PubMed]

82. Qu, D. Studies of the activated carbons used in double-layer supercapacitors. *J. Power Sources* **2002**, *109*, 403–411. [CrossRef]
83. Khiar, A.S.A.; Arof, A.K. Conductivity studies of starch-based polymer electrolytes. *Ionics* **2010**, *16*, 123–129. [CrossRef]

![coatings logo] *coatings*

MDPI

Article

Effect of Sensitization on the Electrochemical Properties of Nanostructured NiO

Matteo Bonomo [1,*], **Daniele Gatti** [1], **Claudia Barolo** [2,3] **and Danilo Dini** [1,*]

[1]   Department of Chemistry, University of Rome "La Sapienza", p.le A. Moro 5, 00185 Rome, Italy;
      daniele.gatti127@gmail.com
[2]   Department of Chemistry, NIS Interdepartmental Centre and INSTM Reference Centre, University of Turin,
      via Pietro Giuria 7, 10125 Torino, Italy; claudia.barolo@unito.it
[3]   ICxT Interdepartmental Centre, Lungo Dora Siena 100, 10153 Torino, Italy
*   Correspondence: matteo.bonomo@uniroma1.it (M.B.); danilo.dini@uniroma1.it (D.D.);
     Tel.: +39-06-4991-3335 (D.D.)

Received: 7 February 2018; Accepted: 25 June 2018; Published: 29 June 2018

Abstract: Screen-printed NiO electrodes were sensitized with 11 different dyes and the respective electrochemical properties were analyzed in a three-electrode cell with the techniques of cyclic voltammetry and electrochemical impedance spectroscopy. The dye sensitizers of NiO were organic molecules of different types (e.g., squaraines, coumarins, and derivatives of triphenyl-amines and erythrosine B), which were previously employed as sensitizers of the same oxide in dye-sensitized solar cells of p-type (p-DSCs). Depending on the nature of the sensitizer, diverse types of interactions occurred between the immobilized sensitizer and the screen-printed NiO electrode at rest and under polarization. The impedance data recorded at open circuit potential were interpreted in terms of two different equivalent circuits, depending on the eventual presence of the dye sensitizer on the mesoporous electrode. The fitting parameter of the charge transfer resistance through the electrode/electrolyte interface varied in accordance to the differences of the passivation action exerted by the various dyes against the electrochemical oxidation of NiO. Moreover, it has been observed that the resistive term R_{CT} associated with the process of dark electron transfer between the dye and NiO substrate is strictly correlated to the overall efficiency of the photoconversion (η) of the corresponding p-DSC, which employs the same dye-sensitized electrode as photocathode.

Keywords: nickel oxide; organic sensitizers; dye-sensitized solar cells

---

1. Introduction

p-type nickel oxide (NiO) [1,2]; is widely employed as electrodic material in photoelectrochemical cells [3–9], electrochromic windows [10–14], and charge storage systems [15–17], among others [18–20]. The wide range of NiO applicability derives from the fact that NiO can constitute a functional material in both bulk [21] and nanostructured [22,23] versions. When nanostructured NiO is employed in the configuration of thin film (thickness, $l < 10 \ \mu m$) it displays photoelectrochemical activity in the presence of opportune redox mediators (e.g., the redox couple I^-/I_3^-) [24]. Moreover, nanostructured films of NiO present solid-state electroactivity [25–27] due to the verification of a series of reversible electrochemical processes that switch the properties of electrical transport [28] and optical absorption [29] of the oxide itself. The optical [10–14,30,31], magnetic [32,33], electrochemical [34], and photoelectrochemical [35,36] properties of NiO (either in the bulk state or in the nanostructured version) are considerably altered when the surface of the oxide is dye-sensitized [37]. Such a type of electrode modification consists mainly in the impartation of additional optical absorption [38–49] and photoelectroactivity to NiO in the NIR-Vis range, i.e., in a spectral range of lower energies with respect to the intrinsic optical absorption of pristine NiO (typically in the near ultraviolet (UV)) [50].

Dye sensitization of nanostructured *p*-type NiO is finalized principally to the realization of devices, such as dye-sensitized solar cells of *p*-type (*p*-DSCs) [51–59], light-fueled electrolyzers for hydrogen generation [5,60–65], and photoelectrochemical reactors for carbon dioxide reduction [66,67]. Because of these finalities, the study of dye-sensitized electrodes is usually aimed at the analysis of the light absorption properties and at the determination of the efficiency of photoelectrochemical conversion in the process of interest. In this context of research, a less considered (but not less important) aspect is represented by the analysis of the electrochemical properties of the electrode in the dye-sensitized state when the system is in dark conditions (i.e., the evaluation of the electrochemical properties of a photoelectrode in the "blank" state) [68]. In the particular case of dye-sensitized NiO for *p*-DSC purposes, the observation of a series of effects imparted by the dye sensitizer to the oxide substrate has been previously reported, even in absence of illumination [68–70]. These effects mostly consisted of the passivation of the NiO surface towards the oxidation of the NiO substrate itself [69–71], the observation of additional redox processes based on the immobilized dye (representing the actual redox-active species) [72], and redox processes following the synergy between the dye and oxide due to the spontaneous exchange of electronic charge between the dye and the oxide in the unbiased state [72]. The present contribution reports a study on the electrochemical and photoelectrochemical properties of NiO thin films prepared via screen printing [36,73,74] when the NiO electrode is either in the pristine state or in the dye-sensitized version. In particular, we considered the sensitization of nanostructured NiO with a series of organic dyes, the optical absorption properties of which span the visible and (near-infrared) NIR ranges. The structures of the nine dyes here employed as sensitizers of NiO are reported in the Supplementary Materials (Figures S1–S3). The colorants here considered are P1 [75–78], Fast Green (FG) [79,80], erythrosine B (ERY) [3,81–83], the series of the differently substituted squaraines DS_35, DS_44, and DS_46 [84,85], and the series of squaraines VG_1, VG_10, and VG_11, which differ to the extent of electronic conjugation [86].

2. Materials and Methods

Mesoporous NiO thin films have been deposited via screen printing of a paste containing NiO nanoparticles in accordance to the procedure reported in [36]. The scheme of preparation of the paste is reported in Table S1. The paste was successively screen printed onto Fluorine doped Tin Oxide (FTO)-coated glass and annealed at 450 °C in an oven, and a non-stoichiometric film of porous nickel oxide (l = 2.5 μm) was obtained [36]. The resulting film of NiO was nanostructured (Figure 1) and was used successively as a photocathode of a *p*-DSC in both pristine and sensitized states. The morphology of the NiO film is the same for all the employed cathodes as proved by SEM analyses. NiO was sensitized with the series of commercial dyes, P1, Erythrosine B (ERYB), and Fast Green (FG), and the series of squaraines prepared in laboratory scale, DS_35, DS_44, DS_46, VG_1, VG_10, and VG_11. The procedures of the syntheses of the six squaraines here considered have been reported elsewhere [87,88]. The sensitization procedure consisted of the dipping of the electrode in a solution of the given sensitizer. The solvent and the dipping time varied with regard to the colorant employed. When P1 was the sensitizer, the electrode was dipped in a 0.3 mM solution of P1 in acetonitrile for 16 h. The solutions of NiO sensitization with ERYB and FG had ethanol as the solvent, and for both dyes, the concentration was 0.3 mM [70,81,83,89]. The corresponding dipping time was 16 h. When the squaraine was employed, the sensitization was performed for 2 h in a 0.3 mM solution of each dye in ethanol. The shorter duration of the dipping time was used to avoid molecular aggregation [86].

Cyclic voltammetry (CV) measurements have been carried out in a three-electrodes cell. The potential was cyclically varied from open circuit voltage toward more cathodic potential to 1.2 V and further back to −0.27 V. More than 10 cycles have been recorded for each device in order to highlight any modification of the electrode behavior due to potential variation. The analyses of three different cells (sensitized with the same dye) led to identical results. Electrochemical impedance spectroscopy (EIS) measurements have been carried out in a three-electrodes configuration cell. A constant perturbation of ±20 mV from open-circuit voltage has been supplied by a potentiostat

(AUTOLAB PGSTAT12® from Metrohm, Herisau, Switzerland) during all the measurements, whereas the frequencies have been modulated from 100 KHz to 1 Hz: applied frequencies lower than 1 Hz led to unreliable (and non-reproducible) experimental points. Bode's plot was used to report the experimental data. In both CV and EIS experiments, the working electrode (WE) was a film of NiO deposited onto a FTO-coated glass, the counter electrode (CE) was a platinum rod, and the reference electrode was an Ag/AgCl electrode (+0.222 mV vs. NHE (Normal Hydrogen Electrode)). All the potentials reported throughout this work are referred to it. In all the experiments, the electrolyte solution was LiClO$_4$ 0.2 M in ACN (Acetonitrile). All figures have been plotted with the application Kaleidagraph 3.6, and their photoelectrochemical characterization has been already described in previous works [36,85,86].

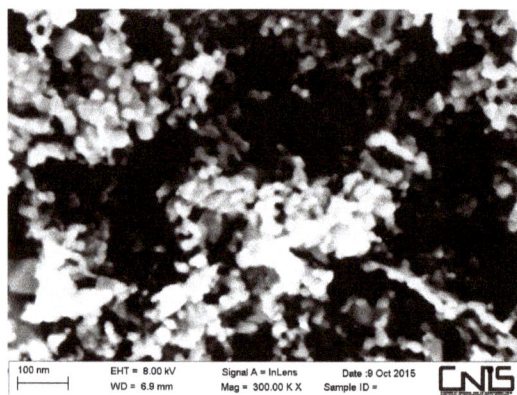

Figure 1. SEM picture of the mesoporous NiO film (thickness, l = 2 μm) here employed as the working electrode and dye-sensitized solar cells of *p*-type (*p*-DSC) photocathode.

3. Results and Discussion

3.1. Analysis of the Voltammetric Data

After sensitization of screen-printed NiO, the open-circuit voltage (V_{OC}) of the three-electrode cell with the dye-sensitized NiO as the working electrode was recorded. The corresponding values are displayed in Table 1.

The bare NiO in the native nanostructured version presents an excess of positive charge on its surface due to the presence of defective Ni(III) centers [26,91]. Such surface-localized species are responsible for the anchoring of the dye sensitizer [68] due to the hydrolysis reaction between the electron deficient site of Ni(III) and the carboxylic group of the colorant [92]. Such a process of sensitization diminishes the positive charge exposed to the electrolyte in passing from the sensitized NiO to the bare NiO with a consequent decrease of the potential difference across the double layer [71]. As a consequence of that, in dark conditions, the open-circuit voltage (OCV$_{dark}$) of the three-electrode cell with sensitized NiO as the working electrode will be lower with respect to the value displayed by the cell with the bare NiO electrode. All observed values of the OCV$_{dark}$ for the three-electrode cells having sensitized NiO as the working electrode (WE) (see Table 1) are consistent with the given depiction of the mechanism of sensitization. Because dye anchoring is supposed to annihilate the positively charged sites on the surface of the defective nanostructured NiO, the extent of the OCV$_{dark}$ decrease is related to the degree of surface passivation and, as such, represents a measure of the efficacy with which a single molecule covers the NiO electrode surface. In fact, the combined analysis of the dye loading and the OCV$_{dark}$ data can be of some usefulness to estimate the apparent volume of a dye when in the surface-immobilized state [86].

Table 1. Second column from left: list of open-circuit potential values V_{OC} of the three-electrode cells differing for the nature of the sensitizer anchored on the NiO-working electrode. In the third column from left, the values of $V_{OC,20}$ refer to the open-circuit value of the cell after electrochemical cycling. The number of consecutive electrochemical cycles was 20 (*vide infra*). The given value was recorded when the open-circuit potential was steady for at least 30 min after the conduction of 20 voltammetric cycles within the potential range $-0.3 \text{ V} \leq E_{appl} \leq 1.2 \text{ V}$ vs. Ag/AgCl. The first column on the right lists the values of dye loading on the NiO film determined via the desorption method described in [90]. For the quantitative determination of dye loading in the case of the NiO sample sensitized with Fast Green (FG), the desorption method could not be adopted because of the scarce effect of dye desorption with the ordinary bleaching agents having medium basic strength.

Dye	V_{OC} (mV) vs. Ag/AgCl	Dye Loading (10^{-8} mol cm^{-2})
–	360	–
P1	140	20 [a]
FG	−55	NA
ERY	265	1.23
VG_1	150	1.63
VG_10	30	1.73
VG_11	120	2.35
DS44	130	1.12
DS35	120	0.87
DS46	85	0.35

[a] from ref. [55]. ERY: erythrosine B.

Within the family of DS squaraines, all three members possess the same extent of electronic conjugation but differ in the length of the alkyl chain [44]. Because the difference (OCV_{dark}(NiO) − OCV_{dark}(NiO-DS)) increases from DS_44 (230 mV) to DS_46 (275 mV), with DS_35 displaying the intermediate value (240 mV), we evince that DS_46, with its longer substituent (dodecyl group) and the lowest surface concentration (0.35×10^{-8} mol cm^2), is the sensitizer with the strongest ability to passivate the surface charge of NiO among the three symmetric squaraines of the DS family. In conclusion, the differences in the passivation action of the DS dyes reside mostly on the bulkiness of the substituent given the equality of extension of the conjugated skeleton (Figure S2). Therefore, after sensitization of NiO with DS squaraines, the alkyl chains also cover those electron-deficient sites of the surface, which have not been involved in the process of dye sensitization. In doing so, the alkyl chains of immobilized DS molecules prevent the further sensitization of the NiO surface despite the availability of free sites of anchoring. Within the group of VG squaraines, the structural differences among the three members consist of the enlargement of the structure in passing from VG_1 to VG_10 and VG_11 due to the additional presence of a condensed additional phenyl ring in VG_10 and VG_11. Such a variation brings about an extension of the conjugated moiety in VG_10 and VG_11 in comparison to VG_1, with consequences on the main property of the electronic polarizability. Data in Table 1 show that VG_1-sensitized NiO produces the cell with the highest value of OCV_{dark}(NiO) within the group of VG dyes. Such a finding is consistent with the previous considerations of the factors controlling the passivating effect of a dye when in the anchored state: VG_1 represents the dye with the smallest volume with respect to VG_10 and VG_11 due to the lack of two condensed benzene rings (Figure S3). For this reason, the passivation of the excess charge on the NiO surface with VG_1 will be less efficacious in comparison with VG_10 and VG_11, and the corresponding value of OCV_{dark}(VG_1-NiO) will be larger (150 mV) than OCV_{dark}(VG_10-NiO) and OCV_{dark}(VG_11-NiO) with 30 and 120 mV, respectively (Table 1). When the OCV_{dark} values of the cells with VG_10- and VG_11-sensitized NiO WEs are compared, a strong diminution of OCV_{dark} is observed from VG_11 to VG_10 (120 mV vs. 30 mV, Table 1). Accordingly, two important points have to be evidenced when VG_10 and VG_11 are confronted: (1) the presence of two strong electron-withdrawing groups in VG_1 (i.e., the cyano groups) which replace an atom of oxygen in the four terms ring; (2) a minor conformational freedom of VG_11 versus VG_10 due to the limitation of the rotation of the benzo-isoindoline moieties around the bond that links them to the squaric ring as imposed by the presence of the bulkier cyano groups in VG_11. The remarkably low value of OCV$_{dark}$(VG_10-NiO) within the VG series could be due to the concomitant action of several factors: (a) enhanced electron donor properties of the benzo-isoindoline moiety towards the electron-deficient sites of NiO with respect to the isoindoline unit of VG_1; (b) despite the analogous extension of the electronically conjugated network, VG_10 lacks two CN groups (i.e., electron-withdrawing groups), which disfavor the retro-donation of negative charge from the dye molecule to the positive centers localized on the NiO surface when VG_11 is the sensitizer. For this reason, it is expected to observe a larger diminution of OCV$_{dark}$ with respect to the bare NiO in passing from the cell with a VG_10-sensitized electrode to the one with a VG_11-sensitizer. In other words, VG_10 compensates more efficiently the electron deficiency of the NiO surface by virtue of its more favorable properties of electron donation with respect to VG_1 and VG_11. A comparison of the trends of OCV$_{dark}$ values for the VG and DS families of squaraines is rendered quite complicated by the differences in the number of anchoring groups (four in DSs and two in VGs), the extension of the conjugated moiety, and the presence of the amino-phenyl moiety in sole DSs. For this reason, we do not attempt an explanation of the differences in the OCV$_{dark}$ values for the groups of VGs and DSs. A similar shift of open-circuit potential is also observed when screen-printed NiO film is sensitized with coumarin 153 and coumarin 343 (see Table S2 and Figure S4 in the Supplementary Materials). When NiO is electrochemically cycled in non-aqueous solvent within the potential range in which the oxide undergoes the series of solid state oxidation processes [27], the following occurs:

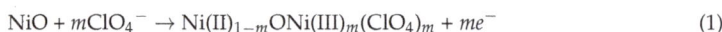

$$NiO + mClO_4^- \rightarrow Ni(II)_{1-m}ONi(III)_m(ClO_4)_m + me^- \tag{1}$$

$$Ni(II)_{1-m}ONi(III)_m(ClO_4)_m + mClO_4^- \rightarrow Ni(II)_{1-m}ONi(IV)_m(ClO_4)_{2m} + me^- \tag{2}$$

Two broad peaks appear in the relative voltammogram of NiO (black trace in Figure 2 and Figure S5 in the Supplementary Materials). The oxidation peak observed at the lower applied potential corresponds to the process of Equation (1), whereas the oxidation occurring at the higher potential refers to the redox process of Equation (2) [3].

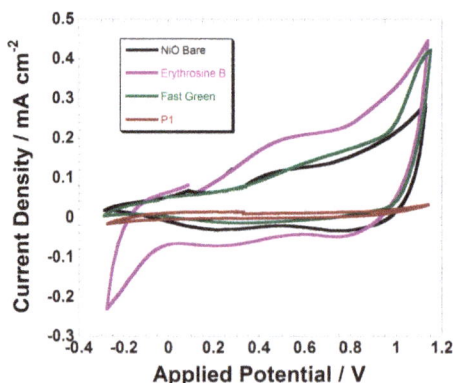

Figure 2. Variation of the voltammogram of screen-printed NiO electrodes in passing from the bare state (black trace) to the sensitized ones with ERY, FG, and P1 (see Figure S1 for the corresponding structures). Scan rate: 100 mV s^{-1}.

The adoption of P1 as a sensitizing agent has a clear effect of NiO surface passivation towards the redox processes of Equations (1) and (2) (brown-reddish trace in Figure 2) in accordance to previously reported data, which referred to the P1-sensitized NiO electrodes deposited via rapid discharge sintering (RDS) [75]. This is not the case of ERY-sensitized NiO (violet trace in Figure 2), because the presence of the sensitizer amplifies the current wave associated with the oxidation of NiO with respect to the electrochemical response of the bare oxide (black trace in Figure 2). Moreover, the NiO electrode sensitized with ERY does not display any additional redox peak or a potential shift of the oxidation peaks with respect to pristine NiO. This combination of findings evidences that the ERY sensitizer does not alter the nature of the redox processes NiO undergoes (Equations (1) and (2)). The verification of a decrease of the open-circuit potential going from the bare NiO to the ERY-sensitized version indicates that ERY mainly acts as an electron-donor towards NiO with consequent increases of the number of surface sites prone to electrochemical oxidation. Thus, a process of the following type can be predicted:

$$h^+{}_{(NiO)} + ERY \rightarrow [e^- \text{-} h^+]_{(NiO)} + (ERY)^+ \tag{3}$$

where the holes $h^+{}_{(NiO)}$ localized on the NiO surface represent the defective Ni(III) sites of the pristine oxide [23,26]. This is equivalent to say that chemisorbed ERY induces the chemical conversion Ni(III) $\rightarrow$ Ni(II). As a consequence of that, an increase in the oxidative current associated with the process of Equation (1) should be expected. In some precedent cases, it has been also verified that ERY exerts a passivating action on the surface of (Rapid Discharge Sintering) RDS-NiO [83] similar to what we have observed here with P1 (brown-reddish trace in Figure 2). Such discrepant behaviors of ERY as a NiO sensitizer have to be ascribed to the differences in the surface compositions of the differently prepared NiO samples. In fact, the extent of the charge transfer from ERY to NiO depends on several characteristics of the NiO substrate, such as the surface concentration of defective Ni(III) sites, surface concentration of the sites of dye anchoring (such "species" do not necessarily coincide with the defective Ni(III) centers) [86], degree of NiO surface hydration [26], porosity [55], and extent of dye coverage.

The FG sensitizer shows an apparent effect of NiO surface passivation (green trace in Figure 2 and brown-reddish trace in the left frame of Figure 3) as far as the process of NiO oxidation at lower potential is concerned (Equation (1)). Moreover, FG introduces an additional peak of oxidation due to the electroactivity of the dye itself (see the comparison of the two traces in the left frame of Figure 3).

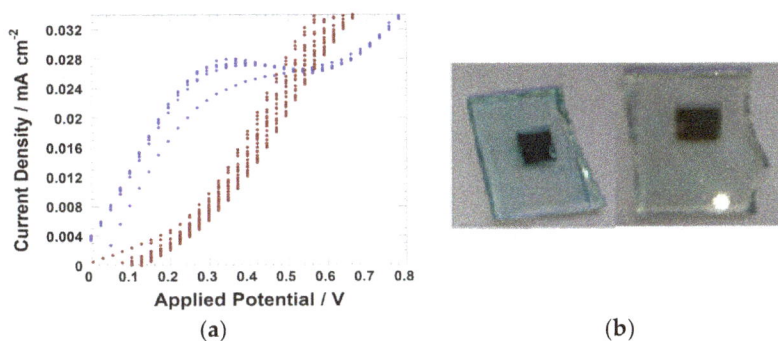

Figure 3. (a) Zooming of the voltammograms of the bare NiO (blue trace) and FG-sensitized NiO (brown-reddish dots); (b) Photograph of FG-sensitized NiO prior the treatment of electrochemical cycling (colored sample on the left) and after electrochemical cycling (bleached sample on the right).

The peak of FG oxidation overlaps with the oxidation process of NiO reported in Equation (2). After one cycle, the NiO electrode became decolored (photograph in the right frame of Figure 3). The oxidation of immobilized FG then brings about the detachment of the dye from the surface of NiO. This effect is probably due to the scarce electrochemical stability of the sulfonic group as a linker of FG [92]. DS squaraines in the immobilized state have also shown an analogous behavior upon continuous electrochemical cycling (Figure 4) [72].

The first cycle of DS_46-sensitized NiO in Figure 4 is characterized by the presence of two irreversible current peaks (Figure 4a) that do not appear in the successive scans (Figure 4b). We attribute this sequence of events to the combined oxidation of DS_46 and NiO during the first cycle with successive peeling-off of the dye from the substrate. In the first cycle, the process of NiO oxidation (Figures 2 and S5) is completely masked by the much ampler wave of current associated with DS_46 oxidation. The amplitude of the oxidation peaks of DS_46 is ca. 10 times larger than the amplitude of the peaks characteristic of NiO oxidation. In the successive cycles, the current peaks we attributed to the oxidation of DS_46 were no longer present. Moreover, the voltammograms of the successive cycles recall the ones of the bare NiO within the same potential range (comparison between the voltammetric profiles in the left frame of Figure 4 and the black trace of Figure 2). The hypothesis of the detachment of DS_46 consequent to its surface oxidation is supported by the observation of electrode decoloration after the treatment of electrochemical cycling (Figures 5 and S6). An analogous electrochemical behavior has been observed when screen-printed NiO was sensitized with the other squaraines DS_35, DS_44, VG_1, VG_10, and VG_11 (see Figures S7–S11 in the Supplementary Materials). The electrochemical behavior of the screen-printed NiO sensitized with the series of squaraines considered here is consistent with the findings reported in previous works about the electrochemical and photoelectrochemical properties of various NiO samples sensitized with the commercial squaraine SQ2 [72,93,94].

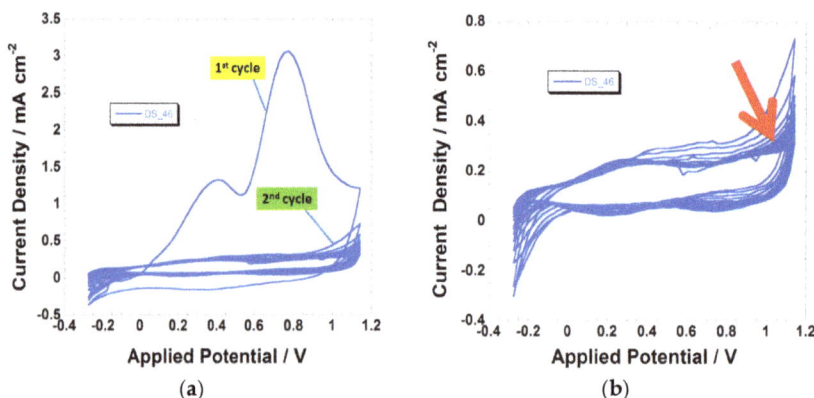

Figure 4. (a) First 20 voltammograms of the DS_46-sensitized NiO electrode at the scan rate of 100 mV s^{-1}. The first two cycles are indicated; (b) Stabilization of the voltammogram of the DS_46-sensitized electrode after the first cycle, evidenced in the left frame. The red arrow indicates the sense of variation of the current profile upon the increase of the number of cycles. See Figure S2 for the visualization of the dye structure.

The voltammograms of coumarin-sensitized NiO electrodes are shown in Figure S12 (see Supplementary Materials). Different from squaraines, surface-immobilized coumarins do not introduce redox peaks associated with the redox activity of the dye. Moreover, the observation of a scarce effect of NiO surface passivation, as well as a poor effect of NiO sensitization indicate that coumarins do not anchor in large amounts onto the mesoporous NiO surface prepared via screen printing.

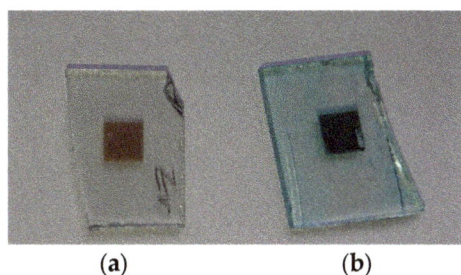

Figure 5. Photographs of DS_46-sensitized NiO film before (**a**) and after (**b**) the electrochemical cycling of Figure 4.

3.2. Analysis of the Impedance Spectra

The electrochemical impedance spectra of the three-electrode cells having the bare or sensitized NiO as the working electrode have been recorded at open-circuit potential (V_{OC}). The corresponding values of V_{OC} for the cells with differently sensitized NiO electrodes are those reported in the second column of Table 1 (*vide supra*). After the assembly of the cell, its value of V_{OC} was first measured. The impedance spectrum of the cell was recorded immediately after the determination of V_{OC} with the cell potential kept at the given value of V_{OC}.

The impedance analysis was not conducted on the cells after the treatment of electrochemical cycling, because it has been observed that electrochemical cycling generally introduces irreversible modifications of the sensitized electrode (i.e., partial or complete desorption of the dye). It is worth mentioning that the electrochemical cycling did not provoke any structural modification in the electrode. To confirm that, we reported some SEM images (see Figure S6 Page: 8 and Table S3) of pristine NiO, sensitized NiO, and cycled-sensitized NiO. No meaningful structural variations may be highlighted. Interestingly, EDX (Energy Dispersion X-ray) analyses showed different elemental distributions for the three electrodes: compared to the bare NiO, the sensitized electrode has a higher carbon content due to the organic molecule chemisorbed onto the NiO surface. After cycling, the carbon content decreases, but it is still higher compared to the pristine electrode. The latter evidence is consistent with the dye degradation evidenced during the CV experiments: the sensitizer occurred in some oxidation reactions (and it did not still adsorb visible light), but it did not completely detach from NiO, and it partially contributes to the electrode passivation (see above). As previously shown, dye desorption/degradation has been verified through the continuous changes of the shape of the voltammograms with cycling (Figure 4 and Figures S7–S12 in the Supplementary Materials) and through the non-negligible variation of the open-circuit potential of the cell in passing from the pristine state to the cycled one (V_{OC} vs. $V_{OC,20}$ in Table 1). Figures 6–8 present the impedance spectra of the differently sensitized cells in the form of Bode plots.

The impedance data presented in Figures 6–8 were fitted with the response calculated for the equivalent circuits displayed in Figures 9 and 10. Two distinct models were considered depending on the eventual presence of the dye sensitizer on the electrode surface. The model of Figure 9 is characterized by the parallel connection of the charge transfer resistance through the bare electrode/electrolyte interface ($R_{NiO/el}$) to the capacitive element $CPE_{NiO/el}$. The latter circuital element is associated with the charge distribution representing the electrochemical double layer at the bare NiO/electrolyte interface. The second model of equivalent circuit (Figure 10) accounts for the presence of the immobilized sensitizer and considers accordingly the additional interface NiO/dye created by the combination of the electrode with the dye sensitizer. The consideration of the NiO/dye interface as the relevant electrical element of the modified electrode has led to the insertion of the parallel terms R_{ct} (=resistance of the charge transfer between NiO substrate and dye) and CPE_{ct} (=element of constant phase, which refers to the charge distribution at the NiO/dye interface during the charge transfer) in the model of Figure 10. The fit of the impedance spectrum of the cell with the bare NiO as the working electrode was realized when $R_{\text{NiO/el}} = 4517 \pm 53\ \Omega$ and the double-layer capacitance at the NiO/electrolyte interface $C_{\text{NiO/el}}$ was $8.06 \pm 0.80\ \mu\text{F}$. Table 2 lists the fitting values of the electrical parameters $R_{\text{NiO-D/el}}$, R_{ct} (Figure 10) and $C_{\text{NiO-D/el}}$. The latter capacitive element refers to the charge distribution at the interface formed by the sensitized NiO electrode in contact with the electrolyte. The simulated impedance spectra are shown in the Supplementary Materials (Figures S13–S22).

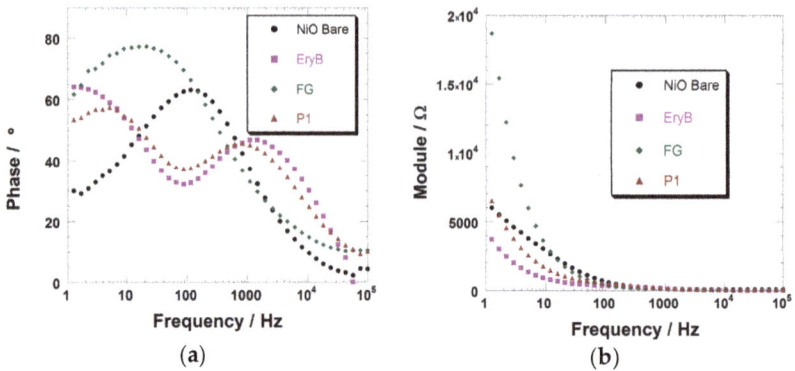

Figure 6. (**a**) Variation of the phase of the electrochemical impedance with the frequency of the potential stimulus for the cells with the bare, ERY-, P1-, and FG-sensitized NiO as the working electrode; (**b**) Variation of the impedance modulus with the frequency of stimulus for the four cells analyzed in the left frame.

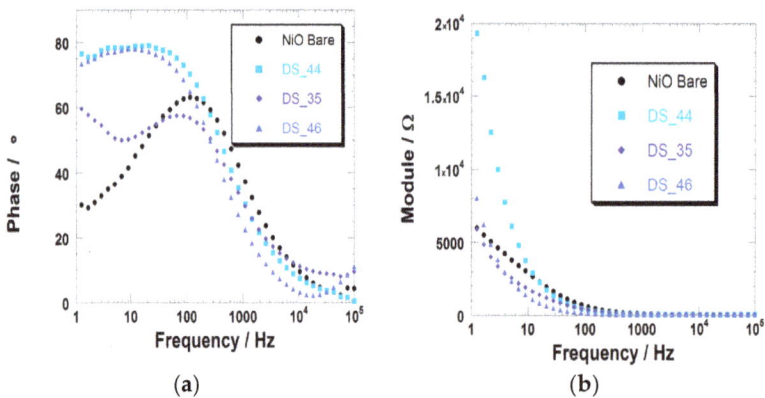

Figure 7. (**a**) Variation of the phase of the electrochemical impedance with the frequency of the potential stimulus for the cells with the bare, DS_35-, DS_44-, and DS_46-sensitized NiO as the working electrode; (**b**) Variation of the impedance modulus with the frequency of stimulus for the four cells analyzed in the left frame.

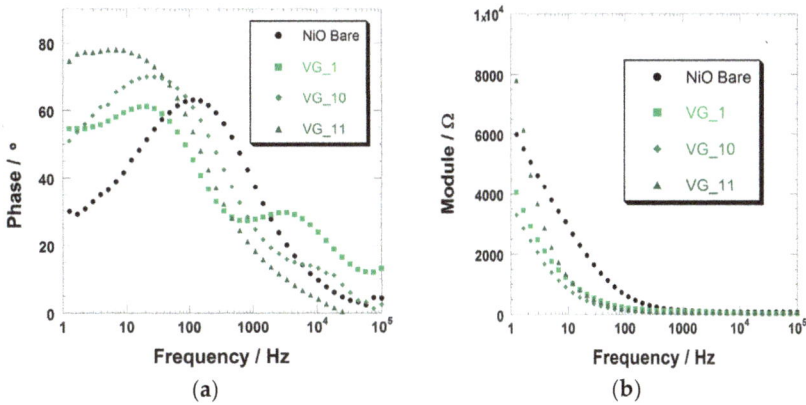

Figure 8. (**a**) Variation of the phase of the electrochemical impedance with the frequency of the potential stimulus for the cells with bare, VG_1-, VG_10-, and VG_11-sensitized NiO as the working electrode. (**b**) Variation of the impedance modulus with the frequency of stimulus for the four cells analyzed in the left frame.

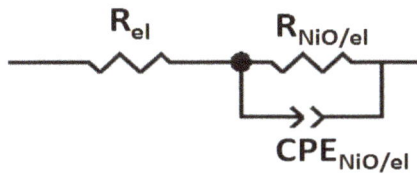

Figure 9. Model of equivalent circuit for the fitting of the electrochemical impedance spectroscopy (EIS) data recorded at the potential of the open circuit of three-electrode cell with the bare NiO as the working electrode. R_{el}: the sum of all the resistances due to the circuital elements except the working electrode. $R_{NiO/el}$: resistance of the charge transfer through the interface created by the bare NiO and the electrolyte; $CPE_{NiO/el}$: element of the constant phase, which refers to the charge distribution during the process of the charge transfer through the interface created by the bare NiO and the electrolyte.

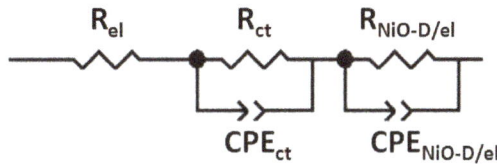

Figure 10. Model of the equivalent circuit for the fitting of the EIS data recorded at the potential of open circuit of three-electrode cell with dye-sensitized NiO as the working electrode. R_{el}: the sum of all the resistances due to the circuital elements except the working electrode; R_{ct}: resistance of the charge transfer between the NiO substrate and the dye; CPE_{ct}: element of the constant phase, which refers to the charge distribution at the NiO/dye interface during the process of the charge transfer between the NiO and the dye; $R_{NiO-D/el}$: resistance of the charge transfer through the interface created by the sensitized electrode and electrolyte; $CPE_{NiO-D/el}$: element of the constant phase, which refers to the charge distribution during the process of the charge transfer through the interface created by the sensitized electrode and the electrolyte.

In the Bode's plot, phase diagrams usually exhibit a number of peaks corresponding to the number of RC (Resistor-Capacitor) elements. This is true when investigated phenomena occur in different time scales. In the present case, this happened for just some of the sensitized electrodes (e.g., P1, ERYB, and VG_11). When FG or the dyes from the DS series are employed as the sensitizer, the two phenomena (i.e., the charge transfer from the NiO to the dye and the charge transfer from the sensitized electrode to the electrolyte) occurs in a quite similar time scale, and the two peaks tend to merge into (a broadened) one, straightforwardly. It is worth mentioning that a different time scale between the two reactions is typical of well-performing sensitizers (see Table 3 below).

Impedance data confirm that the dye sensitizer generally acts as a passivating agent of the NiO surface, because the resistance of the charge transfer through the electrode/electrolyte interface is systematically higher for the sensitized electrode ($R_{NiO-D/el}$ > 14 kΩ, Table 2, second column from left) with respect to the pristine NiO ($R_{NiO/el}$ < 5 kΩ). Because the characteristic electrochemical processes of NiO consist of the simultaneous injection of the electronic charges and ion uptake (Equations (1) and (2)), the presence of the dye sensitizer with its steric hindrance impedes the approach of the charge, compensating ions towards the surface of NiO during its oxidation. As a confirmation of that, the largest values of the charge transfer resistance are found when the NiO is sensitized by the dyes with the largest skeleton or the longer pending groups within the same series. These are the cases of VG_11 within the VG series (Figure S3) and of DS_46 within the DS series (Figure S2).

Table 2. List of the fitting values of $R_{NiO-D/el}$, R_{ct}, and $C_{NiO-D/el}$ for the simulation of the impedance spectra of the cells with sensitized NiO electrodes at the condition of open-circuit potential. The meaning of the terms $R_{NiO-D/el}$ and R_{ct} is the same as reported in the caption of Figure 10. $C_{NiO-D/el}$ is the capacitance of the double layer at the sensitized NiO/electrolyte interface.

Dye	$R_{NiO-D/el}$ (Ω)	R_{ct} (Ω)	$C_{NiO-D/el}$ (µF)	R_{rec} (Ω)
P1	58,212 ± 456	292.6 ± 2.9	6.52 ± 0.70	85.2
ERY	23,611 ± 221	455.5 ± 4.9	8.62 ± 0.69	63.1
FG	37,938 ± 412	4023 ± 53	6.91 ± 0.76	58.6
DS_44	15,230 ± 144	1328 ± 18	6.72 ± 0.86	52.3
DS_35	29,035 ± 258	2899 ± 36	7.66 ± 0.68	72.9
DS_46	69,673 ± 665	3204 ± 39	7.97 ± 0.76	96.7
VG_1	14,447 ± 153	602.3 ± 10.3	6.08 ± 1.16	15.8
VG_10	22,059 ± 196	424.2 ± 12.0	6.30 ± 1.15	60.3
VG_11	71,463 ± 423	187.9 ± 2.3	8.59 ± 0.68	71.5

Analogous considerations hold when the resistances of FG- and P1-sensitized NiO electrodes are compared to that of ERY-sensitized NiO (Figure S1). Dye loading (Table 1, first column from the right) does not appear as crucial as the size of the dye in controlling the interfacial resistance $R_{NiO-D/el}$. This is probably because the variations of dye loading are not large enough to generate substantial differences of passivating action within the set of dye sensitizers here considered. With the exception of NiO sensitization with ERY and VG_11, dye chemisorption generally leads to a decrease of the double-layer capacitance of the electrode/electrolyte interface $C_{NiO-D/el}$ with respect to the bare NiO. Such a diminution of $C_{NiO-D/el}$ is consistent with the effect of masking of the NiO surface charge, which is produced by the sensitizer as previously discussed during the analysis of the trend of V_{OC} (Table 1).

The dark resistance of the charge transfer between the NiO substrate and the dye sensitizer (the R_{ct} term defined in Figure 10) refers to the process of the charge injection in the NiO with the dye in the ground state, which acts as mediator. We show here that such a resistive term (that is determined in the absence of illumination in a three-electrode cell with a dye-sensitized working electrode) can represent a valid parameter for the evaluation of the efficacy with which the dye in the optically excited state transfers electronic charge to the semiconducting substrate during the operation of an illuminated dye-sensitized solar cell [95]. Table 3 reports the fundamental photoelectrochemical parameters of the *p*-DSCs, employing the same dye-sensitized electrodes of the experiments shown in Figures 6–8 when the redox shuttle is the couple I^-/I_3^- and the counter electrode is platinized FTO. R_{rec} is the recombination resistance at the NiO/electrolyte interface in complete device configuration.

It is found that the sensitizers producing the largest efficiencies of conversion η (>0.030%, Table 3) in the corresponding *p*-DSCs employing the NiO cathodes display generally the lowest values of the dark parameter R_{CT} within the same series of compounds. This is particularly evident in the cases of P1 and ERY and in the cases of the DS_44 and VG_11 sensitizers within their respective families of squaraines. The existence of a correlation between the trends of η and R_{CT} within the same class of compounds indicates that when the electron transfer (*et*) between the NiO and the ground state of the sensitizer is efficient, the process of *et* between the NiO substrate and the excited state of the sensitizer is as efficient as the one conducted in dark conditions, despite the involvement of different electronic states of the sensitizer (and probably of different energy levels in the electronic structure of the mesoporous, semiconducting NiO) [96] in these two processes.

Table 3. Values of the relevant parameters' open-circuit photopotential (V_{OC}), short-circuit current density (J_{SC}), fill factor (FF), and overall conversion efficiency (η), which characterize the photoelectrochemical performance of a *p*-DSC with a sensitized NiO photocathode. For the assembly of the DSC and the determination of the photoelectrochemical parameters, we adopted the experimental conditions reported in [97]. Data of the *p*-DSCs employing FG-sensitized NiO photocathodes could not be determined for the scarce time stability of the relative photoelectrochemical response.

Dye	V_{OC} (mV)	J_{SC} (mA cm^{-2})	FF (%)	η (%)	Ref.
–	96	−0.261	40.76	0.01	–
ERY	88	−1.019	36.02	0.032	[86]
P1	125.6	−1.188	32.9	0.049	–
VG_1	87	−0.577	37.2	0.018	[86]
VG_10	102	−0.435	40.87	0.018	[86]
VG_11	93	−1.16	36.12	0.043	[86]
DS_44	101	−0.991	37.12	0.037	[85]
DS_35	95	−0.821	37.94	0.03	[85]
DS_46	94	−0.503	36.46	0.017	[85]

4. Conclusions

In the present work, the electrochemical characterization of screen-printed NiO electrodes sensitized with the commercial benchmarks ERY, P1, and FG and with two series of squaraines, differing according to the extent of electronic conjugation (VG series) and the size of peripheral substituents (DS series), has been considered. The presence of the chemisorbed sensitizers P1 and FG on the screen-printed NiO induces an effect of passivation of the oxide surface against NiO oxidation. Such an effect manifests itself through the decrease of the intensity of the characteristic peaks of NiO oxidation when the oxide passes from the bare to the P1- or FG-sensitized state. In such cases, the electronic interactions between the dye sensitizer and the NiO surface can be considered weak, because no process of charge transfer between the dye and the oxide occurs in dark conditions or for small polarizations of the sensitized NiO electrode. Squaraines and, to a lesser extent, FG undergo a process of electrochemical oxidation in the immobilized state when the applied potential falls in the range of NiO electroactivity. Such a redox process brings about the disruption of the electronic conjugation in the dye skeleton and its successive detachment from the substrate. This was verified by the decoloration of the electrochemically cycled NiO. Because the solid-state electrochemical process of NiO oxidation consists of the dark injection of holes (i.e., the formation of Ni(III) centers, which render nickel oxide a non-stoichiometric system), such a finding warns us about the employment of squaraines (and FG) as dye sensitizers of NiO in *p*-type dye-sensitized solar cells: their use must be particularly judicious. The occurrence of the oxidation for NiO and squaraines within the same range of applied potential indicated that the simultaneous presence of holes in the NiO and the oxidized dye in the immobilized state is deleterious for the chemical stability of the linker anchoring the dye to the NiO surface. Consequently, the diffusion of the photoinjected holes towards the NiO bulk has to occur faster than the onset of the degrading process of linker rupture, which starts from the excess of positive charge at the dye/electrode junction. Different to all other sensitizers here examined, ERY produced the opposite effect of dark current enhancement corresponding with the electrochemical oxidation processes of NiO. Such a finding has been attributed to the occurrence of a spontaneous process of electron transfer from the ERY sensitizer to the defective Ni(III) centers of the NiO surface in the absence of any external polarization. The consequence of the resulting conversion Ni(III) → Ni(II) would be the increase of the surface sites of NiO, which undergo electrochemical oxidation. The electrochemical impedance spectra of the differently sensitized electrodes have been analyzed and modeled with two distinct equivalent circuits depending on the version of the NiO electrode (i.e., pristine or sensitized). Within a set of dye, we could find a correlation between the trend of the dark charge transfer resistance at the electrode/dye junction and the trend of the overall efficiency of photoconversion in the corresponding dye-sensitized solar cells employing the same sensitized electrode as a photocathode. The photoelectrochemical cells displaying the highest efficiencies of solar conversion were those that employed sensitized NiO electrodes with the lowest values of charge transfer resistance through the dye/NiO junction in the absence of illumination. This finding would indicate that the electronic communication between the NiO substrate and the dye sensitizer is the most important factor of control of the electrochemical and photoelectrochemical processes occurring at this type of modified semiconductor.

Supplementary Materials: The following are available online at http://www.mdpi.com/2079-6412/8/7/232/s1, Figure S1: Structure of the commercial dyes, Figure S2: Structure of the squaraines (DS-series), Figure S3: Structure of the squaraines (VG-series), Figure S4: Structures of coumarin dyes, Figure S5: Voltammogram of bare NiO, Figure S6: SEM images of different electrodes, Figure S7–S12: CV curves, Figure S13–S22: Bode plots; Table S1: Procedure of the preparation of the paste P3, Table S2: Variation of VOC for the cells with coumarin-sensitized NiO, Table S3: EDX elemental analyses.

Author Contributions: Conceptualization, Danilo Dini and Matteo Bonomo; Methodology, All Authors; Validation, Danilo Dini, Matteo Bonomo and Claudia Barolo; Formal Analysis, Daniele Gatti and Matteo Bonomo; Investigation, Daniele Gatti and Matteo Bonomo; Resources, Matteo Bonomo and Claudia Barolo; Data Curation, Matteo Bonomo and Daniele Gatti; Writing-Original Draft Preparation, Matteo Bonomo and Danilo Dini; Writing-Review & Editing, Matteo Bonomo and Danilo Dini; Supervision, Danilo Dini, Matteo Bonomo and Claudia Barolo; Project Administration, Danilo Dini and Claudia Barolo.

Funding: This research received no external funding.

Conflicts of Interest: The authors declare no conflict of interest.

References

1. Passerini, S.; Scrosati, B. Characterization of Nonstoichiometric Nickel Oxide Thin-Film Electrodes. *J. Electrochem. Soc.* **1994**, *141*, 889–895. [CrossRef]
2. Wei, L.; Jiang, L.; Yuan, S.; Ren, X.; Zhao, Y.; Wang, Z.; Zhang, M.; Shi, L.; Li, D. Valence Band Edge Shifts and Charge-transfer Dynamics in Li-Doped NiO Based p-type DSSCs. *Electrochim. Acta* **2016**, *188*, 309–316. [CrossRef]
3. He, J.; Lindström, H.; Hagfeldt, A.; Lindquist, S.-E. Dye-Sensitized Nanostructured p-Type Nickel Oxide Film as a Photocathode for a Solar Cell. *J. Phys. Chem. B* **1999**, *103*, 8940–8943. [CrossRef]
4. Sahara, G.; Abe, R.; Higashi, M.; Morikawa, T.; Maeda, K.; Ueda, Y.; Ishitani, O. Photoelectrochemical CO$_2$ reduction using a Ru(II)-Re(I) multinuclear metal complex on a p-type semiconducting NiO electrode. *Chem. Commun.* **2015**, *51*, 10722–10725. [CrossRef] [PubMed]
5. Gross, M.A.; Creissen, C.E.; Orchard, K.L.; Reisner, E. Photoelectrochemical hydrogen production in water using a layer-by-layer assembly of a Ru dye and Ni catalyst on NiO. *Chem. Sci.* **2016**, *7*, 5537–5546. [CrossRef]
6. Dini, D. Nanostructured Metal Oxide Thin Films as Photoactive Cathodes of p-Type Dye-Sensitised Solar Cells. *Phys. Chem. Commun.* **2016**, *3*, 14–51.
7. Shan, B.; Sherman, B.D.; Klug, C.M.; Nayak, A.; Marquard, S.L.; Liu, Q.; Bullock, R.M.; Meyer, T.J. Modulating Hole Transport in Multilayered Photocathodes with Derivatized p-Type Nickel Oxide and Molecular Assemblies for Solar-Driven Water Splitting. *J. Phys. Chem. Lett.* **2017**, 4374–4379. [CrossRef] [PubMed]

8. Li, L.; Dai, H.; Luo, D.; Wang, S.; Sun, X. Nickel Oxide Nanosheets for Enhanced Photoelectrochemical Water Splitting by Hematite (α-Fe$_2$O$_3$) Nanowire Arrays. *Energy Technol.* **2016**, *639798*, 758–763. [CrossRef]

9. Yao, K.; Li, F.; He, Q.; Wang, X.; Jiang, Y.; Huang, H.; Jen, A.K.Y. A copper-doped nickel oxide bilayer for enhancing efficiency and stability of hysteresis-free inverted mesoporous perovskite solar cells. *Nano Energy* **2017**, *40*, 155–162. [CrossRef]

10. Mihelčič, M.; Šurca Vuk, A.; Jerman, I.; Orel, B.; Švegl, F.; Moulki, H.; Faure, C.; Campet, G.; Rougier, A. Comparison of electrochromic properties of Ni$_{1-x}$O in lithium and lithium-free aprotic electrolytes: From Ni$_{1-x}$O pigment coatings to flexible electrochromic devices. *Sol. Energy Mater. Sol. Cells* **2014**, *120*, 116–130. [CrossRef]

11. Da Rocha, M.; Rougier, A. Electrochromism of non-stoichiometric NiO thin film: As single layer and in full device. *Appl. Phys. A Mater. Sci. Process.* **2016**, *122*. [CrossRef]

12. Wen, R.T.; Granqvist, C.G.; Niklasson, G.A. Anodic electrochromism for energy-efficient windows: Cation/anion-based surface processes and effects of crystal facets in nickel oxide thin films. *Adv. Funct. Mater.* **2015**, *25*, 3359–3370. [CrossRef]

13. Zhao, C.C.; Chen, C.; Du, F.L.; Wang, J.M. Template synthesis of NiO ultrathin nanosheets using polystyrene nanospheres and their electrochromic properties. *RSC Adv.* **2015**, *5*, 38533–38537. [CrossRef]

14. Estrada, W.; Andersson, A.M.; Granqvist, C.G.; Gorenstein, A.; Decker, F. Infrared reflectance spectroscopy of electrochromic NiO$_x$H$_y$ films made by reactive DC sputtering. *J. Mater. Res.* **1991**, *6*, 1715–1719. [CrossRef]

15. Choi, S.H.; Kang, Y.C. Ultrafast synthesis of yolk-shell and cubic NiO nanopowders and application in lithium ion batteries. *ACS Appl. Mater. Interfaces* **2014**, *6*, 2312–2316. [CrossRef] [PubMed]

16. Gu, L.; Xie, W.; Bai, S.; Liu, B.; Xue, S.; Li, Q.; He, D. Facile fabrication of binder-free NiO electrodes with high rate capacity for lithium-ion batteries. *Appl. Surf. Sci.* **2016**, *368*, 298–302. [CrossRef]

17. Yue, G.H.; Zhao, Y.C.; Wang, C.G.; Zhang, X.X.; Zhang, X.Q.; Xie, Q.S. Flower-like nickel oxide nanocomposites anode materials for excellent performance lithium-ion batteries. *Electrochim. Acta* **2015**, *152*, 315–322. [CrossRef]

18. Wang, C.; Wang, T.; Wang, B.; Zhou, X.; Cheng, X.; Sun, P.; Zheng, J.; Lu, G. Design of α-Fe$_2$O$_3$ nanorods functionalized tubular NiO nanostructure for discriminating toluene molecules. *Sci. Rep.* **2016**, *6*, 26432. [CrossRef] [PubMed]

19. Wang, C.; Cui, X.; Liu, J.; Zhou, X.; Cheng, X.; Sun, P.; Hu, X.; Li, X.; Zheng, J.; Lu, G. Design of Superior Ethanol Gas Sensor Based on Al-Doped NiO Nanorod-Flowers. *ACS Sens.* **2016**, *1*, 131–136. [CrossRef]

20. Bai, G.; Dai, H.; Deng, J.; Liu, Y.; Ji, K. Porous NiO nanoflowers and nanourchins: Highly active catalysts for toluene combustion. *Catal. Commun.* **2012**, *27*, 148–153. [CrossRef]

21. Yang, J.; Lai, Y.; Chen, J.S. Effect of heat treatment on the properties of non-stoichiometric p-type nickel oxide films deposited by reactive sputtering. *Thin Solid Films* **2005**, *488*, 242–246. [CrossRef]

22. Cavallo, C.; Di Pascasio, F.; Latini, A.; Bonomo, M.; Dini, D. Nanostructured Semiconductor Materials for Dye-Sensitized Solar Cells. *J. Nanomater.* **2017**, *2017*, 5323164. [CrossRef]

23. D'Amario, L.; Jiang, R.; Cappel, U.B.; Gibson, E.A.; Boschloo, G.; Rensmo, H.; Sun, L.; Hammarström, L.; Tian, H. Chemical and Physical Reduction of High Valence Ni States in Mesoporous NiO Film for Solar Cell Application. *ACS Appl. Mater. Interfaces* **2017**, *9*, 33470–33477. [CrossRef] [PubMed]

24. Boschloo, G.; Hagfeldt, A. Characteristics of the iodide/triiodide redox mediator in dye-sensitized solar cells. *Acc. Chem. Res.* **2009**, *42*, 1819–1826. [CrossRef] [PubMed]

25. Awais, M.; Dini, D.; Don MacElroy, J.M.; Halpin, Y.; Vos, J.G.; Dowling, D.P. Electrochemical characterization of NiO electrodes deposited via a scalable powder microblasting technique. *J. Electroanal. Chem.* **2013**, *689*, 185–192. [CrossRef]

26. Marrani, A.G.; Novelli, V.; Sheehan, S.; Dowling, D.P.; Dini, D. Probing the redox states at the surface of electroactive nanoporous nio thin films. *ACS Appl. Mater. Interfaces* **2014**, *6*, 143–152. [CrossRef] [PubMed]

27. Bonomo, M.; Marrani, A.G.; Novelli, V.; Awais, M.; Dowling, D.P.; Vos, J.G.; Dini, D. Surface properties of nanostructured NiO undergoing electrochemical oxidation in 3-methoxy-propionitrile. *Appl. Surf. Sci.* **2017**, *403*, 441–447. [CrossRef]

28. Nakaoka, K.; Ueyama, J.; Ogura, K. Semiconductor and electrochromic properties of electrochemically deposited nickel oxide films. *J. Electroanal. Chem.* **2004**, *571*, 93–99. [CrossRef]

29. Surca, A.; Orel, B.; Pihlar, B.; Bukovec, P. Optical, spectroelectrochemical and structural properties of sol-gel derived Ni-oxide electrochromic film. *J. Electroanal. Chem.* **1996**, *408*, 83–100. [CrossRef]

30.   Moulki, H.; Faure, C.; Mihelčič, M.; Šurca Vuk, A.; Švegl, F.; Orel, B.; Campet, G.; Alfredsson, M.; Chadwick, A.V.; Gianolio, D.; et al. Electrochromic performances of nonstoichiometric NiO thin films. *Thin Solid Films* **2014**, *553*, 63–66. [CrossRef]

31.   Xia, X.H.; Tu, J.P.; Zhang, J.; Wang, X.L.; Zhang, W.K.; Huang, H. Electrochromic properties of porous NiO thin films prepared by a chemical bath deposition. *Sol. Energy Mater. Sol. Cells* **2008**, *92*, 628–633. [CrossRef]

32.   Balaev, D.A.; Dubrovskiy, A.A.; Krasikov, A.A.; Popkov, S.I.; Balaev, A.D.; Shaikhutdinov, K.A.; Kirillov, V.L.; Mart'yanov, O.N. Magnetic properties of NiO nano particles: Contributions of the antiferromagnetic and ferromagnetic subsystems in different magnetic field ranges up to 250 kOe. *Phys. Solid State* **2017**, *59*, 1547–1552. [CrossRef]

33.   Zhang, X.K.; Yuan, J.J.; Xie, Y.M.; Yu, Y.; Yu, H.J.; Zhu, X.R.; Kuang, F.G.; Shen, H. Magnetic nature of surface and exchange bias effect in NiO nanosheets. *Appl. Phys. Lett.* **2016**, *109*. [CrossRef]

34.   Cai, G.; Wang, X.; Cui, M.; Darmawan, P.; Wang, J.; Eh, A.L.S.; Lee, P.S. Electrochromo-supercapacitor based on direct growth of NiO nanoparticles. *Nano Energy* **2015**, *12*, 258–267. [CrossRef]

35.   Liu, Q.; Wei, L.; Yuan, S.; Ren, X.; Zhao, Y.; Wang, Z.; Zhang, M.; Shi, L.; Li, D.; Li, A. Influence of interface properties on charge density, band edge shifts and kinetics of the photoelectrochemical process in p-type NiO photocathodes. *RSC Adv.* **2015**, *5*, 71778–71784. [CrossRef]

36.   Bonomo, M.; Naponiello, G.; Venditti, I.; Zardetto, V.; Carlo, A.D.; Dini, D. Electrochemical and Photoelectrochemical Properties of Screen-Printed Nickel Oxide Thin Films Obtained from Precursor Pastes with Different Compositions. *J. Electrochem. Soc.* **2017**, *164*, H137–H147. [CrossRef]

37.   Piccinin, S.; Rocca, D.; Pastore, M. Role of Solvent in the Energy Level Alignment of Dye-Sensitized NiO Interfaces. *J. Phys. Chem. C* **2017**, *121*, 22286–22294. [CrossRef]

38.   Pham, T.T.T.; Saha, S.K.; Provost, D.; Farré, Y.; Raissi, M.; Pellegrin, Y.; Blart, E.; Vedraine, S.; Ratier, B.; Aldakov, D.; et al. Toward Efficient Solid-State p-Type Dye-Sensitized Solar Cells: The Dye Matters. *J. Phys. Chem. C* **2017**, *121*, 129–139. [CrossRef]

39.   Ameline, D.; Diring, S.; Farre, Y.; Pellegrin, Y.; Naponiello, G.; Blart, E.; Charrier, B.; Dini, D.; Jacquemin, D.; Odobel, F. Isoindigo derivatives for application in p-type dye sensitized solar cells. *RSC Adv.* **2015**, *5*, 85530–85539. [CrossRef]

40.   Nattestad, A.; Mozer, A.J.; Fischer, M.K.R.; Cheng, Y.-B.; Mishra, A.; Bäuerle, P.; Bach, U. Highly efficient photocathodes for dye-sensitized tandem solar cells. *Nat. Mater.* **2009**, *9*, 31–35. [CrossRef] [PubMed]

41.   Odobel, F.; Pellegrin, Y. Recent advances in the sensitization of wide-band-gap nanostructured p-type semiconductors. Photovoltaic and photocatalytic applications. *J. Phys. Chem. Lett.* **2013**, *4*, 2551–2564. [CrossRef]

42.   Wood, C.J.; Cheng, M.; Clark, C.A.; Horvath, R.; Clark, I.P.; Hamilton, M.L.; Towrie, M.; George, M.W.; Sun, L.; Yang, X.; et al. Red-absorbing cationic acceptor dyes for photocathodes in tandem solar cells. *J. Phys. Chem. C* **2014**. [CrossRef]

43.   Nikolaou, V.; Charisiadis, A.; Charalambidis, G. Recent advances and insights in dye-sensitized NiO photocathodes for photovoltaic devices. *J. Mater. Chem. A Mater. Energy Sustain.* **2017**, *121*, 21077–21113. [CrossRef]

44.   Bonomo, M.; Saccone, D.; Magistris, C.; Di Carlo, A.; Barolo, C.; Dini, D. Effect of alkyl chain length on the sensitizing action of substituted non symmetric squaraines for p-type dye-sensitized solar cells. *ChemElectroChem* **2017**, *4*, 2385–2397. [CrossRef]

45.   Farré, Y.; Raissi, M.; Fihey, A.; Pellegrin, Y.; Blart, E.; Jacquemin, D.; Odobel, F. A Blue Diketopyrrolopyrrole Sensitizer with High Efficiency in Nickel-Oxide-based Dye-Sensitized Solar Cells. *ChemSusChem* **2017**, *10*, 2618–2625. [CrossRef] [PubMed]

46.   Naik, P.; Planchat, A.; Pellegrin, Y.; Odobel, F.; Vasudeva Adhikari, A. Exploring the application of new carbazole based dyes as effective p-type photosensitizers in dye-sensitized solar cells. *Sol. Energy* **2017**, *157*, 1064–1073. [CrossRef]

47.   Farré, Y.; Raissi, M.; Fihey, A.; Pellegrin, Y.; Blart, E.; Jacquemin, D.; Odobel, F. Synthesis and properties of new benzothiadiazole-based push-pull dyes for p-type dye sensitized solar cells. *Dyes Pigments* **2017**, *148*, 154–166. [CrossRef]

48.   Sinopoli, A.; Wood, C.J.; Gibson, E.A.; Elliott, P.I.P. New cyclometalated iridium(III) dye chromophore complexes for p-type dye-sensitised solar cells. *Dyes Pigments* **2017**, *140*, 269–277. [CrossRef]

49. Bonomo, M.; Sabuzi, F.; Di Carlo, A.; Conte, V.; Dini, D.; Galloni, P. KuQuinones as sensitizers of NiO based p-type dye-sensitized solar cells. *New J. Chem.* **2017**, *41*, 2769–2779. [CrossRef]

50. Awais, M.; Rahman, M.; Don MacElroy, J.M.; Coburn, N.; Dini, D.; Vos, J.G.; Dowling, D.P. Deposition and characterization of NiO$_x$ coatings by magnetron sputtering for application in dye-sensitized solar cells. *Surf. Coat. Technol.* **2010**, *204*, 2729–2736. [CrossRef]

51. Nattestad, A.; Perera, I.; Spiccia, L. Developments in and prospects for photocathodic and tandem dye-sensitized solar cells. *J. Photochem. Photobiol. C Photochem. Rev.* **2016**, *28*, 44–71. [CrossRef]

52. Flynn, C.J.; Oh, E.E.; McCullough, S.M.; Call, R.W.; Donley, C.L.; Lopez, R.; Cahoon, J.F. Hierarchically-structured NiO nanoplatelets as mesoscale p-type photocathodes for dye-sensitized solar cells. *J. Phys. Chem. C* **2014**, *118*, 14177–14184. [CrossRef]

53. Dini, D.; Halpin, Y.; Vos, J.G.; Gibson, E.A. The influence of the preparation method of NiO$_x$ photocathodes on the efficiency of p-type dye-sensitized solar cells. *Coord. Chem. Rev.* **2015**, *304–305*, 179–201. [CrossRef]

54. Wood, C.J.; Summers, G.H.; Gibson, E.A. Increased photocurrent in a tandem dye-sensitized solar cell by modifications in push–pull dye-design. *Chem. Commun.* **2015**, *51*, 3915–3918. [CrossRef] [PubMed]

55. Wood, C.J.; Summers, G.H.; Clark, C.A.; Kaeffer, N.; Braeutigam, M.; Carbone, L.R.; D'Amario, L.; Fan, K.; Farré, Y.; Narbey, S.; et al. A comprehensive comparison of dye-sensitized NiO photocathodes for solar energy conversion. *Phys. Chem. Chem. Phys.* **2016**, *18*, 10727–10738. [CrossRef] [PubMed]

56. Farrè, Y.; Zhang, L.; Pellegrin, Y.; Planchat, A.; Blart, E.; Boujtita, M.; Hammarstrom, L.; Jacquemin, D.; Odobel, F. Second Generation of Diketopyrrolopyrrole Dyes for NiO-Based Dye-Sensitized Solar Cells. *J. Phys. Chem. C* **2016**, *120*, 7923–7940. [CrossRef]

57. Brisse, R.; Faddoul, R.; Bourgeteau, T.; Tondelier, D.; Leroy, J.; Campidelli, S.; Berthelot, T.; Geffroy, B.; Jousselme, B. Inkjet printing NiO-based p-Type dye-sensitized solar cells. *ACS Appl. Mater. Interfaces* **2017**, *9*, 2369–2377. [CrossRef] [PubMed]

58. Liu, Q.; Wei, L.; Yuan, S.; Ren, X.; Zhao, Y.; Wang, Z. The effect of Ni (CH$_3$COO)$_2$ post-treatment on the charge dynamics in p-type NiO dye-sensitized solar cells. *J. Mater. Sci.* **2015**, 6668–6676. [CrossRef]

59. Kong, W.; Li, S.; Chen, Z.; Wei, C.; Li, W.; Li, T.; Yan, Y.; Jia, X.; Xu, B.; Zhang, W. p-Type Dye-Sensitized Solar Cells with a CdSeS Quantum-Dot-Sensitized NiO Photocathode for Outstanding Short-Circuit Current. *Part. Part. Syst. Charact.* **2015**, *32*, 1078–1082. [CrossRef]

60. Willkomm, J.; Orchard, K.L.; Reynal, A.; Pastor, E.; Durrant, J.R.; Reisner, E. Dye-sensitised semiconductors modified with molecular catalysts for light-driven H$_2$ production. *Chem. Soc. Rev.* **2016**, *45*, 9–23. [CrossRef] [PubMed]

61. Kamire, R.J.; Majewski, M.B.; Hoffeditz, W.L.; Phelan, B.T.; Farha, O.K.; Hupp, J.T.; Wasielewski, M.R. Photodriven hydrogen evolution by molecular catalysts using Al$_2$O$_3$-protected perylene-3,4-dicarboximide on NiO electrodes. *Chem. Sci.* **2017**, *8*, 541–549. [CrossRef] [PubMed]

62. Meng, P.; Wang, M.; Yang, Y.; Zhang, S.; Sun, L. CdSe quantum dots/molecular cobalt catalyst co-grafted open porous NiO film as a photocathode for visible light driven H$_2$ evolution from neutral water. *J. Mater. Chem. A* **2015**, *3*, 18852–18859. [CrossRef]

63. Hoogeveen, D.A.; Fournier, M.; Bonke, S.A.; Fang, X.Y.; Mozer, A.J.; Mishra, A.; Bäuerle, P.; Simonov, A.N.; Spiccia, L. Photo-electrocatalytic hydrogen generation at dye-sensitised electrodes functionalised with a heterogeneous metal catalyst. *Electrochim. Acta* **2016**, *219*, 773–780. [CrossRef]

64. Antila, L.J.; Ghamgosar, P.; Maji, S.; Tian, H.; Ott, S.; Hammarström, L. Dynamics and Photochemical H$_2$ Evolution of Dye–NiO Photocathodes with a Biomimetic FeFe-Catalyst. *ACS Energy Lett.* **2016**, *1*, 1106–1111. [CrossRef]

65. Massin, J.; Lyu, S.; Pavone, M.; Muñoz-García, A.B.; Kauffmann, B.; Toupance, T.; Chavarot-Kerlidou, M.; Artero, V.; Olivier, C. Design and synthesis of novel organometallic dyes for NiO sensitization and photo-electrochemical applications. *Dalt. Trans.* **2016**, *45*, 12539–12547. [CrossRef] [PubMed]

66. Bonomo, M.; Dini, D. Nanostructured p-type semiconductor electrodes and photoelectrochemistry of their reduction processes. *Energies* **2016**, *9*, 373. [CrossRef]

67. Tian, H. Molecular Catalyst Immobilized Photocathodes for Water/Proton and Carbon Dioxide Reduction. *ChemSusChem* **2015**. [CrossRef] [PubMed]

68. Bonomo, M.; Dini, D.; Marrani, A.G.; Zanoni, R. X-ray photoelectron spectroscopy investigation of nanoporous NiO electrodes sensitized with Erythrosine B. *Colloids Surf. A Physicochem. Eng. Asp.* **2017**. [CrossRef]

69. Awais, M.; Dowling, D.P.; Decker, F.; Dini, D. Electrochemical characterization of nanoporous nickel oxide thin films spray-deposited onto indium-doped tin oxide for solar conversion scopes. *Adv. Condens. Matter Phys.* **2015**, *2015*, 186375. [CrossRef]

70. Awais, M.; Dowling, D.D.; Decker, F.; Dini, D. Photoelectrochemical properties of mesoporous NiO$_x$ deposited on technical FTO via nanopowder sintering in conventional and plasma atmospheres. *Springerplus* **2015**, *4*, 564–588. [CrossRef] [PubMed]

71. Gregg, B.A. Interfacial processes in the dye-sensitized solar cell. *Coord. Chem. Rev.* **2004**, *248*. [CrossRef]

72. Sheehan, S.; Naponiello, G.; Odobel, F.; Dowling, D.P.; Di Carlo, A.; Dini, D. Comparison of the photoelectrochemical properties of RDS NiO thin films for p-type DSCs with different organic and organometallic dye-sensitizers and evidence of a direct correlation between cell efficiency and charge recombination. *J. Solid State Electrochem.* **2015**, *19*, 975–986. [CrossRef]

73. Bonomo, M.; Naponiello, G.; Carlo, A.D.; Dini, D. Characterization of Screen-Printed Nickel Oxide Electrodes for p-type Dye-Sensitized Solar Cells. *J. Mater. Sci. Nanotechnol.* **2016**, *4*, 201. [CrossRef]

74. Sakurai, K.; Fujihara, S. Fabrication of Nanostructured NiO Thick Films by Facile Printing Method and their Dye-Sensitized Solar Cell Performance. *Key Eng. Mater.* **2010**, *445*, 74–77. [CrossRef]

75. Gibson, E.A.; Awais, M.; Dini, D.; Dowling, D.P.; Pryce, M.T.; Vos, J.G.; Boschloo, G.; Hagfeldt, A. Dye sensitised solar cells with nickel oxide photocathodes prepared via scalable microwave sintering. *Phys. Chem. Chem. Phys.* **2013**, *15*, 2411–2420. [CrossRef] [PubMed]

76. Li, N.; Gibson, E.A.; Qin, P.; Boschloo, G.; Gorlov, M.; Hagfeldt, A.; Sun, L. Dou ble-layered NiO photocathodes for p-Type DSSCs with record IPCE. *Adv. Mater.* **2010**, *22*, 1759–1762. [CrossRef] [PubMed]

77. Qin, P.; Zhu, H.; Edvinsson, T.; Boschloo, G.; Hagfeldt, A.; Sun, L. Design of an Organic Chromophore for P-Type Dye-Sensitized Solar Cells. *J. Am. Chem. Soc.* **2008**, *130*, 8570–8571. [CrossRef] [PubMed]

78. Bonomo, M.; Congiu, M.; De Marco, M.L.; Dowling, D.P.; Di Carlo, A.; Graeff, C.F.O.; Dini, D. Limits on the use of cobalt sulfide as anode of p-type dye-sensitized solar cells Dedicated to Professor Roberto Federici on the occasion of his retirement. *J. Phys. D Appl. Phys.* **2017**, *50*. [CrossRef]

79. Perera, V.P.S.; Pitigala, P.; Jayaweera, P.V.V.; Bandaranayake, K.M.P.; Tennakone, K. Dye-sensitized solid-state photovoltaic cells based on dye multilayer-semiconductor nanostructures. *J. Phys. Chem. B* **2003**, *107*, 13758–13761. [CrossRef]

80. Perera, V.P.S.; Pitigala, P.K.D.D.P.; Senevirathne, M.K.I.; Tennakone, K. A solar cell sensitized with three different dyes. *Sol. Energy Mater. Sol. Cells* **2005**, *85*, 91–98. [CrossRef]

81. Awais, M.; Rahman, M.; Don MacElroy, J.M.; Dini, D.; Vos, J.G.; Dowling, D.P. Application of a novel microwave plasma treatment for the sintering of nickel oxide coatings for use in dye-sensitized solar cells. *Surf. Coat. Technol.* **2011**, *205*, S245–S249. [CrossRef]

82. Awais, M.; Dowling, D.D.; Rahman, M.; Vos, J.G.; Decker, F.; Dini, D. Spray-deposited NiO$_x$ films on ITO substrates as photoactive electrodes for p-type dye-sensitized solar cells. *J. Appl. Electrochem.* **2013**, *43*, 191–197. [CrossRef]

83. Awais, M.; Gibson, E.; Vos, J.G.; Dowling, D.P.; Hagfeldt, A.; Dini, D. Fabrication of Efficient NiO Photocathodes Prepared via RDS with Novel Routes of Substrate Processing for p-Type Dye-Sensitized Solar Cells. *ChemElectroChem* **2014**, *1*, 384–391. [CrossRef]

84. Langmar, O.; Saccone, D.; Amat, A.; Fantacci, S.; Viscardi, G.; Barolo, C.; Costa, R.D.; Guldi, D.M. p-type Squaraines Designing to Control Charge Injection and Recombination Processes in NiO based p-type Dye-Sensitized Solar Cells. *ChemSusChem* **2017**. [CrossRef] [PubMed]

85. Bonomo, M.; Magistris, C.; Buscaino, R.; Fin, A.; Barolo, C.; Dini, D. Effect of Sodium Hydroxide Pretreatment of NiO$_x$ Cathodes on the Performance of Squaraine-Sensitized p-Type Dye-Sensitized Solar Cells. *ChemistrySelect* **2018**, *3*, 1066–1075. [CrossRef]

86. Bonomo, M.; Barbero, N.; Matteocci, F.; Di Carlo, A.; Barolo, C.; Dini, D. Beneficial Effect of Electron-Withdrawing Groups on the Sensitizing Action of Squaraines for p-Type Dye-Sensitized Solar Cells. *J. Phys. Chem. C* **2016**, *120*, 16340–16353. [CrossRef]

87. Barbero, N.; Magistris, C.; Park, J.; Saccone, D.; Quagliotto, P.; Buscaino, R.; Medana, C.; Barolo, C.; Viscardi, G. Microwave-Assisted Synthesis of Near-Infrared Fluorescent Indole-Based Squaraines. *Org. Lett.* **2015**, *17*, 3306–3309. [CrossRef] [PubMed]

88. Galliano, S.; Novelli, V.; Barbero, N.; Smarra, A.; Viscardi, G.; Borrelli, R.; Sauvage, F.; Barolo, C. Dicyanovinyl and cyano-ester benzoindolenine squaraine dyes: The effect of the central functionalization on dye-sensitized solar cell performance. *Energies* **2016**, *9*, 486. [CrossRef]

89. Novelli, V.; Awais, M.; Dowling, D.P.; Dini, D. Electrochemical Characterization of Rapid Discharge Sintering (RDS) NiO Cathodes for Dye-Sensitized Solar Cells of p-Type. *Am. J. Anal. Chem.* **2015**, *6*, 176–187. [CrossRef]

90. Venditti, I.; Barbero, N.; Vittoria Russo, M.; Di Carlo, A.; Decker, F.; Fratoddi, I.; Barolo, C.; Dini, D. Electrodeposited ZnO with squaraine sentisizers as photoactive anode of DSCs. *Mater. Res. Express* **2014**, *1*, 15040. [CrossRef]

91. Bonomo, M.; Dini, D.; Marrani, A.G. Adsorption Behavior of I_3^- and I^- Ions at a Nanoporous NiO/Acetonitrile Interface Studied by X-ray Photoelectron Spectroscopy. *Langmuir* **2016**, *32*. [CrossRef] [PubMed]

92. Galoppini, E. Linkers for anchoring sensitizers to semiconductor nanoparticles. *Coord. Chem. Rev.* **2004**, *248*, 1283–1297. [CrossRef]

93. Chang, C.H.; Chen, Y.C.; Hsu, C.Y.; Chou, H.H.; Lin, J.T. Squaraine-arylamine sensitizers for highly efficient p-type dye-sensitized solar cells. *Org. Lett.* **2012**, *14*, 4726–4729. [CrossRef] [PubMed]

94. Jiang, J.-Q.; Sun, C.-L.; Shi, Z.-F.; Zhang, H.-L. Squaraines as light-capturing materials in photovoltaic cells. *RSC Adv.* **2014**, *4*, 32987–32996. [CrossRef]

95. Hagfeldt, A.; Boschloo, G.; Sun, L.; Kloo, L.; Pettersson, H. Dye-Sensitized Solar Cells. *Chem. Rev.* **2010**, *110*, 6595–6663. [CrossRef] [PubMed]

96. Hagfeldt, A.; Graetzel, M. Light-Induced Redox Reactions in Nanocrystalline Systems. *Chem. Rev.* **1995**, *95*, 49–68. [CrossRef]

97. Bonomo, M.; Carella, A.; Centore, R.; Di Carlo, A.; Dini, D. First Examples of Pyran Based Colorants as Sensitizing Agents of p-Type Dye-Sensitized Solar Cells. *J. Electrochem. Soc.* **2017**, *164*, F1412–F1418. [CrossRef]

coatings

MDPI

Article

A DFT-Based Model on the Adsorption Behavior of H₂O, H⁺, Cl⁻, and OH⁻ on Clean and Cr-Doped Fe(110) Planes

Jun Hu [1,2,*], Chaoming Wang [1], Shijun He [1], Jianbo Zhu [1], Liping Wei [1] and Shunli Zheng [2,*]

[1] School of Chemical Engineering, Northwest University, Xi'an 710069, China, cmwang158@163.com (C.W.); heshijun0717@163.com (S.H.); jianzhu@nwu.edu.cn (J.Z.); weiliping@nwu.edu.cn (L.W.)
[2] School of Materials Science and Engineering, Nanyang Technological University, 50 Nanyang Avenue, Singapore 639798, Singapore
* Correspondence: hujun32456@163.com.cn (J.H.); zhengsl@ntu.edu.sg (S.Z.); Tel.: +86-136-6928-4868 (J.H.)

Received: 1 December 2017; Accepted: 23 January 2018; Published: 29 January 2018

Abstract: The impact of four typical adsorbates, namely H₂O, H⁺, Cl⁻, and OH⁻, on three different planes, namely, Fe(110), Cr(110) and Cr-doped Fe(110), was investigated by using a density functional theory (DFT)-based model. It is verified by the adsorption mechanism of the abovementioned four adsorbates that the Cr-doped Fe(110) plane is the most stable facet out of the three. As confirmed by the adsorption energy and electronic structure, Cr doping will greatly enhance the electron donor ability of neighboring Fe atoms, which in turn prompts the adsorption of the positively charged H⁺. Meanwhile, the affinity of Cr to negatively charged adsorbates (e.g., Cl⁻ and O of H₂O, OH⁻) is improved due to the weakening of its electron donor ability. On the other hand, the strong bond between surface atoms and the adsorbates can also weaken the bond between metal atoms, which results in a structure deformation and charge redistribution among the native crystal structure. In this way, the crystal becomes more vulnerable to corrosion.

Keywords: density functional theory; electron transfer; electronic structures; bond population; density of states

---

1. Introduction

The interaction between small molecules and metal surface is crucial in applied researches such as interfacial catalysis, anisotropic synthesis, anti-corrosion, and so forth [1]. For example, the interfacial reactions involving H₂O is significantly affected by the adsorption of H₂O molecules on the metal surface [2,3]. In addition, the adsorption of other small molecules can also greatly influence the catalytic efficiency of relative metallic catalysts, but this has been less investigated [4]. Therefore, it is paramount to understand the interaction between metal surface and various small molecules [5–7].

Quantum chemical simulation can be used to intuitively understand the geometric structure and electronic structure of metals after adsorption of H₂O. During the past few decades, considerable academic efforts have been devoted to the investigation of ion adsorption and activation over metallic iron surfaces. For example, Rafael et al. [8] studied the adsorption mechanism of a single H₂O molecule on the Fe(100) surface, which exhibited the highest adsorption energy for 0.11 and 0.25 coverages when adsorption took place on the top site of Fe. Zhao et al. [9] reported that H₂O monomers would preferentially bind to the top site and lie nearly flat on the Fe(110) surface. In addition to H₂O molecules, adsorption of H⁺ and Cl⁻ in acidic solution also plays an important role in anti-corrosion and surface stability. Bozso et al. [10] reported the adsorption energies of H⁺ on Fe(110), (100), and (111) planes, which were 26, 24, and 21 kcal/mol, respectively. Keisuke et al. [11] suggested that the small Fe clusters could form chemical bonds with gaseous H, and further absorb 10 H atoms in total by two types of bonding mechanisms. The total magnetic moment would decrease from 4 to 0 μ_B through hydrogenation while the distance of Fe–Fe bond elongates from 2.00 to 2.79 Å. Despite the great deal of research efforts on the adsorption behavior on Fe surface, only a few have investigated the adsorption of Cl^- on surfaces of Fe and FeCr alloys, which is theoretically associated with the initiation of pitting corrosion and crack growth [12]. The comparison of different adsorbates on Fe and FeCr alloy surfaces may lead to an in-depth understanding on the anti-corrosion effect of Cr atoms.

In this study, adsorption models of aggressive species, i.e. H_2O, Cl^-, H^+, and OH^-, on Fe(110) and Cr-doped Fe(110) facets were simulated by the quantum chemistry calculation based on density functional theory (DFT). In this comprehensive investigation which considers the adsorption energies, surface energies, bond length and population, charge transferring and density of states (DOS), we expect to reveal the impact of Cr doping on the surface proprieties of Fe.

2. Theoretical Calculation Methods

The $DMol^3$ module of the Materials Studio software 8.0 (Accelrys Inc., San Diego, CA, USA) was employed for the quantum chemistry calculations. In DFT calculations, the Generalized Gradient Approximation (GGA) in the form of the Perdew–Burke–Ernzerhof (PBE) method was used for the exchange–correlation function [13–17]. The spin value was set to be unrestricted, and the formal spin was expressed as the initial value [18]. The energy convergence tolerance was higher than 1.0×10^{-5} Ha. Structures were relaxed using a geometry optimization method until the forces on all atoms were less than 0.002 Ha/Å to satisfy the convergence criterion. The double-numerical quality basis was employed, which was set with Double Numerical plus Polarization (DNP) functions. The Effective Core Potential (ECP) was used to handle the core electrons of the metallic atoms. A thermal smearing was adopted at 0.005 hartree. A Monkhorst–Pack grid of $4 \times 4 \times 1$ k point sampling in the surface Brillouin zone were used for bulk and surface calculations [19]. The k-point separation was 0.05 Å. The electron configuration of the valence was set as follows: Fe-$3d^6 4s^2$, Cr-$3d^5 4s^1$, O-$2s^2 2p^4$, H-$1s^1$, and Cl-$3s^2 3p^5$. During the calculations, the Fe(110) plane was chosen as the ideal model system to investigate the structure, stability, and adsorption properties because it is the most stable surface under practical conditions. A 2×2 supercell of the Fe(110) lattice structure including 4 atomic layers was constructed, and the two bottom layers were fixed. The vacuum thickness of the slab was set to 15 Å to avoid interactions between the adsorbed layer and the top layer of the bulky phase. The atomic layers and vacuum thickness were verified with small errors, as shown in Figure 1. Furthermore, the Fe(110)Cr surface was created by replacing one Fe atom in the surface center with Cr with 1/4 mL coverage. During the calculation, a top site (above the Fe atom of the central plane), a bridge site (above the location between the two Fe atoms), an hcp site (above an acute triangle of the Fe atoms of the plane), and an fcc site (above an acute triangle of the Fe atoms of the plane) takes the adsorption process into account, and the most stable adsorption site was determined by the minimum adsorption energy of the system with other pivotal properties. The bond length of Fe–Fe atoms was optimized to be 2.9627 Å on bulk Fe, which was consistent with the reported value of cubic α-Fe (JCPDS 6-0696, a = 2.8664 Å) [20].

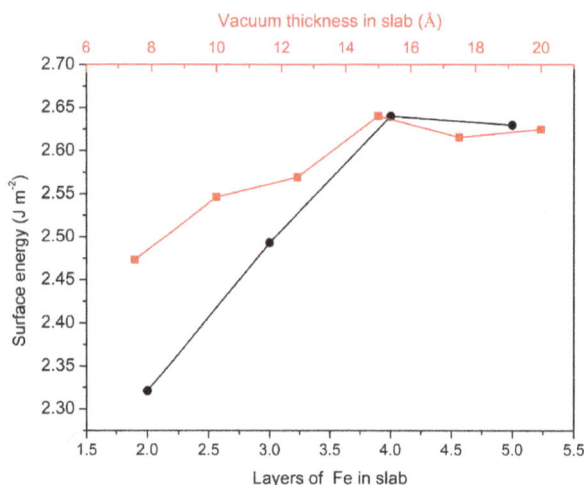

Figure 1. The surface energy of the Fe(110) slab as a function of Fe layers and vacuum using the Generalized Gradient Approximation in the form of the Perdew–Burke–Ernzerhof method (PBE-GGA) method.

3. Results and Discussion

3.1. Surface Energies Calculation

The energy of clean surface can be defined as Equation (1) [21,22].

$$\sigma_{clean} = \frac{1}{2S}(E_{slab}^N - N \times E_{bulk}) \tag{1}$$

Different adsorbate species were added to the top site for the energy calculation of the adsorbed surface. The surface energy after adsorption of different species were calculated according to the Equation (2):

$$\sigma_{clean} + \sigma_{ads} = \frac{1}{S}(E_{slab}^N - N \times E_{bulk} - nE_{species}) \tag{2}$$

where E_{slab}^N is the total electronic energy of the slab, E_{bulk} is the electronic energy of a single metal atom in the bulk, $E_{species}$ is the optimized energy of certain species, which can be derived from the reference species including H_2O (-76.4016541 Ha), H_2 (-1.1166113 Ha), and Cl_2 (-920.121142 Ha). For example, $E_{OH^-} = E_{H_2O} - 0.5H_2$. N is the total number of atoms in the slab, and n is the total number of species on the surface. S is the surface area, and $1/2$ is applied here due to the two equal surfaces in the slab model. It should be noted that the upper and bottom surfaces must be equivalent when the slab model is created. It is clearly shown in the equations that those surfaces of low surface free energy will be more stable and vice versa. Figure 2 compared the surface energies of three different metal facets, which are calculated based on the adsorption of different species on their most stable sites.

As shown in Figure 2, the surface energy of the clean Fe(110) facet was smaller than those of the Fe(110)Cr and Cr(110) facets. The high surface energies of those surfaces indicate it is difficult to exist in reality. So both the Fe and Cr surfaces would be easy to oxidize, especially for the Cr surface. The surface energies were greatly reduced upon adsorption of H_2O, H^+, Cl^-, and OH^- species, which suggested it is energy favorable for the process of adsorption in aqueous solution. Furthermore, the Fe(110)Cr facet showed the smallest surface energy regardless of which species was absorbed, indicating that Cr doping can strengthen adsorption.

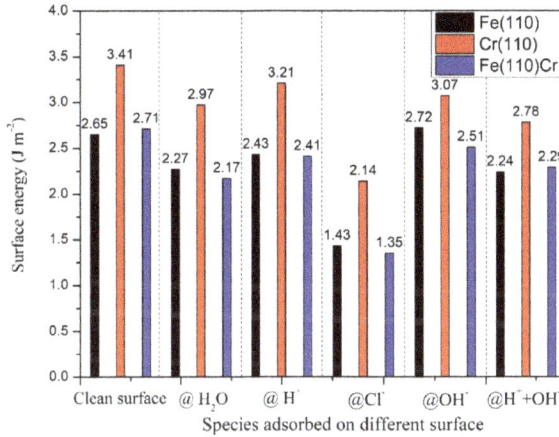

Figure 2. The calculation of surface energy for three different metal facets, based on the adsorption of different species. "@" signs stand for bond, close contact, and adsorption state on the facet.

3.2. The Most Stable Adsorption Structure of Relevant Particles

The adsorption energy (E_{ads}) between a plane and adsorbates is the key parameter in evaluating the stability of that plane [23], which can be calculated as Equation (3):

$$E_{ads} = E_{species-surface} - E_{surface} - E_{species} \qquad (3)$$

where $E_{surface}$ and $E_{species - surface}$ represent the energies of metal planes before and after adsorption, respectively [24]. Theoretically, a more negative value of E_{ads} indicates a more stable adsorption between the adsorbate and the plane. The adsorption energies of four types of species binding on three facets are shown in Figure 3.

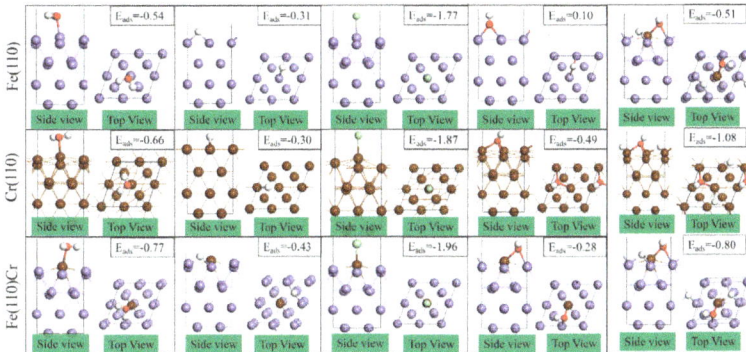

Figure 3. The geometric structures for stable adsorption of H_2O, H^+, Cl^-, and OH^- on Fe(110), Cr(110), and Fe(110)Cr planes, respectively. In the slab, purple spheres stand for Fe atoms, brown ones for Cr atoms, red ones for O atoms, green ones for Cl atoms, and white ones for H atoms. This color scheme applies to the following figures as well. The unit of E_{ads} is eV.

In the modeling, the top, bridge, hcp, and fcc sites take adsorption into account. It was found that the most stable sites for the Fe(110), Cr(110), and Fe(110)Cr surfaces were top, top, and hcp respectively, as calculated from all adsorbates taken into account. The adsorption structures were consistent with the literature [25,26]. The geometric structures for the adsorption of each ion on different surfaces are shown in Figure 3. It can be inferred that, at the most stable site for adsorption, the alignment of H_2O molecules is parallel to the crystal surface of Fe(110)Cr, which agrees well with previous studies [27,28]. Furthermore, regarding the adsorption of H^+, its adsorption energy on the most stable site of the Fe(110)Cr facet is more negative than those of the other two surfaces. However, the adsorption energies of H_2O and Cl^- on the Fe(110)Cr were increased. These results preliminarily illustrated that Cr doping would greatly impact the electronic property of the neighboring Fe atoms since H^+ was adsorbed on the neighboring Fe atoms while H_2O and Cl^- were adsorbed on Cr atoms. The Cr doping changed the charge of neighboring Fe atoms from 0 to -0.022 e, as shown in Figure 4, and rendered them preferential for the adsorption of positively charged hydrogen. Furthermore, the highest occupied molecular orbital (HOMO) region of Cr-doped surface was mainly distributed within the Fe atom, while the lowest unoccupied molecular orbital (LUMO) region was mainly distributed within the Cr atom. Basically, the energy of the HOMO orbital is directly related to the ionization potential and characterizes the susceptibility of the molecules attacked by electrophiles. Cr doping would greatly enhance the electron donor ability of neighboring Fe atoms, which in turn would prompt the adsorption of the positively charged H^+. Meanwhile, the affinity of Cr to negatively charged adsorbates (e.g., Cl^- and O of H_2O, OH^-) was improved due to the weakening of its electron donor ability.

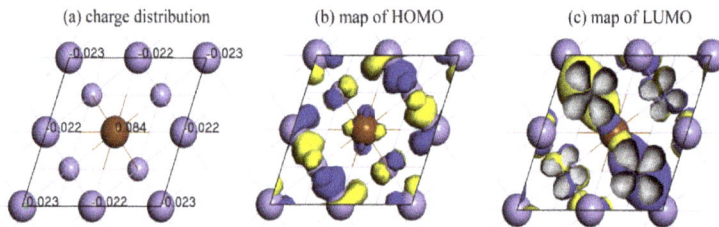

(a) charge distribution   (b) map of HOMO   (c) map of LUMO

Figure 4. (**a**) The charge distribution on the Fe(110)Cr facets. The map of HOMO (**b**) and LUMO (**c**) in Fe(110)Cr facets, respectively. The unit of charge is e.

3.3. Density of States

DOS and partial density of states (PDOS), as shown in Figure 5, provided a more comprehensive explanation of the electronic structure [29].

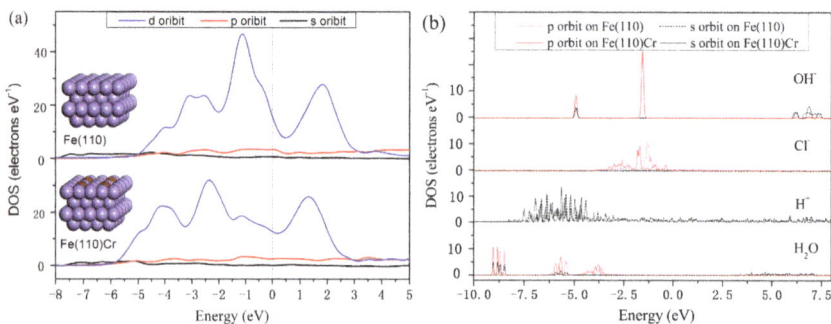

Figure 5. (**a**) PDOS of the Fe(110) and Fe(110)Cr facets. (**b**) Total DOS of H_2O, H^+, Cl^-, and OH^- species before and after their adsorption on Fe(110) and Fe(110)Cr facets. Each adsorbate takes the total DOS calculation into account. The data before adsorption were calculated with optimization, and the data after absorption were calculated by deleting metal atoms in the model without optimization.

As indicated in Figure 5a, the conduction of metal mainly originated from the excitation of electrons from d orbit. After Cr doping, the middle peak (from −2 eV to 0 eV) of the original Fe(110) facet showed a negative shift as well as a decrease in intensity. Figure 5b shows the PDOS of H_2O, H^+, Cl^-, and OH^- on Fe(110) and Cr-doped Fe(110) planes. It is clearly shown that the PDOS peaks of H_2O, Cl^-, and H^+ on the Fe(110)Cr plane were shifted to a deeper energy level compared with those of the clean Fe(110) plane, which indicates a stronger bond between the Fe(110)Cr surface and the adsorbates.

3.4. The Effect Adsorbed Species on Surface

The average slab bond lengths after adsorption of H_2O, Cl^-, and H^+ on the most stable sites are presented in Figure 6.

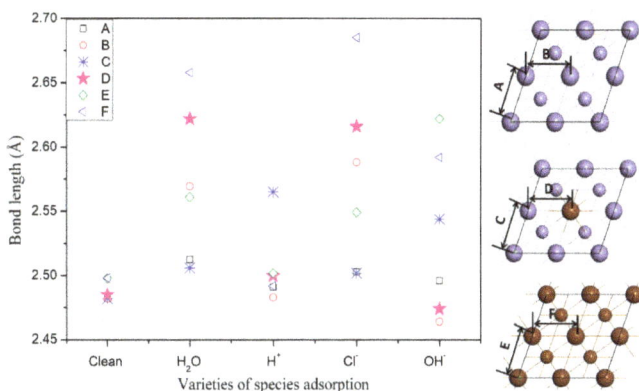

Figure 6. The change in slab bond length after adsorption of H_2O, H^+, Cl^-, and OH^- on Fe(110) and Fe(110)Cr surfaces.

As shown in Figure 6, initially the Fe–Fe bond length on the clean Fe(110) surface was equal to that on the clean Fe(110)Cr surface. By comparing the adsorption of the four species on these three facets, it was revealed that all of the bond lengths on the three facets were greatly changed by adsorption, which confirms that the bond strength of facets can be affected by the adsorption of these four species. By comparing the bond lengths of D (as indicated in Figure 6) on the Cr-doped Fe(110) surface after adsorption of different species, it can be said that the adsorption of H_2O resulted in the most significant increase in bond length. The impact of the four species on surfaces was sorted as $H_2O > Cl^- > H^+ > OH^-$. We can conclude that, generally, the adsorbed H_2O and Cl^- will induce a greater structure deformation on surfaces, which will weaken the bond between metal atoms and render the crystal vulnerable to corrosion.

4. Conclusions

The adsorption behavior of four typical adsorbates, namely, H_2O, H^+, Cl^-, and OH^-, on three different planes, namely, Fe(110), Cr(110), and Cr-doped Fe(110), were investigated using a DFT-based model. The surface energy study suggests that the Cr-doped Fe(110) surface is more stable than Fe(110) and Cr(110) facets upon adsorption of these four typical adsorbates. As confirmed by adsorption energy and electronic structure, Cr doping will greatly enhance the electron donor ability of neighboring Fe atoms, which in turn prompts the adsorption of the positively charged H^+. Meanwhile, the affinity of Cr to negatively charged adsorbates (e.g., Cl^- and O of H_2O, OH^-) is improved due to the weakening of its electron donor ability. On the other hand, the strong bond between surface atoms and the adsorbates can also weaken the bond between metal atoms, which results in a structure deformation and charge redistribution among the native crystal structure. In this way, the crystal becomes more vulnerable to corrosion.

Acknowledgments: Financial support from the National Natural Science Foundation of China (No. 21676216, 51606153), the China Postdoctoral Science Foundation (No. 2014M550507; 2015T81046), and the Innovative Projects of Northwest University (YZZ17140) is greatly acknowledged. This research project was also supported by the Center for High Performance Computing at Northwestern Polytechnical University and Haixia Ma, Northwest University, China.

Author Contributions: Jun Hu conceived and designed the simulations; Chaoming Wang and Shijun He performed the simulations; Jianbo Zhu and Liping Wei analyzed the data; Shunli Zheng wrote the paper.

Conflicts of Interest: The authors declare no conflict of interest.

References

1. Duy, D.V.; Aleksey, G.L.; Truong, K.N.; Thoi, T.N. Nitrogen trapping ability of hydrogen-induced vacancy and the effect on the formation of AlN in aluminum. *Coatings* **2017**, *7*, 79. [CrossRef]

2. Menzel, D. Water on a metal surface. *Science* **2002**, *295*, 58–59. [CrossRef] [PubMed]

3. Roman, O.; Elena, V.; Joachim, S. Water adsorption and O-defect formation on Fe_2O_3(0001) surfaces. *Phys. Chem. Chem. Phys.* **2016**, *18*, 25560–25568.

4. Magali, B.; Nathalie, T.; Joseph, M. Adsorption energy of small molecules on core–shell Fe@Au nanoparticles: Tuning by shell thickness. *Phys. Chem. Chem. Phys.* **2016**, *18*, 9112–9123.

5. Liu, X.; Kang, C.; Qiao, H.; Ren, Y.; Tan, X.; Sun, S. Theoretical studies of the adsorption and migration behavior of boron atoms on hydrogen-terminated diamond (001) surface. *Coatings* **2017**, *7*, 57. [CrossRef]

6. Lakshmi, R.V.; Bharathidasan, T.; Basu, J. Superhydrophobicity of AA2024 by a simple solution immersion technique. *Surf. Innov.* **2015**, *1*, 241–247. [CrossRef]

7. Wang, X.; Yin, X.; Nalaskowski, J.; Du, H.; Miller, J.D. Molecular features of water films created with bubbles at silica surfaces. *Surf. Innov.* **2015**, *3*, 20–26. [CrossRef]

8. Freitas, R.R.Q.; Rivelino, R.; de Brito Mota, F.; de Castilho, C.M.C. Dissociative adsorption and aggregation of water on the Fe(100) Surface: A DFT Study. *J. Phys. Chem. C* **2012**, *116*, 20306–20314. [CrossRef]

9. Zhao, W.; Wang, J.D.; Liu, F.B.; Chen, D.R. First principles investigation of water adsorption on Fe (110) crystal surface containing N. *Adv. Tribol.* **2008**, *1*, 654–657.

10. Bozsom, F.; Ertlm, G.; Grunze, M.; Weiss, M. Chemisorption of hydrogen on iron surfaces. *Appl. Surf. Sci.* **1977**, *1*, 103–119. [CrossRef]

11. Keisuke, T.; Shigehito, I.; Somei, O. Chemisorption of hydrogen on Fe clusters through hybrid bonding mechanisms. *Appl. Phys. Lett.* **2013**, *102*, 11308.

12. Sun, C.; Hui, R.; Qu, W.; Yick, S. Progress in corrosion resistant materials for supercritical water reactors. *Corros. Sci.* **2009**, *51*, 2508–2523. [CrossRef]

13. Wang, H.; Nie, X.; Guo, X.; Song, C. A computational study of adsorption and activation of CO_2 and H_2 over Fe(100) surface. *J. CO2 Utilization* **2016**, *15*, 107–114. [CrossRef]

14. Cremaschi, P.; Yang, H.; Whitten, J.L. Ab initio chemisorption studies of H on Fe(110). *Surf. Sci.* **1995**, *330*, 255–264. [CrossRef]

15. Wang, W.; Wang, G.; Shao, M. First-principles modeling of direct versus oxygen-assisted water dissociation on Fe(100) surfaces. *Catalysts* **2016**, *6*, 29. [CrossRef]

16. Hu, J.; Zhao, X.; Chen, W.; Su, H.; Chen, Z. Theoretical insight into the mechanism of photoelectrochemical oxygen evolution reaction on $BiVO_4$ anode with oxygen vacancy. *J. Phys. Chem. C* **2017**, *121*, 18702–18709. [CrossRef]

17. Horányi, G. Investigation of the specific adsorption of HSO_4^- (SO_4^{2-}) and Cl^- ions on Co and Fe by radiotracer technique in the course of corrosion of the metals in perchlorate media. *Corros. Sci.* **2004**, *46*, 1741–1749. [CrossRef]

18. Ghiasi, M.; Kamalinahad, S.; Arabieh, M.; Zahedi, M. Carbonic anhydrase inhibitors: A quantum mechanical study of interaction between some antiepileptic drugs with active center of carbonic anhydrase enzyme. *Comput. Theor. Chem.* **2012**, *992*, 59–69. [CrossRef]

19. Zhang, H. A DFT study on direct hydrogenation of amide catalyzed by a PNN Ru (II) pincer complex. *Comput. Theor. Chem.* **2015**, *1066*, 1–6. [CrossRef]

20. Kohn, W.; Sham, L.J. Self-consistent equations including exchange and correlation effects. *Phys. Rev.* **1965**, *140*, A1133. [CrossRef]
21. David, H.; Franz, D.F.; Dieter, V. Structure and surface energy of Au_{55} nanoparticles: An *ab initio* study. *Comput. Mater. Sci.* **2017**, *134*, 137–144.
22. Ji, D.P.; Zhu, Q.X.; Wang, S.Q. Detailed first-principles studies on surface energy and work function of hexagonal metals. *Surf. Sci.* **2016**, *651*, 137–146. [CrossRef]
23. Pople, J.A.; Gill, P.M.W.; Johnson, B.G. Kohn—Sham density-functional theory within a finite basis set. *Chem. Phys. Lett.* **1992**, *199*, 557–560. [CrossRef]
24. Lanzani, G.; Laasonen, K. SO_2 and its fragments on a Cu(110) surface. *Surf. Sci.* **2008**, *602*, 321–344. [CrossRef]
25. Henderson, M.A. The interaction of water with solid surfaces: Fundamental aspects revisited. *Surf. Sci. Rep.* **2002**, *46*, 1–308. [CrossRef]
26. Hodgson, A.; Haq, S. Water adsorption and the wetting of metal surfaces. *Surf. Sci. Rep.* **2009**, *64*, 381–451. [CrossRef]
27. Xie, W.; Peng, L.; Peng, D.; Gu, F.L.; Liu, J. Processes of H_2 adsorption on Fe(110) surface: A density functional theory study. *Appl. Surf. Sci.* **2014**, *296*, 47–52. [CrossRef]
28. Michaelides, A.; Ranea, V.A.; de Andres, P.L.; King, D.A. General model for water monomer adsorption on close-packed transition and noble metal surfaces. *Phys. Rev. Lett.* **2003**, *90*. [CrossRef] [PubMed]
29. Tang, J.; Xie, K.; Liu, Y.; Chen, X.; Ma, W.; Liu, Z. Density functional theory study of leaching efficiency of acids on Si(110) surface with adsorbed boron. *Hydrometallurgy* **2015**, *151*, 84–90. [CrossRef]

coatings

MDPI

Review

Strategies of Anode Materials Design towards Improved Photoelectrochemical Water Splitting Efficiency

Jun Hu [1], Shuo Zhao [1], Xin Zhao [2],* and Zhong Chen [2],*

[1] School of Chemical Engineering, Northwest University, Xi'an 710069, China; hujun@nwu.edu.cn (J.H.); zhao981017@foxmail.com (S.Z.)

[2] School of Materials Science and Engineering, Nanyang Technological University, 50 Nanyang Avenue, Singapore 639798, Singapore

* Correspondence: xinzhao@ntu.edu.sg (X.Z.); ASZChen@ntu.edu.sg (Z.C.); Tel.: +65-6790-4256 (Z.C.)

Received: 4 April 2019; Accepted: 6 May 2019; Published: 9 May 2019

Abstract: This review presents the latest processes for designing anode materials to improve the efficiency of water photolysis. Based on different contributions towards the solar-to-hydrogen efficiency, we mainly review the strategies to enhance the light absorption, facilitate the charge separation, and enhance the surface charge injection. Although great achievements have been obtained, the challenges faced in the development of anode materials for solar energy to make water splitting remain significant. In this review, the major challenges to improve the conversion efficiency of photoelectrochemical water splitting reactions are presented. We hope that this review helps researchers in or coming to the field to better appreciate the state-of-the-art, and to make a better choice when they embark on new research in photocatalytic water splitting.

Keywords: photoelectrochemical; anode materials; surface; water splitting

1. Introduction

Nowadays, it is becoming the world's biggest challenge to meet the increasing energy demand by developing new sustainable substitution energy and reducing carbon dioxide emission. Overall, water splitting using sunlight is identified as a promising technology for production of renewable fuels and chemicals without causing environmental pollution.

In a photoelectrochemical cell for solar water-splitting, there are three main elements; namely, an anode, a cathode, and an electrolyte. At the anode, the oxygen evolution reaction (OER) will happen, while through the hydrogen evolution reaction (HER), hydrogen forms at the cathode, as shown in reactions R1 and R2 [1].

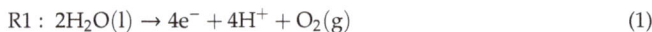

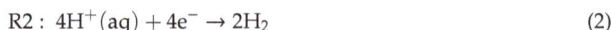

$$R1: \ 2H_2O(l) \rightarrow 4e^- + 4H^+ + O_2(g) \tag{1}$$

$$R2: \ 4H^+(aq) + 4e^- \rightarrow 2H_2 \tag{2}$$

The two half-reactions constitute the overall water splitting process ($2H_2O \rightarrow 2H_2 + O_2$), with the thermodynamic threshold energy $E_0 = 1.23$ eV. Although both reactions are important for the overall water splitting efficiency, the process is mainly hindered by the hypokinetic four-electron water oxidation. Based on electron transfer process, OER can be divided into four reaction paths with 1.23 eV in the surface site (denoted by the * symbol) of anode materials for thermodynamic threshold energy, as shown in reactions R3 to R6 [2].

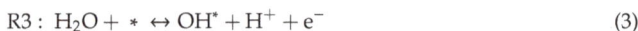

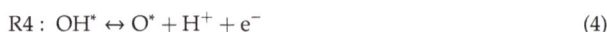

$$R3: \ H_2O + \ * \ \leftrightarrow OH^* + H^+ + e^- \tag{3}$$

$$R4: \ OH^* \leftrightarrow O^* + H^+ + e^- \tag{4}$$

$$R5 : O^* + H_2O \leftrightarrow OOH^* + H^+ + e^- \tag{5}$$

$$R6 : OOH^* \leftrightarrow O_2 + H^+ + e^- \tag{6}$$

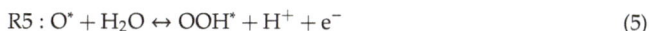

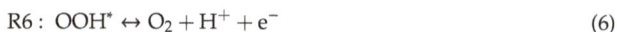

After the Honda–Fujishima effect was discovered in the early 1970s [3], intensive studies were carried out for photoelectrochemical (PEC) water splitting on semiconducting materials. The semiconducting anode materials will greatly determine the efficiency of absorbing light, suitable over-potential, and stability. It is challenging to develop efficient new anode materials or improve the ability of existing anode materials. Consequently, extensive efforts were implemented to improve the efficiency of anode materials; for example, TiO_2, α-Fe_2O_3, WO_3, and $BiVO_4$ [4]. However, there is still lack of cheap and stable photoanode materials with sufficient efficiency for water oxidation. Although significant progress has been made by developing new materials, nanostructures, and other novel concepts over the past decades, efficiencies are still below our expectations. Therefore, to assist future development, we wish to summarize the strategies to develop anode materials for improved efficiency in PEC water splitting in a concise manner. It is up to the readers to decide how different strategies could be applied in their specific applications, or to apply them synergistically in some situations.

The solar-to-hydrogen (STH) efficiency is an important metric for benchmarking and performance evaluation. The STH for a PEC system constructed by photoelectrodes can be expressed by [5]:

$$STH = \frac{J_P \times 1.23}{I_0} \times 100\% \tag{7}$$

where J_P is the photocurrent observed from the experiment, I_0 is the power intensity of the sunlight. The number 1.23 in the equation is the approximated value of the threshold energy for water splitting which requires the bandgap energy of any possible photoanode lager than 1.23 eV. For a photoelectrode, J_P is determined by the photogeneration absorption efficiency (η_{abs}), the product of the injection efficiency (η_{sep}), and the injection efficiency (η_{inj}) of the photogenerated carrier to the reactant [6], which can be described by:

$$J_P = J_0 \times \eta_{abs} \times \eta_{sep} \times \eta_{inj} \tag{8}$$

where J_0 is the theoretical photocurrent when 100% of the photons in the solar spectrum with energies exceeding the bandgap are absorbed and converted. Theoretical STH and solar photocurrent of a photoelectrode under AM 1.5 G irradiation (100 mW·cm^{-2}) are displayed in Figure 1 [6].

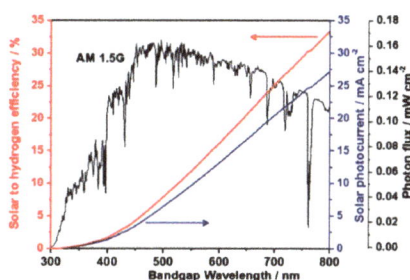

Figure 1. Dependence of theoretical solar-to-hydrogen (STH) efficiency and solar photocurrent density of photoelectrodes on their bandgap absorption edges. Reprinted with permission from ref. [5]. Copyright 2013. The Royal Society of Chemistry.

Thus, to achieve efficient PEC water splitting, several key criteria must be met, as shown in Figure 2: (1) wide range of light absorption of a material that can utilize more energy from sunlight; (2) effective generation of the charges and their transportation from the internal electrode to the photoelectric surface; (3) fast reaction of photogenerated carriers on the surface. Therefore, we will summarize recent progress of improving those efficiencies separately.

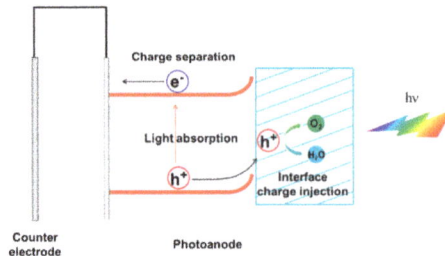

Figure 2. Schematic illustration of the three efficiencies of photoelectrochemical (PEC) water splitting for enhancing performance.

In this review, we summarize recent progress in the development of anode materials for photoelectrochemical water splitting, which are mainly based on our research work. This is carried out through a review of the strategies to enhance the light absorption, facilitate the charge separation, and enhance the surface charge injection. Finally, we share our critical views on the remaining challenges in this field.

2. Strategies to Enhance the Light Absorption

Generally, light absorption of a photoelectrode depends on its bandgap. For a known photoelectrode with a fixed bandgap, the film thickness and structure would affect its light absorption efficiency. Some reported strategies used to enhance the light absorption are summarized below.

2.1. Nanostructure Formation

Nanostructure is a good way to capture the scattered light if the light is not absorbed immediately after it reaches the surface. Nanostructure formation is a well-known strategy for silicon solar cells to reduce reflection loss and increase the energy conversion efficiency [7]. Similarly, nanostructure construction is proven effective in improving the light capture in PEC electrodes [8]. It has been reported that multiple light scattering in engineered tapered nanostructures enhances light absorption. (Figure 3) [9]. With a nanoporous Mo:BiVO$_4$ layer on the engineering conical nanophotonic structure combined with a solar cell, a STH efficiency of 6.2% was realized. Zhou et al. reported a hierarchical BiVO$_4$ photoelectrode, which consists of small nanoparticles and voids, could induce efficient light harvesting due to multiple light scattering [6]. Kim reported high-efficiency light absorption of GaN truncated nanocones. The truncated nanocones can trap the light within the nanostructure; thus, light loss, which is caused by surface reflection, is reduced. The TiO$_2$ solar cell prepared by using a 1D nanowire array can have a light conversion efficiency of up to 5.02%, which is much higher than that of a simple TiO$_2$ powder [10].

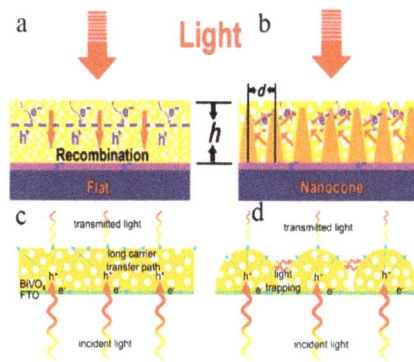

Figure 3. Optical absorption and electron transport on nanoporous $BiVO_4$ with (**a**) the flat substrate and (**b**) the conductive nanocone substrate. Reprinted with permission from ref. [9]. Copyright 2016. Science. (**c**) flat $BiVO_4$ film and (**d**) hierarchical $BiVO_4$ nanostructured film. Reprinted with permission from ref. [10]. Copyright 2008. American Chemical Society.

2.2. Band Engineering

To reduce the band gap of semiconductors, one commonly adopted and promising strategy is to add a foreign element (doping). Doping of foreign elements may include adding cation(s), anion(s), or adding cation(s) and anion(s) at same time.

Addition of cations usually induces a shallow valence band in the wide band gap through the hybrid valence orbital [11]. In metal oxide photocatalysts, incorporation of metal ions will introduce a donor level by hybridizing the O 2p valence orbitals with a metal ion having high-lying valence orbitals, such as Bi(III), Ag(I), Sn(II), Pb(II), or Cu(I). The crystal structure and the amount of metal incorporated greatly affect the contribution of the valence band. For example, in a MoO_3 electrode, Li^+, Na^+, and Mg^{2+} can greatly reduce the gap of the band [12,13]. Thus, modulating the band structure by cation doping is considered to be effective at the atomic level.

Anion doping has been employed to shift the absorption edge from the UV to the visible region; indeed, it is one of the strategies to enhance the light absorption. A lot of attention has been paid to nonmetal doping into TiO_2 to tune the light absorption, such as N doping as well as C, F, S, B doping [14]. For example, N and B doping greatly expands the light absorption to 700 nm [15]. Also, O incorporation in MoS_2 is able to improve the hybridization between Mo d-orbital and S p-orbital, resulting in a much smaller band gap [16]. Similarly, replacement of O by N in Ta_2O_5 induces a narrowed bandgap; for example, it was observed that the absorption was extended from 320 to 500 nm by N-doping (Figure 4) [17]. Simultaneously adding cations and anions can greatly change the band structure for enhancing the light absorption; e.g., the band gap of GaN is about 3.4 eV and the band gap of ZnO is 3.2 eV (Figure 5). However, the solid solution of GaN and ZnO narrowed the band gap to around 2.58 eV with visible light response [18]. This strategy provides us with a powerful method for adjusting the bandgap of photoelectrode materials through band engineering.

Another strategy for enlarging the light absorption is by intrinsic doping, such as introducing oxygen vacancy in oxides. The typical example is black TiO_2, in which lots of Ti^{3+} species are generated accompanied with the oxygen vacancy, being self-doping [19,20]. Doping and presence of lattice vacancies introduce midgap states which form the band tails and narrow the band gap. The light absorption of TiO_2 was enlarged to around 1000 nm from the 400 nm of white TiO_2. However, it is noted that the enlarged light absorption has little contribution to the enhanced performance, because the there is little visible light response for the black TiO_2 (Figure 6) [21]. The enhanced performance is possibly from the better conductivity.

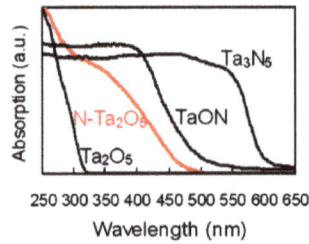

Figure 4. Diffuse reflectance absorption spectrum of N-Ta$_2$O$_5$ power (8.9 at. % N), nondoped Ta$_2$O$_5$, TaON, and Ta$_3$N$_5$ in the ultraviolet and visible light region. Reprinted with permission from ref. [17]. Copyright 2018. American Institute of Physics.

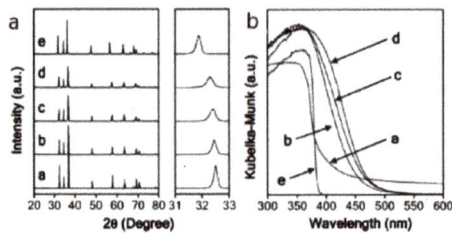

Figure 5. (**a**) Power X-ray diffraction pattern and (**b**) ultraviolet light visible diffuse reflectance spectra: a, b, c, d, e stand for GaN (ref), GaN:ZnO (Zn 3.4 at. %), GaN:ZnO (Zn 6.4 at. %), GaN:ZnO (Zn 13.3 at. %), and ZnO. Reprinted with permission from ref. [18]. Copyright 2018. American Chemical Society.

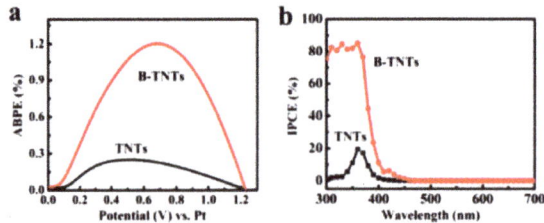

Figure 6. (**a**) Applied bias photon-to-current efficiency (ABPE) of the TiO$_2$ nanotubes (TNTs) and black titania nanotube (B-TNTs) as a function of applied potential; (**b**) incident-photon-to-current-conversion efficiency (IPCE) spectra in the region of 300–700 nm at 0.23 V vs. Ag/AgCl. Reprinted with permission from ref. [21]. Copyright 2014. The Royal Society of Chemistry.

2.3. Dual Absorber

Wang et al. reported extraordinary PEC performance by using dual BiVO$_4$ photoanodes made of two BiVO$_4$ which has the same transparency (Figure 7) [22]. The second electrode can use the transmitted light unabsorbed by the first electrode. It was found that the enhanced performance was mainly because of the higher light utilization rate in the airway scope of band edge (400–520 nm). A similar strategy was employed by Kim et al. with a different absorber; BiVO$_4$ in the front, and Fe$_2$O$_3$ behind BiVO$_4$ (Figure 8) [23]. This can further utilize the solar light, because Fe$_2$O$_3$ has a smaller bandgap than BiVO$_4$; thus, the transmitted light can be utilized by Fe$_2$O$_3$. This has increased the efficiency of hydrogen conversion to 7.7%. Wang et al. reported PEC water splitting using a hematite/Si nanowire dual-absorber system, which yielded a lower onset potential [24]. In their work, hematite was coated on the surface of Si nanowire, and the transmitted light through hematite was utilized by Si. The generated electrons of hematite combine with the holes generated on Si. Thus, holes generated on hematite are used for water oxidation, and at the same time, electrons generated on Si are used for reduction reaction. The match of the band edge positions of the two absorbers is important. A double absorber series battery was reported by Sivula et al., in which the light transmitted through the photoelectrode material is utilized by the dye-sensitized solar cells (DSSC) that provides the bias needed by the photoelectrode [25]. This concept is very useful in tandem PEC cells, which do not need additional bias and can increase the solar conversion efficiency. Besides, Chang and Ye also added Co_3O_4 and carbon quantum dot (CQDs) components on the surface of $BiVO_4$ to enlarger light absorption range [26,27]. For instance, CQDs decorated $BiVO_4$ photoanode cocatalyzed by a bi-layered Ni-FeOOH oxygen evolution catalyst have obtained an outstanding photocurrent density of 5.99 mA·cm^{-2} at 1.23 V vs. reversible hydrogen electrode (RHE) with an enlarged light absorption range. Noble metal nanoparticles can also enhance the absorption light scattering/trapping, such as Au nanoparticles coated $BiVO_4$ and hematite [28,29]. However, the plasmonic effect also involves direct electron transfer mechanism (DET) and plasmon induced resonance electron transfer (PIRET) to enhance the performance except the light absorption. Currently, the full understanding of the involved physical mechanisms remains elusive, and requires more work to be clarifies [30].

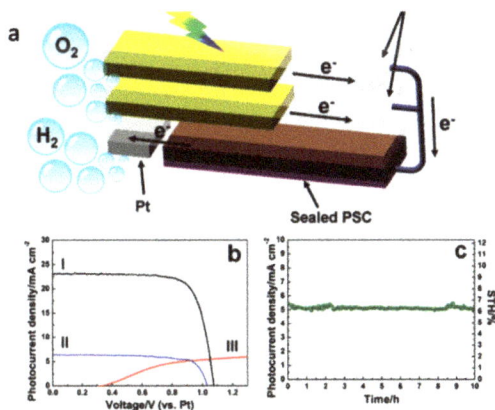

Figure 7. (**a**) The $BiVO_4$-FeOOH/NiOOH dual photoanodes coupled with s sealed perovskite solar cell; (**b**) I: *J–V* curve of perovskite solar cell, II: *J–V* curve of perovskite solar cell behind the BVO-FeOOH/NiOOH dual photoanodes, III: *J–V* curve of BVO-FeOOH/NiOOH under AM 1.5 G illumination; (**c**) Stability of unassisted water splitting (0 V vs. the counter electrode Pt). Reprinted with permission from ref. [22]. Copyright 2018. Wiley.

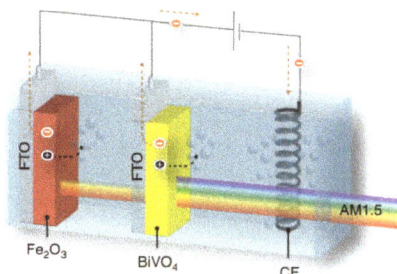

Figure 8. Wavelength-selective solar light absorption by hetero-type dual photoanode. Reprinted with permission from ref. [23]. Copyright 2018. Nature.

3. Strategies to Improve the Charge Separation

3.1. Doping

Doping is proved to be an effective way to improve the conductivity of semiconducting compounds and thus increase the charge separation efficiency [5]. Zhao et al. reported that extrinsic Ti doping and oxygen vacancy generation (intrinsic doping) have greatly enhanced the PEC performance of hematite (Figure 9) [31]. In another example, incorporation of Mo^{6+} into a partial site of V^{5+} in $BiVO_4$ not only has changed the crystal symmetry of $BiVO_4$ and introduced some polarons, but also provided more free carriers (electrons) which facilitate the charge transport [32,33].

Exploration of new materials with suitable band position and narrow bandgap for water splitting is still a vital issue. Since doping is proven as an effective method to ameliorate the efficiency through improved conductivity, this method is usually combined with the design and synthesis of new PEC electrode materials. Recently, the performance of a new promising material, Zn_2FeO_4, for PEC water splitting was improved by Ti doping (Figure 10) [34]. Lumley et al. reported that the performance of $Cu_{11}V_6O_{26}$ was enhanced after doping by Mo or W [35]. Jo et al. found the 0.5% PO_4-doped $BiVO_4$ can lower the charge transfer resistance of $BiVO_4$ remarkably, about 30 times higher than that of $BiVO_4$ before doping [36].

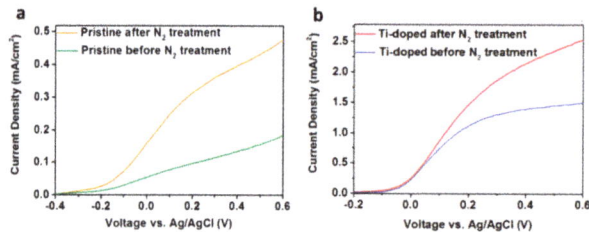

Figure 9. The photocurrents of (**a**) pristine and (**b**) Ti-doped hematite before and after N_2 treatment under AM 1.5 G solar simulator illumination. The N_2 treatment is carried out in nitrogen gas at 600 °C for 2 h. Reprinted with permission from ref. [31]. Copyright 2017. The Royal Society of Chemistry.

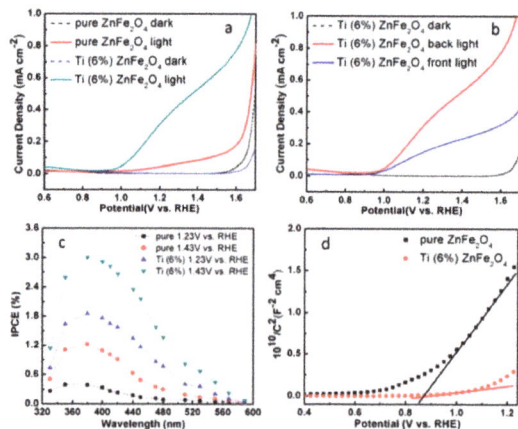

Figure 10. The photocurrents of (**a**) $ZnFe_2O_4$ photoanodes and Ti-doped $ZnFe_2O_4$ photoanodes in 1 M NaOH aqueous solution under AM 1.5 G illumination (100 mW·cm^{-2}); (**c**) IPCE spectrum of Ti-doped $ZnFe_2O_4$ and pure $ZnFe_2O_4$ in range from 330 to 589 nm at different applied potential; (**d**) Mott–Schottky plots of pure Ti-doped $ZnFe_2O_4$ and $ZnFe_2O_4$ photoanodes in 1 M NaOH, the AC amplitude is 5 mV and the frequency is 1000 Hz. Reprinted with permission from ref. [34]. Copyright 2018. American Chemical Society.

One special doping is gradient doping, which means that the dopant concentration varies from surface to the bulk. In such a case, the band bending not only occurs on the surface, but also in the bulk, which provides an electric field throughout the bulk material. It is easier for the carriers to separate and transport to the surface in a gradient electrode. Yang et al. investigated how gradient doping affects a GaAs photocathode and found that both the escape probability and the diffusion length increased when compared with uniform doping [37]. Abdi et al. synthesized a doped W gradually decreased $BiVO_4$ from 1% to 0% between the interface of the FTO/semiconductor and semiconductor/electrolyte (Figure 11) [38]. Based on their experiment, the efficiency of charge separation increases to ~60%, while the efficiency of charge separation is ~38% for equally doped $BiVO_4$ (Figure 12).

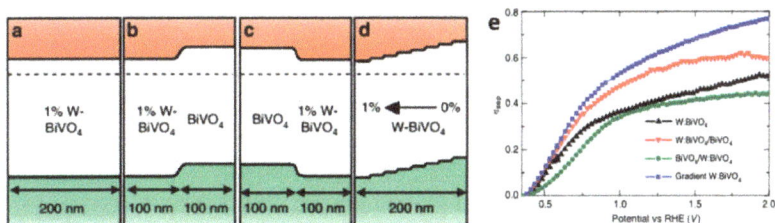

Figure 11. (**a**) 1% W-doped $BiVO_4$; (**b**) W:$BiVO_4$ homojunction; (**c**) W:$BiVO_4$ reverse homojunction; and (**d**) gradient-doped W:$BiVO_4$; (**e**) Carrier-separation efficiency (η_{sep}) in different applied potential for 1% W-doped $BiVO_4$ (black triangle), W:$BiVO_4$ homojunction (red inverted triangle), W:$BiVO_4$ reverse homojunction (green circle), and gradient-doped W:$BiVO_4$ (blue square). Reprinted with permission from ref. [38]. Copyright 2018. Nature.

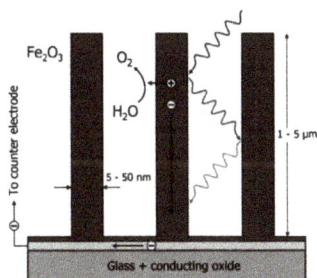

Figure 12. Illustration of the morphology of an α-Fe_2O_3 photo-anode for water splitting. The small diameter of nanowires ensures short hole diffusion path lengths. Reprinted with permission from ref. [39]. Copyright 2008. The Royal Society of Chemistry.

3.2. Nanostructure to Shorten the Diffusion Length

Besides conductivity improvement, the minority carrier's diffusivity is another key factor influencing the charge separation efficiency. Nanostructure provides a shorter diffusion distance for the minority carrier; therefore, it has been widely applied to improve the charge separation as shown in Figure 12 [39]. Kim et al. found nanoporous morphology of $BiVO_4$ can effectively suppress recombination, which yields an electron-hole separation efficiency as high as 90% at 1.23 V vs. RHE [40]. Zhao et al. also found enhancement of the performance by nanostructure compared with dense structure (Figure 13) and proved that the main contribution of the observed enhancement is the improved charge separation efficiency, which is caused by the easy diffusion of minority carriers to the interface [41]. Zhu et al. reported a nanorod-structured $ZnFe_2O_4$ with a new benchmark solar photocurrent of 1.0 mA·cm^{-2} at 1.23 V and 1.7 mA·cm^{-2} at 1.6 V vs. RHE (Figure 14) [42]. In general, if the average particle diameter of the nanopores is shorter than holes diffusion length, carrier recombination in bulk can be effectively inhibited. Table 1 shows holes diffusion length of some typical materials for OER.

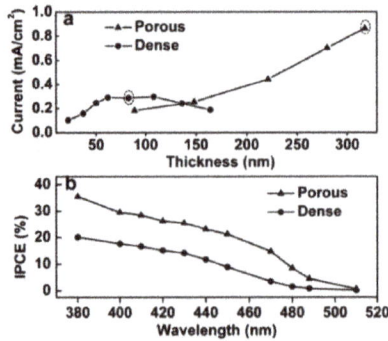

Figure 13. (**a**) The photocurrent at 1.23 V RHE of the porous and the dense Mo doped $BiVO_4$ films with different thickness in 0.5 M Na_2SO_4 aqueous solution under front (electrolyte-$BiVO_4$ surface side) illumination with a 500 W Xe lamp equipped with a 420 nm cut-off filter at a scan rate of 30 mV·s^{-1}; (**b**) IPCE at 1.23 V RHE of the porous and the dense $BiVO_4$ films in 0.5 M Na_2SO_4 aqueous solution under front illumination. Reprinted with permission from [41]. Copyright 2018. Wiley.

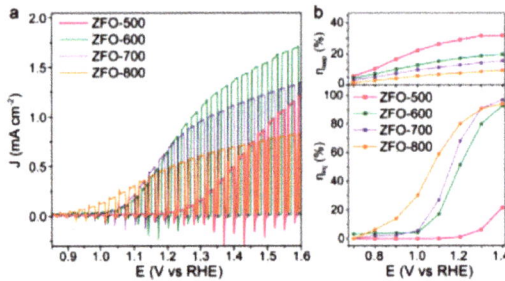

Figure 14. The performance of $ZnFe_2O_4$ (ZFO) photoanodes. (**a**) Linear scanning *J–V* curves in 1 m NaOH under 100 mW·cm^{-2} illumination; (**b**) Calculated charge separation efficiency, η_{sep}, and minority charge carrier injection efficiency, η_{inj}. Reprinted with permission from ref. [42]. Copyright 2018. Wiley.

Table 1. The hole diffusion length of different materials.

Materials	Hole Diffusion Length	Ref.
$BiVO_4$	100 nm	[43]
α-Fe_2O_3	2–4 nm	[44]
TiO_2	70 nm	[45]
WO_3	150 nm	[46]

3.3. Heterojunction

A heterojunction comprises, but is not limited to, two materials, whereby electron–hole pairs are produced by one or more semiconductor photocatalysts and then transfer to the other material, which facilitates the space charge separation [47]. The two materials should meet some requirements. Taking a photoanode as an example, the valence band of A is more positive than B; at the same time, the conduction band of photocatalyst B should be more negative than A. This would facilitate the electron and hole transfer from one to the other (Figure 15). A lot of work has reported the heterojunction for enhanced PEC performance, such as WO_3/$BiVO_4$ [48,49], Cu_2O/g-C_3N_4 [50], Cu_2O/TiO_2, p-Cu_2O/n-TaON [51]. It is also found that heterojunction can be formed between different facets of the same crystal, e.g., $BiVO_4$ (010) and $BiVO_4$ (110) surfaces. The electrons and holes can accumulate on different surfaces of $BiVO_4$ (Figure 16) [52], which causes the space separation of electron–hole. Wang et al. found improved hydrogen production in a TiO_2 photocatalyst by forming {001}-{010} "quasi" heterojunctions [53]. Another case of heterojunctions is α-Fe_2O_3-TiO_2 heterojunction. Holes generated in TiO_2 will be transferred to α-Fe_2O_3 with electrons preferentially accumulating in TiO_2 at the same time. Under specific conditions, hybrid composites such as Fe_2TiO_5, Fe_3TiO_4, and $FeTiO_3$ form at the interface between hematite and titania that, despite some controversial results, exhibit remarkable performances as photoanode materials in PEC devices [54].

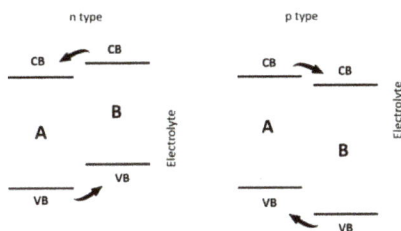

Figure 15. The band structure for the different heterojunctions. CB stands for conduct band and VB stands for valance band.

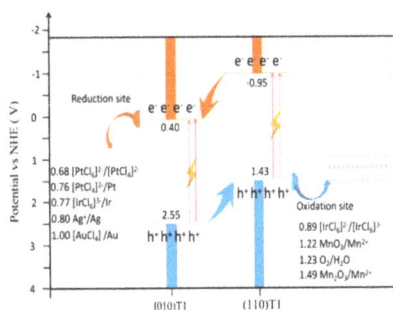

Figure 16. Schematic diagram of the heterojunction on between $BiVO_4$ (010) and $BiVO_4$ (110) surfaces. Reprinted with permission from ref. [52]. Copyright 2018. American Chemical Society.

3.4. Internal Electric Field to Improve the Charge Separation

Recent studies indicate that the internal electric field in a photocatalyst can improve the charge separation [55]. The internal electric field provides a driving force for the separation of photoinduced charge carriers. Ferroelectric materials are a big family for having internal electric field due to their spontaneous electric polarization property among various photocatalytic materials. The spontaneous polarization in ferroelectric materials are due to the noncentrosymmetric property and the polarization occurs because the positive and negative charges have different centers of symmetry [56]. Perovskite oxides (ABO_3) are among the most extensively studied ferroelectrics, such as $BaTiO_3$ [57] and $BiFeO_3$ [58]. Rohrer et al. found that $Ba_xSr_{1-x}TiO_3$ solid solution by replacing Ba with Sr can enlarge the space-charge region and improve the activity [59]. Zhang et al. also found that incorporation of carbon into the Bi_3O_4Cl lattice can increase internal electric field to boost bulk charge separation [60]. Hao et al. found that double perovskite ferroelectric $Bi_2FeMo_xNi_{1-x}O_6$ thin films also possess strong ferroelectric self-polarization, which has played a crucial part in improving the photovoltaic effect [61]. Another case of charge separation is α-Fe_2O_3-TiO_2 heterojunction. Some research works found that the Fe_2O_3-TiO_2 heterojunction showed enhancement in the photocurrent [62,63]. However, somehow it is not consistent with our general understanding on the heterojunction, because hematite has a more positive conduction band potential than TiO_2, which prohibits the electron transfer from hematite to

TiO_2. Further study shows that this kind of heterojunction is through iron titanates (e.g., Fe_2TiO_5, Fe_3TiO_4, and $FeTiO_3$) formed at the interface [54].

4. Strategies to Enhance the Surface Charge Injection

4.1. Catalyst Loading

In PEC water splitting, the catalyst loading play three vital roles for improving both the activity and reliability: (1) to reduce the overpotential or activation energy; (2) to assist charge separation at the semiconductor/electrolyte interface; (3) to suppress the photocorrosion and improve the durability of the working device [64]. Transition metal or transition metal oxides are usually used as catalyst loading to achieve a higher activity and decent reaction rates [65]. Recently, many (Fe, Co, Ni)-based catalyst loading have been widely investigated. For example, cobalt phosphate (CoPi) is the star cocatalyst to lower the overpotential and enhance the performance of water oxidation (Figure 17) [66]. Hydro-oxides are also one kind of effective catalyst loading, such as FeOOH, NiOOH, and CoOOH [67,68]. Kim et al. employed FeOOH/NiOOH dual-layer catalyst loading and achieved maximum ABPE of 1.75% due to reduction of the surface recombination [40]. Zhang et al. extended the catalyst loading to metal organic framework (MOF) materials, through applying in situ synthesized poly[-Co_2(benzimidazole)$_4$] nanoparticles on the surface of $BiVO_4$ to boost surface reaction kinetics (Figure 18) [69]. In this respect, the bilayered Ni-FeOOH oxygen evolution catalyst layer has been reported as the best catalysts for $BiVO_4$ to date [70].

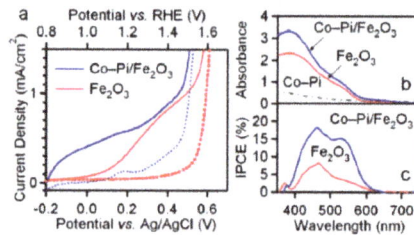

Figure 17. (**a**) Dark (dashed) and photocurrent (solid) densities for Fe_2O_3 (red) and Co–Pi/Fe_2O_3 (blue) photoanodes, collected using simulated AM 1.5 illumination (1 sun, backside) at a scan rate of $50 \ mV \cdot s^{-1}$; (**b**) Electronic absorption and (**c**) IPCE spectra for Fe_2O_3 and Co–Pi/Fe_2O_3 at 1.23 and 1 V vs. RHE, respectively. The absorption spectrum of Co–Pi on FTO without Fe_2O_3 is included in (**b**), but no photocurrent was detected for these anodes. Reprinted with permission from ref. [66]. Copyright 2018. American Chemical Society.

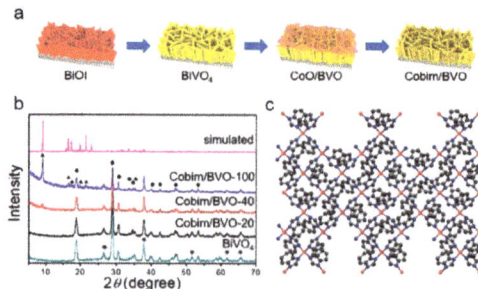

Figure 18. (**a**) Synthesis process of [Co_2(bim)$_4$]-modified $BiVO_4$ photoanodes; (**b**) XRD patterns of $BiVO_4$, Cobim/BiVO-20, Cobim/BiVO-40, Cobim/BiVO-100, and simulated [[Co_2(bim)$_4$], where the symbols asterisk, dot, and triangle correspond to the peaks from FTO, $BiVO_4$, and [Co_2(bim)$_4$], respectively; (**c**) Crystal structure of [Co_2(bim)$_4$], where gray, blue, and red balls represent carbon, nitrogen, and cobalt, respectively. Reprinted with permission from ref. [69]. Copyright 2018. American Chemical Society.

4.2. Surface Treatment and Surface Passivation

Point defects are usually found on the surface of photoelectrode. In some cases, defects play a positive role in the surface catalysis. However, in some other cases, defects serve as recombination centers. Thus, it is important to suppress the recombination on these surface centers. Recently, it was found that the surface treatment can improve the PEC performance. The mechanism for surface treatment and surface passivation is illustrated in Figure 19 [71].

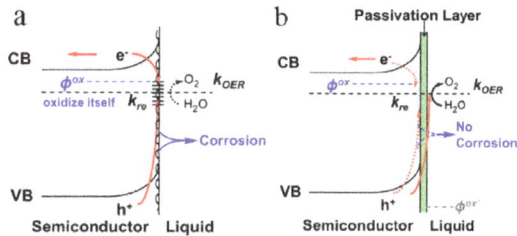

Figure 19. Band structure of an n-type semiconductor photoelectrode experiencing corrosion (**a**) when contacting an aqueous electrolyte; (**b**) with passivation layer without corrosion. ϕ_{ox} is the corrosion potential of the n-type semiconductor; and ϕ_{red} is the corrosion potential of the p-type semiconductor. Corrosion may reduce the light absorption and/or generate more surface defect states. Reprinted with permission from ref. [71] Copyright 2014. The Royal Society of Chemistry.

Surface defect states will lead to high charge recombination rate and water oxidation with low efficiency, caused by the photogenerated holes. Surface passivation and treatment can passivate defect states, forcefully suppressing surface recombination for improving water oxidation efficiency. For example, Luo et al. found that electrochemical cyclic voltammetry surface pretreatment on Mo-doped $BiVO_4$ can greatly enhances the photocurrent of Mo-doped $BiVO_4$ [72]. This is mainly because the pretreatment that can remove the surface recombination center MoO_x which is segregated during the preparation of the photoelectrode. After the preprocessing, MoO_x on the surface is dissolved into the electrolyte (Figure 20).

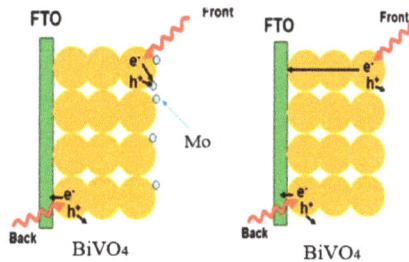

Figure 20. Transfer of photogenerated electrons in Mo-doped $BiVO_4$ photoelectrodes illuminated from the front-side and the back-side. Reprinted with permission from ref. [72]. Copyright 2018. American Chemical Society.

Further, Wang et al. found that the electrochemical treatment is not only suitable for doped $BiVO_4$, but also applicable to pristine $BiVO_4$ and other materials, such as TiO_2, Fe_2O_3 [73]. Besides the surface treatment, a thin passivation layer is also proved to be effective to enhance the surface catalysis. Originally, passivation layers were applied to semiconductor photoelectrodes to improve their photochemical or chemical stability. It was reported that even an extremely thin layer on the photoelectrodes can change the surface properties [74]. Hisatomi investigated the oxide surface layer on the onset potential of hematite photoanode, such as Al_2O_3, Ga_2O_3, and In_2O_3, which can greatly weaken the original potential for oxidation of water [75]. Formal et al. covered an extremely thin layer (0.1 to 2 nm) of alumina on nanostructured hematite photoanodes. The photocurrent was shifted cathodically by as much as 100 mV. Further investigation found that the cathodic shift is owing to the passivation of surface trapping states caused by the Al_2O_3 overlayer [76]. Besides, IrO_2 and CoO_x thin layers on TaON photoanodes were found to improve the durability by reducing the self-oxidation of the electrode [77,78]. In general, a suitable structure of the passivation layer for photoelectrodes can either improve the reaction kinetics, or increase the chemical or physical protection against corrosion, or both.

4.3. Active Sites

The chemical reactions for water splitting occur on the electrode surface through its active sites. The active sites, therefore, affect the performance of the surface catalysis. The low catalytic activities are mainly attributed to the relatively lower activity and smaller number of reactive sites. The activity of active sites can be reflected by the overpotential at the anode. Experimentally, it might be extremely challenging to identify the atomic surface active sites. Therefore, calculations using density functional theory (DFT) become important to reveal the overpotential at different sites, as well as to understand the atomic/molecular mechanism. There are some methods to calculate the overpotential based on different reaction process [79,80]. The well acceptable reaction process is the four intermediate steps reaction path, as shown in R3 to R6.

The overpotential tends to zero if $\Delta G_{R3}^{\theta} = \Delta G_{R4}^{\theta} = \Delta G_{R5}^{\theta} = \Delta G_{R6}^{\theta} = 1.23$ eV and this would define the ideal catalyst, as shown in Figure 21 [81].

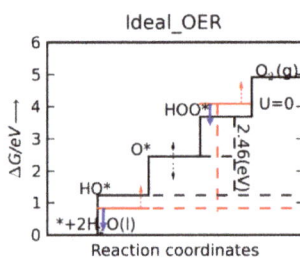

Figure 21. The free energy diagram for the OER ideal catalyst. The black continuous line is the OER ideal catalyst, where the over-potential is 0 eV because all free energies of the intermediate steps equal to 1.23 V. Red lines show the actual HOO* and HO* levels for the catalysts, which have to be moved down and up, respectively, in order to reduce the over-potential. Unfortunately, a metal oxide catalyst usually changes these two levels in the same direction, as indicated by the red (move up) and blue arrows (move down). O* level is positioned between the HOO* and HO* level. Reprinted with permission from ref. [81]. Copyright 2012. The Royal Society of Chemistry.

Until now, the atomic level interaction between catalytic activity and active sites is still an outstanding problem, owing to the large difference between idealized models and real catalysts. Thus, it is necessary to identify the pivotal factors that influence the activity of reactive sites, in order to promote the catalytic properties. It is generally considered that the low coordinated steps, edges, terraces, kinks, and/or corner atoms are often the favorable catalytic reaction sites [82,83]. For example, Sun et al. suggested that the low-coordinated Co^{3+} atoms of porous atomically-thick Co_3O_4 sheets serve as the catalytically active sites [84]. Currently, there are no quantitative models to describe the relationship between the coordinated number and activity. The main reason for this difficulty is that there might be other key parameters that have not been separately verified. For example, some results show that the surface state of the atom and position of neighbor atom will greatly affect the activity of the reactive sites. Hu et al. found that oxygen vacancy in $BiVO_4$ will induce more active sites on a $BiVO_4$ surface, which makes the vacancy sites active for water oxidation (Figure 22) [85]. Similarly, a Ti 3d defect state in the band gap of titania also plays an important role in the surface catalysis of defect contained TiO_2 [86]. Zhao et al. also revealed that W doping and W–Ti codoping into $BiVO_4$ have led to more active sites on the surface and reduced charge transfer resistance (Figure 23) [87,88]. Theoretical calculation revealed that Ti sites become active after the codoping.

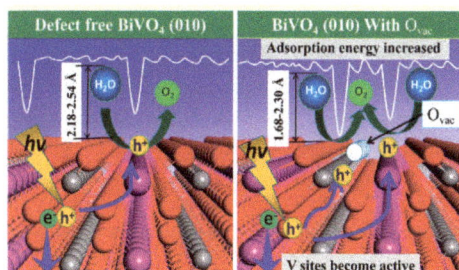

Figure 22. PEC water splitting process on a $BiVO_4$ (010) facet with or without O_{vac}. O_{vac} will greatly affect the adsorption energy. Reprinted with permission from ref. [85]. Copyright 2017. American Chemical Society.

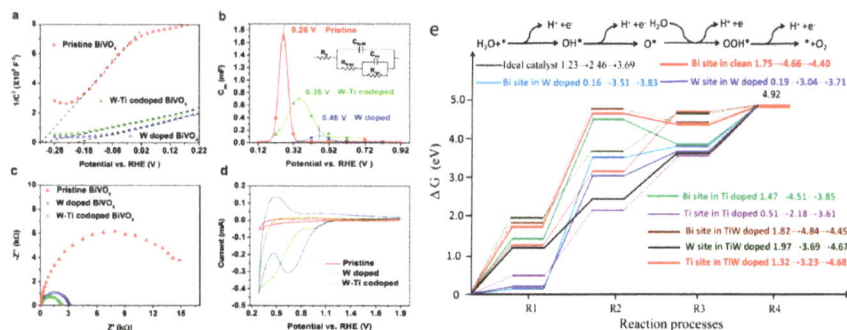

Figure 23. Electrochemistry results in 0.5 M Na_2SO_4 aqueous solution under 1.5 G illumination (a) Mott-Schottky plots of pristine, W doped $BiVO_4$ and W–Ti codoped $BiVO_4$ at the frequency of 1 kHz; (b) C_{ss} determined from EIS (rectangle points). Inset is the equivalent circuits employed to fit the experimental EIS data; (c) EIS of pristine, W doped, and W–Ti codoped $BiVO_4$ photoelectrodes at the applied potential of 1.23 V RHE; (d) Cyclic voltammetry scan of pristine, W doped and W–Ti codoped $BiVO_4$ in the dark at a scan rate of 100 mV·s^{-1} after holding the electrode at the potential of 1.8 V RHE for 120 s; (e) Free-energy profiles of OER on different sites of $BiVO_4$ surfaces. Reprinted with permission from ref. [88]. Copyright 2018. The Royal Society of Chemistry.

5. Conclusions and Outlook

In this review, we have highlighted the latest studies directed at developing metal oxide-based photoanodes, including $BiVO_4$, α-Fe_2O_3, TiO_2, and WO_3 nanostructures. We have summarized the strategies to enhance the light absorption, charge separation, and surface charge injection of this class of n-type semiconductor oxide materials. Although great achievements have been obtained, significant challenges remain before commercially viable solar-driven water splitting can be realized. Among them, one challenge is to balance the contributions from factors affecting light absorption, charge separation, and charge injection. Sometimes, their contributions contradict each other; e.g., a thicker layer increases absorption efficiency, but reduces the charge separation efficiency due to a longer distance to be travelled by the charge carrier. Another challenge lies with the lack of thorough understanding of the effect of the surface-state, due to the complex roles it plays in the water splitting process. Surface states may affect the amount of stored charges, act as surface recombination centers, and pin the Fermi-level. Most of the existing models can only explain one of these potential effects and there is no commonly used model to describe the full impact of surface state. A big challenge is to combine all the advantages and improve the overall efficiency for water splitting. Usually, it is inevitable to introduce some other problems when improving one factor. For example, nanostructuring is mostly employed to enhance the charge separation and light absorption; however, more surface states will be formed which sometimes are unfavorable to the interfacial charge injection. Thus, catalyst loading is required to passivate the surface states. Then, the challenge becomes finding a suitable catalyst loading to passivate and to control its thickness through a proper deposition technology.

Recently, researchers are focusing on the formation of ultrathin layers of cocatalyst on the surface, which will provide very efficient charge injection while avoiding the recombination in the surface layer and at the interface. However, it is very difficult to precisely control the formation of a uniform cover layer of the cocatalyst on the surface with reasonable cost of production. New low-cost processing technology and materials need to be developed. Another related issue is the stability and durability from the point of view of practical application. Newly developed materials should be coated with good adhesion while having good chemical and photochemical stability. More work is needed to build a more comprehensive model to guide the design and optimization of the electrode materials for solar water splitting.

Funding: This research was funded by Ministry of Education of Singapore (RG15/16), the National Natural Science Foundation of China (Nos. 21676216, 21576224) and China Postdoctoral Science Foundation (Nos. 2014M550507; 2015T81046).

Conflicts of Interest: The authors declare no conflict of interest.

References

1. van de Krol, R.; Grätzel, M. *Photoelectrochemical Hydrogen Production*; Springer: Berlin, Germany, 2012; pp. 13–69.
2. Inoue, H.; Shimada, T.; Kou, Y.; Nabetani, Y.; Masui, D.; Takagi, S.; Tachibana, H. The water oxidation bottleneck in artificial photosynthesis: How can we get through it? an alternative route involving a two-electron process. *ChemSusChem* **2011**, *4*, 173–179. [CrossRef]
3. Fujishima, A.; Honda, K. Electrochemical photolysis of water at a semiconductor electrode. *Nature* **1972**, *238*, 37–38. [CrossRef] [PubMed]
4. Sivula, K.; Van De Krol, R. Semiconducting materials for photoelectrochemical energy conversion. *Nat. Rev. Mater.* **2016**, *1*, 15010. [CrossRef]
5. Li, Z.; Luo, W.; Zhang, M.; Feng, J.; Zou, Z. Photoelectrochemical cells for solar hydrogen production: Current state of promising photoelectrodes, methods to improve their properties, and outlook. *Energy Environ. Sci.* **2013**, *6*, 347–370. [CrossRef]
6. Zhou, Y.; Zhang, L.; Lin, L.; Wygant, B.R.; Liu, Y.; Zhu, Y.; Zheng, Y.; Mullins, C.B.; Zhao, Y.; Zhang, X. Highly efficient photoelectrochemical water splitting from hierarchical WO$_3$/BiVO$_4$ nanoporous sphere arrays. *Nano Lett.* **2017**, *17*, 8012–8017. [CrossRef] [PubMed]
7. Zhao, J.; Wang, A.; Green, M.A.; Ferrazza, F. 19.8% efficient "honeycomb" textured multicrystalline and 24.4% monocrystalline silicon solar cells. *Appl. Phys. Lett.* **1998**, *73*, 1991–1993. [CrossRef]
8. Zhao, X.; Chen, Z. Enhanced photoelectrochemical water splitting performance using morphology-controlled BiVO$_4$ with W doping. *Beilstein J. Nanotechnol.* **2017**, *8*, 2640–2647. [CrossRef] [PubMed]
9. Qiu, Y.; Liu, W.; Chen, W.; Zhou, G.; Hsu, P.-C.; Zhang, R.; Liang, Z.; Fan, S.; Zhang, Y.; Cui, Y. Efficient solar-driven water splitting by nanocone BiVO$_4$-perovskite tandem cells. *Sci. Adv.* **2016**, *2*, e1501764. [CrossRef]
10. Feng, X.; Shankar, K.; Varghese, O.K.; Paulose, M.; Latempa, T.J.; Grimes, C.A. Vertically aligned single crystal TiO$_2$ nanowire arrays grown directly on transparent conducting oxide coated glass: Synthesis details and applications. *Nano Lett.* **2008**, *8*, 3781–3786. [CrossRef] [PubMed]
11. Lou, S.N.; Amal, R.; Scott, J.; Ng, Y.H. Concentration-mediated band gap reduction of Bi$_2$MoO$_6$ photoanodes prepared by Bi^{3+} cation insertions into anodized MoO$_3$ thin films: Structural, optical, and photoelectrochemical properties. *ACS Appl. Energy Mater.* **2018**, *1*, 3955–3964. [CrossRef]
12. Mai, L.; Yang, F.; Zhao, Y.; Xu, X.; Xu, L.; Hu, B.; Luo, Y.; Liu, H. Molybdenum oxide nanowires: Synthesis & properties. *Mater. Today* **2011**, *14*, 346–353. [CrossRef]

13. Lou, S.N.; Ng, Y.H.; Ng, C.; Scott, J.; Amal, R. Harvesting, storing and utilising solar energy using MoO_3: Modulating structural distortion through pH adjustment. *ChemSusChem* **2014**, *7*, 1934–1941. [CrossRef]

14. Asahi, R.; Morikawa, T.; Irie, H.; Ohwaki, T. Nitrogen-doped titanium dioxide as visible-light-sensitive photocatalyst: Designs, developments, and prospects. *Chem. Rev.* **2014**, *114*, 9824–9852. [CrossRef]

15. Liu, G.; Yin, L.C.; Wang, J.; Niu, P.; Zhen, C.; Xie, Y.; Cheng, H.M. A red anatase TiO_2 photocatalyst for solar energy conversion. *Energy Environ. Sci.* **2012**, *5*, 9603–9610. [CrossRef]

16. Xie, J.; Zhang, J.; Li, S.; Grote, F.; Zhang, X.; Zhang, H.; Wang, R.; Lei, Y.; Pan, B.; Xie, Y. Controllable disorder engineering in oxygen-incorporated MoS_2 ultrathin nanosheets for efficient hydrogen evolution. *J. Am. Chem. Soc.* **2013**, *135*, 17881–17888. [CrossRef] [PubMed]

17. Morikawa, T.; Saeki, S.; Suzuki, T.; Kajino, T.; Motohiro, T. Dual functional modification by N doping of Ta_2O_5: P-type conduction in visible-light-activated N-doped Ta_2O_5. *Appl. Phys. Lett.* **2010**, *96*, 142111. [CrossRef]

18. Maeda, K.; Takata, T.; Hara, M.; Saito, N.; Inoue, Y.; Kobayashi, H.; Domen, K. GaN: ZnO solid solution as a photocatalyst for visible-light-driven overall water splitting. *J. Am. Chem. Soc.* **2005**, *127*, 8286–8287. [CrossRef] [PubMed]

19. Chen, X.; Liu, L.; Yu, P.Y.; Mao, S.S. Increasing solar absorption for photocatalysis with black hydrogenated titanium dioxide nanocrystals. *Science* **2011**, *331*, 746–750. [CrossRef]

20. Naldoni, A.; Altomare, M.; Zoppellaro, G.; Liu, N.; Kment, Š.; Zbořil, R.; Schmuki, P. Photocatalysis with reduced TiO_2: From black TiO_2 to cocatalyst-free gydrogen production. *ACS Catal.* **2019**, *9*, 345–364. [CrossRef] [PubMed]

21. Cui, H.; Zhao, W.; Yang, C.; Yin, H.; Lin, T.; Shan, Y.; Xie, Y.; Gu, H.; Huang, F. Black TiO_2 nanotube arrays for high-efficiency photoelectrochemical water-splitting. *J. Mater. Chem. A* **2014**, *2*, 8612–8616. [CrossRef]

22. Wang, S.; Chen, P.; Bai, Y.; Yun, J.H.; Liu, G.; Wan, L. New $BiVO_4$ dual photoanodes with enriched oxygen vacancies for efficient solar-driven water splitting. *Adv. Mater.* **2018**, *30*, 1800486. [CrossRef] [PubMed]

23. Kim, J.H.; Jang, J.W.; Jo, Y.H.; Abdi, F.F.; Lee, Y.H.; Van De Krol, R.; Lee, J.S. Hetero-type dual photoanodes for unbiased solar water splitting with extended light harvesting. *Nat. Commun.* **2016**, *7*, 13380. [CrossRef] [PubMed]

24. Mayer, M.T.; Du, C.; Wang, D. Hematite/Si nanowire dual-absorber system for photoelectrochemical water splitting at low applied potentials. *J. Am. Chem. Soc.* **2012**, *134*, 12406–12409. [CrossRef] [PubMed]

25. Brillet, J.; Yum, J.-H.; Cornuz, M.; Hisatomi, T.; Solarska, R.; Augustynski, J.; Graetzel, M.; Sivula, K. Highly efficient water splitting by a dual-absorber tandem cell. *Nat. Photonics* **2012**, *6*, 824. [CrossRef]

26. Chang, X.; Wang, T.; Zhang, P.; Zhang, J.; Li, A.; Gong, J. Enhanced surface reaction kinetics and charge separation of p–n heterojunction $Co_3O_4/BiVO_4$ photoanodes. *J. Am. Chem. Soc.* **2015**, *137*, 8356–8359. [CrossRef] [PubMed]

27. Ye, K.; Wang, Z.; Gu, J.; Xiao, S.; Yuan, Y.; Zhu, Y.; Zhang, Y.; Mai, W.; Yang, S. Carbon quantum dots as a visible light sensitizer to significantly increase the solar water splitting performance of bismuth vanadate photoanodes. *Energy Environ. Sci.* **2017**, *10*, 772–779. [CrossRef]

28. Kim, J.K.; Shi, X.; Jeong, M.J.; Park, J.; Han, H.S.; Kim, S.H.; Guo, Y.; Heinz, T.F.; Fan, S.; Lee, C.L.; et al. Enhancing Mo:$BiVO_4$ Solar Water Splitting with Patterned Au Nanospheres by Plasmon-Induced Energy Transfer. *Adv. Energy Mater.* **2018**, *8*, 1701765. [CrossRef]

29. Thimsen, E.; Formal, F.L.; Grätzel, M.; Warren, S.C. Influence of plasmonic Au nanoparticles on the photoactivity of Fe_2O_3 electrodes for water splitting. *Nano Lett.* **2011**, *11*, 35–43. [CrossRef]

30. Mascaretti, L.; Dutta, A.; Kment, Š.; Shalaev, V.M.; Boltasseva, A.; Zbořil, R.; Naldoni, A. Plasmon-enhanced photoelectrochemical water splitting for efficient renewable energy storage. *Adv. Mater.* **2019**, 1805513. [CrossRef]

31. Zhao, X.; Feng, J.; Chen, S.; Huang, Y.; Sum, T.C.; Chen, Z. New insight into the roles of oxygen vacancies in hematite for solar water splitting. *Phys. Chem. Chem. Phys.* **2017**, *19*, 1074–1082. [CrossRef]

32. Luo, W.; Yang, Z.; Li, Z.; Zhang, J.; Liu, J.; Zhao, Z.; Wang, Z.; Yan, S.; Yu, T.; Zou, Z. Solar hydrogen generation from seawater with a modified $BiVO_4$ photoanode. *Energy Environ. Sci.* **2011**, *4*, 4046–4051. [CrossRef]

33. Berglund, S.P.; Rettie, A.J.E.; Hoang, S.; Mullins, C.B. Incorporation of Mo and W into nanostructured $BiVO_4$ films for efficient photoelectrochemical water oxidation. *Phys. Chem. Chem. Phys.* **2012**, *14*, 7065–7075. [CrossRef]

34. Guo, Y.; Zhang, N.; Wang, X.; Qian, Q.; Zhang, S.; Li, Z.; Zou, Z. A facile spray pyrolysis method to prepare Ti-doped ZnFe$_2$O$_4$ for boosting photoelectrochemical water splitting. *J. Mater. Chem. A* **2017**, *5*, 7571–7577. [CrossRef]

35. Lumley, M.A.; Choi, K.S. Investigation of pristine and (Mo, W)-doped Cu$_{11}$V$_6$O$_{26}$ for use as photoanodes for solar water splitting. *Chem. Mater.* **2017**, *29*, 9472–9479. [CrossRef]

36. Jo, W.J.; Jang, J.W.; Kong, K.; Kang, H.J.; Kim, J.Y.; Jun, H.; Parmar, K.P.S.; Lee, J.S. Phosphate doping into monoclinic BiVO$_4$ for enhanced photoelectrochemical water oxidation activity. *Angew. Chem. Int. Ed.* **2012**, *51*, 3147–3151. [CrossRef]

37. Yang, Z.; Chang, B.; Zou, J.; Qiao, J.; Gao, P.; Zeng, Y.; Li, H. Comparison between gradient-doping GaAs photocathode and uniform-doping GaAs photocathode. *Appl. Opt.* **2007**, *46*, 7035–7039. [CrossRef]

38. Abdi, F.F.; Han, L.; Smets, A.H.; Zeman, M.; Dam, B.; Van De Krol, R. A bismuth vanadate–cuprous oxide tandem cell for overall solar water splitting. *Nat. Commun.* **2013**, *4*, 2195. [CrossRef]

39. Krol, R.; Liang, Y.; Schoonman, J. Solar hydrogen production with nanostructured metal oxides. *J. Mater. Chem.* **2008**, *18*, 2311–2320. [CrossRef]

40. Kim, T.W.; Choi, K.-S. Nanoporous BiVO$_4$ photoanodes with dual-layer oxygen evolution catalysts for solar water splitting. *Science* **2014**, *343*, 990–994. [CrossRef]

41. Zhao, X.; Luo, W.; Feng, J.; Li, M.; Li, Z.; Yu, T.; Zou, Z. Quantitative analysis and visualized evidence for high charge separation efficiency in a solid-liquid bulk heterojunction. *Adv. Energy Mater.* **2014**, *4*, 1301785. [CrossRef]

42. Zhu, X.; Guijarro, N.; Liu, Y.; Schouwink, P.; Wells, R.A.; Formal, F.L.; Sun, S.; Gao, C.; Sivula, K. Spinel structural disorder influences solar-water-splitting performance of ZnFe$_2$O$_4$ nanorod photoanodes. *Adv. Mater.* **2018**, *30*, 1801612. [CrossRef]

43. Iwase, A.; Yoshino, S.; Takayama, T.; Ng, Y.H.; Amal, R.; Kudo, A. Water splitting and CO$_2$ reduction under visible light irradiation using Z-Scheme systems consisting of metal sulfides, CoO$_x$-loaded BiVO$_4$, and a reduced graphene oxide electron mediator. *J. Am. Chem. Soc.* **2016**, *138*, 10260–10264. [CrossRef]

44. Booshehri, A.Y.; Goh, S.C.; Hong, J.D.; Jiang, R.; Xu, R. Effect of depositing silver nanoparticles on BiVO$_4$ in enhancing visible light photocatalytic inactivation of bacteria in water. *J. Mater. Chem. A* **2014**, *2*, 6209–6217. [CrossRef]

45. Zhang, Z.; Wang, P. Optimization of photoelectrochemical water splitting performance on hierarchicalTiO$_2$ nanotube arrays. *Energy Environ. Sci.* **2012**, *5*, 6506–6512. [CrossRef]

46. Butler, M.A. Photoelectrolysis and physical properties of the semiconducting electrode WO$_2$. *J. Appl. Phys.* **1977**, *48*, 1914–1920. [CrossRef]

47. Moniz, S.J.; Shevlin, S.A.; Martin, D.J.; Guo, Z.-X.; Tang, J. Visible-light driven heterojunction photocatalysts for water splitting–a critical review. *Energy Environ. Sci.* **2015**, *8*, 731–759. [CrossRef]

48. Su, J.; Guo, L.; Bao, N.; Grimes, C.A. Nanostructured WO$_3$/BiVO$_4$ heterojunction films for efficient photoelectrochemical water splitting. *Nano Lett.* **2011**, *11*, 1928–1933. [CrossRef]

49. Rao, P.M.; Cai, L.; Liu, C.; Cho, I.S.; Lee, C.H.; Weisse, J.M.; Yang, P.; Zheng, X. Simultaneously efficient light absorption and charge separation in WO$_3$/BiVO$_4$ core/shell nanowire photoanode for photoelectrochemical water oxidation. *Nano Lett.* **2014**, *14*, 1099–1105. [CrossRef]

50. Zhang, S.; Yan, J.; Yang, S.; Xu, Y.; Cai, X.; Li, X.; Zhang, X.; Peng, F.; Fang, Y. Electrodeposition of Cu$_2$O/g-C$_3$N$_4$ heterojunction film on an FTO substrate for enhancing visible light photoelectrochemical water splitting. *Chin. J. Catal.* **2017**, *38*, 365–371. [CrossRef]

51. Hou, J.; Yang, C.; Cheng, H.; Jiao, S.; Takeda, O.; Zhu, H. High-performance p-Cu$_2$O/n-TaON heterojunction nanorod photoanodes passivated with an ultrathin carbon sheath for photoelectrochemical water splitting. *Energy Environ. Sci.* **2014**, *7*, 3758–3768. [CrossRef]

52. Hu, J.; Chen, W.; Zhao, X.; Su, H.; Chen, Z. Anisotropic electronic characteristics, adsorption, and stability of low-index BiVO$_4$ surfaces for photoelectrochemical applications. *ACS Appl. Mater. Interfaces* **2018**, *10*, 5475–5484. [CrossRef]

53. Wang, D.; Kanhere, P.; Li, M.; Tay, Q.; Tang, Y.; Huang, Y.; Sum, T.C.; Mathews, N.; Sritharan, T.; Chen, Z. Improving Photocatalytic H$_2$ Evolution of TiO$_2$ via Formation of {001}−{010} Quasi-Heterojunctions. *J. Phys. Chem. C* **2013**, *117*, 22894–22902. [CrossRef]

54. Kment, S.; Riboni, F.; Pausova, S.; Wang, L.; Wang, L.; Han, H.; Hubicka, Z.; Krysa, J.; Schmuki, P.; Zboril, R. Photoanodes based on TiO_2 and α-Fe_2O_3 for solar water splitting—Superior role of 1D nanoarchitectures and of combined heterostructures. *Chem. Soc. Rev.* **2017**, *46*, 3716–3769. [CrossRef]

55. Huang, X.; Wang, K.; Wang, Y.; Wang, B.; Zhang, L.; Gao, F.; Zhao, Y.; Feng, W.; Zhang, S.; Liu, P. Enhanced charge carrier separation to improve hydrogen production efficiency by ferroelectric spontaneous polarization electric field. *Appl. Catal. B-Environ.* **2018**, *227*, 322–329. [CrossRef]

56. Li, L.; Salvador, P.A.; Rohrer, G.S. Photocatalysts with internal electric fields. *Nanoscale* **2014**, *6*, 24–42. [CrossRef]

57. Giocondi, J.L.; Rohrer, G.S. The influence of the dipolar field effect on the photochemical reactivity of $Sr_2Nb_2O_7$ and $BaTiO_3$ microcrystals. *Top. Catal.* **2008**, *49*, 18–23. [CrossRef]

58. Schultz, A.M.; Zhang, Y.; Salvador, P.A.; Rohrer, G.S. Effect of crystal and domain orientation on the visible-light photochemical reduction of Ag on $BiFeO_3$. *ACS Appl. Mater. Interfaces* **2011**, *3*, 1562–1567. [CrossRef]

59. Bhardwaj, A.; Burbure, N.V.; Gamalski, A.; Rohrer, G.S. Composition dependence of the photochemical reduction of Ag by $Ba_{1-x}Sr_xTiO_3$. *Chem. Mater.* **2010**, *22*, 3527–3534. [CrossRef]

60. Li, J.; Cai, L.; Shang, J.; Yu, Y.; Zhang, L. Giant enhancement of internal electric field boosting bulk charge separation for photocatalysis. *Adv. Mater.* **2016**, *28*, 4059–4064. [CrossRef]

61. Cui, X.; Li, Y.; Sun, N.; Dua, J.; Lia, X.; Yang, H.; Hao, X. Double perovskite $Bi_2FeMo_xNi_{1-x}O_6$ thin films: Novel ferroelectric photovoltaic materials with narrow bandgap and enhanced photovoltaic performance. *Sol. Energy Mater. Sol. C* **2018**, *187*, 9–14. [CrossRef]

62. Han, H.; Riboni, F.; Karlicky, F.; Kment, S.; Goswami, A.; Sudhagar, P.; Yoo, J.; Wang, L.; Tomanec, O.; Petr, M.; et al. α-Fe_2O_3/TiO_2 3D hierarchical nanostructures for enhanced photoelectrochemical water splitting. *Nanoscale* **2017**, *9*, 134–142. [CrossRef] [PubMed]

63. Jeon, T.H.; Choi, W.; Park, H. Photoelectrochemical and photocatalytic behaviors of hematite-decorated Titania nanotube arrays: Energy level mismatch versus surface specific reactivity. *J. Phys. Chem. C* **2011**, *115*, 7134–7142. [CrossRef]

64. Ran, J.; Zhang, J.; Yu, J.; Jaroniec, M.; Qiao, S.Z. Earth-abundant cocatalysts for semiconductor-based photocatalytic water splitting. *Chem. Soc. Rev.* **2014**, *43*, 7787–7812. [CrossRef] [PubMed]

65. Maeda, K.; Domen, K. Photocatalytic water splitting: Recent progress and future challenges. *J. Phys. Chem. Lett.* **2010**, *1*, 2655–2661. [CrossRef]

66. Zhong, D.K.; Sun, J.; Inumaru, H.; Gamelin, D.R. Solar water oxidation by composite catalyst/α-Fe_2O_3 photoanodes. *J. Am. Chem. Soc.* **2009**, *131*, 6086–6087. [CrossRef] [PubMed]

67. Kim, T.W.; Ping, Y.; Galli, G.A.; Choi, K.S. Simultaneous enhancements in photon absorption and charge transport of bismuth vanadate photoanodes for solar water splitting. *Nat. Commun.* **2015**, *6*, 8769. [CrossRef] [PubMed]

68. Tang, F.; Cheng, W.; Su, H.; Zhao, X.; Liu, Q. Smoothing surface trapping states in 3D coral-like $CoOOH$-wrapped-$BiVO_4$ for efficient photoelectrochemical water oxidation. *ACS Appl. Mater. Interfaces* **2018**, *10*, 6228–6234. [CrossRef]

69. Zhang, W.; Li, R.; Zhao, X.; Chen, Z.; Law, A.W.K.; Zhou, K. A cobalt-based metal–organic framework as cocatalyst on $BiVO_4$ photoanode for enhanced photoelectrochemical water oxidation. *ChemSusChem* **2018**, *11*, 2710–2716. [CrossRef]

70. Shi, X.; Choi, I.Y.; Kan, Z.; Kwon, J.; Dong, Y.K.; Lee, J.K.; Sang, H.O.; Kim, J.K.; Park, J.H. Efficient photoelectrochemical hydrogen production from bismuth vanadate-decorated tungsten trioxide helix nanostructures. *Nat. Commun.* **2014**, *5*, 4775. [CrossRef]

71. Liu, R.; Zheng, Z.; Spurgeon, J.; Yang, X. Enhanced photoelectrochemical water-splitting performance of semiconductors by surface passivation layers. *Energy Environ. Sci.* **2014**, *7*, 2504–2517. [CrossRef]

72. Luo, W.; Li, Z.; Yu, T.; Zou, Z. Effects of surface electrochemical pretreatment on the photoelectrochemical performance of Mo-Doped $BiVO_4$. *J. Phys. Chem. C* **2012**, *116*, 5076–5081. [CrossRef]

73. Wang, S.; Chen, P.; Yun, J.H.; Hu, Y.; Wang, L. An electrochemically treated $BiVO_4$ photoanode for efficient photoelectrochemical water splitting. *Angew. Chem. Int. Ed.* **2017**, *56*, 8500–8504. [CrossRef]

74. Riha, S.C.; Klahr, B.M.; Tyo, E.C.; Seifert, S.; Vajda, S.; Pellin, M.J.; Hamann, T.W.; Martinson, A.B.F. Atomic layer deposition of a submonolayer catalyst for the enhanced photoelectrochemical performance of water oxidation with hematite. *ACS Nano* **2013**, *7*, 2396–2405. [CrossRef]

75. Hisatomi, T.; Le Formal, F.; Cornuz, M.; Brillet, J.; Tétreault, N.; Sivula, K.; Grätzel, M. Cathodic shift in onset potential of solar oxygen evolution on hematite by 13-group oxide overlayers. *Energy Environ. Sci.* **2011**, *4*, 2512–2515. [CrossRef]

76. Le Formal, F.; Tetreault, N.; Cornuz, M.; Moehl, T.; Gr€atzel, M.; Sivula, K. Passivating surface states on water splitting hematite photoanodes with alumina overlayers. *Chem. Sci.* **2011**, *2*, 737–743. [CrossRef]

77. Abe, R.; Higashi, M.; Domen, K. Facile fabrication of an efficient oxynitride TaON photoanode for overall water splitting into H_2 and O_2 under visible light irradiation. *J. Am. Chem. Soc.* **2010**, *132*, 11828–11829. [CrossRef]

78. Higashi, M.; Domen, K.; Abe, R. Highly stable water splitting on oxynitride TaON photoanode system under visible light irradiation. *J. Am. Chem. Soc.* **2012**, *134*, 6968–6971. [CrossRef]

79. Yang, J.; Wang, D.; Zhou, X.; Li, C. A theoretical study on the mechanism of photocatalytic oxygen evolution on $BiVO_4$ in aqueous solution. *Chem. Eur. J.* **2013**, *19*, 1320–1326. [CrossRef]

80. Zhao, Z.Y. Single water molecule adsorption and decomposition on the low-index stoichiometric rutile TiO_2 surfaces. *J. Phys. Chem. C* **2014**, *118*, 4287–4295. [CrossRef]

81. Valdés, Á.; Brillet, J.; Grätzel, M.; Gudmundsdóttir, H.; Hansen, H.A.; Jónsson, H.; Klüpfel, P.; Kroes, G.J.; Formal, F.L.; Man, I.C.; et al. Solar hydrogen production with semiconductor metal oxides: New directions in experiment and theory. *Phys. Chem. Chem. Phys.* **2012**, *14*, 49–70. [CrossRef]

82. Sun, Y.F.; Liu, Q.H.; Gao, S.; Cheng, H.; Lei, F.C.; Sun, Z.H.; Jiang, Y.; Su, H.B.; Wei, S.Q.; Xie, Y. Pits confined in ultrathin cerium (IV) oxide for studying catalytic centers in carbon monoxide oxidation. *Nat. Commun.* **2013**, *4*, 2899. [CrossRef]

83. Sun, Y.F.; Lei, F.C.; Gao, S.; Pan, B.C.; Zhou, J.F.; Xie, Y. Atomically thin tin dioxide sheets for efficient catalytic oxidation of carbon monoxide. *Angew. Chem. Int. Ed.* **2013**, *52*, 10569. [CrossRef]

84. Sun, Y.F.; Gao, S.; Lei, F.C.; Liu, J.W.; Liang, L.; Xie, Y. Atomically-thin non-layered cobalt oxide porous sheets for highly efficient oxygen-evolving electrocatalysts. *Chem. Sci.* **2014**, *5*, 3976–3982. [CrossRef]

85. Hu, J.; Zhao, X.; Chen, W.; Su, H.; Chen, Z. Theoretical insight into the mechanism of photoelectrochemical oxygen evolution reaction on $BiVO_4$ anode with oxygen vacancy. *J. Phys. Chem. C* **2017**, *121*, 18702–18709. [CrossRef]

86. Wendt, S.; Sprunger, P.T.; Lira, E.; Madsen, G.K.H.; Li, Z.; Hansen, J.O.; Matthiesen, J.; Blekinge-Rasmussen, A.; Lægsgaard, E.; Hammer, B.; Besenbacher, F. The role of interstitial sites in the Ti_3D defect state in the band gap of Titania. *Science* **2008**, *320*, 1755–1759. [CrossRef] [PubMed]

87. Zhao, X.; Hu, J.; Chen, S.; Chen, Z. An investigation on the role of W doping in $BiVO_4$ photoanodes used for solar water splitting. *Phys. Chem. Chem. Phys.* **2018**, *20*, 13637–13645. [CrossRef] [PubMed]

88. Zhao, X.; Hu, J.; Wu, B.; Banerjee, A.; Chakraborty, S.; Feng, J.; Zhao, Z.; Chen, S.; Ahuja, R.; Sum, T.C.; Chen, Z. Simultaneous enhancement in charge separation and onset potential for water oxidation in a $BiVO_4$ photoanode by W–Ti codoping. *J. Mater. Chem. A* **2018**, *6*, 16965–16974. [CrossRef]

coatings

MDPI

Review

Recent Studies of Semitransparent Solar Cells

Dong Hee Shin and Suk-Ho Choi *

Department of Applied Physics and Institute of Natural Sciences, Kyung Hee University, Yongin 17104, Korea;
sdh0105@hanmail.net
* Correspondence: sukho@khu.ac.kr; Tel.: +82-31-201-2418

Received: 12 August 2018; Accepted: 15 September 2018; Published: 20 September 2018

Abstract: It is necessary to develop semitransparent photovoltaic cell for increasing the energy density from sunlight, useful for harvesting solar energy through the windows and roofs of buildings and vehicles. Current semitransparent photovoltaics are mostly based on Si, but it is difficult to adjust the color transmitted through Si cells intrinsically for enhancing the visual comfort for human. Recent intensive studies on translucent polymer- and perovskite-based photovoltaic cells offer considerable opportunities to escape from Si-oriented photovoltaics because their electrical and optical properties can be easily controlled by adjusting the material composition. Here, we review recent progress in materials fabrication, design of cell structure, and device engineering/characterization for high-performance/semitransparent organic and perovskite solar cells, and discuss major problems to overcome for commercialization of these solar cells.

Keywords: semitransparent; organic; perovskite; polymer; solar cell; transparent conductive electrode; color perception

1. Introduction

Rapid increase in industrialization and world population has sparked a strong demand for sustainable energy sources that can replace fossil fuels. In recent years, searching for environmental-friendly energy sources has been recognized as one of the most challenging issues in the scientific and engineering research fields. Among them, solar energy has received strong attention as one of the most promising candidates because it is clean, infinite, and relatively accessible. However, the main obstacle to overcome for further growth of the solar energy market and the spread of its technologies is the insufficient energy harvesting of the solar irradiation. One way to increase the energy harvesting is to incorporate semitransparent photovoltaic modules (PVMs) into transparent surfaces of high-rise buildings and automobiles or window panels in individual homes.

Currently, Si is a principal material for semitransparent photovoltaic products most available in the solar cell market [1]. A crystalline Si (c-Si) solar cell is usually placed on glass for semitransparency, and is manufactured by employing a technique of partial shading. Due to the opaqueness of c-Si cell, the transparency and electrical output can be adjusted by controlling the space between the cells of the panel. The light is transmitted through the exposed area of semitransparent photovoltaic c-Si panels during the generation of the electricity. Maximum power conversion efficiency (PCE) of c-Si solar cells is currently very close to 26.7% [2], but to achieve this, the active layer should be thick due to the low absorption coefficient of Si, as shown in Table 1 [3,4]. Despite high performance of c-Si PVMs, they are not aesthetically appealing due to the limitation in the color selection, originating from the inherent absorption characteristics of Si. The c-Si PVMs for power generation of buildings are therefore placed on the ceilings/roofs instead of the sidewalls. For the use of semitransparent photovoltaic cells for the sidewall windows of buildings, their solar energy density should be enhanced. On the other hand, the semitransparency of dye-sensitized solar cells (DSSCs) was first demonstrated by using porous TiO_2 nanoparticles (NPs) in 1991 [5]. After that, many researchers have been intensively studying

DSSCs [6–12] due to their potentially-high semitransparency [13]. In general, the TiO_2/fluorine-doped tin oxide (FTO)/glass photo-anode offers high average visible light transmittance (AVT) of >80% at micron-level thicknesses for 10–20 nm TiO_2 NPs, but the mesoporous TiO_2 photo-anode contains a number of grain boundaries, leading to poor charge transport, thereby degrading the device performance [14]. Furthermore, the liquid nature of the materials used in DSSCs still induces several problems due to the leaks [15], which requires solid-state conductors to be substituted for the liquid ones for their practical applications, a tough challenge at the moment.

Table 1. Comparison of thickness and absorption properties between various semiconductor materials used in solar cells.

Photovoltaic Material	Thickness (μm)	Absorption Properties
c-Si	180	Broad band absorption Absorption coefficient: 10^2 cm^{-1}
Organic material	0.1	Confined band absorption Absorption coefficient: 10^5 cm^{-1}
Perovskites	0.3	Broad band absorption Absorption coefficient: 10^5–10^6 cm^{-1}

Semitransparent PVMs employing using organic solar cells (OSCs) and perovskite solar cells (PSCs) are recently highly attractive due to their high absorption coefficient. Organic semiconductors are inherently excitonic with high absorption intensity in the wavelength rang of 250 to 350 nm, originating from the intramolecular charge transfer (ICT) within the organic molecules or polymers. By changing the chemical arrangements, the ICT absorption occurs in near infrared (NIR) region, resulting in considerably-high transparency in the visible region [16,17]. Compared to OSCs with low charge carrier mobility and efficiency, PSCs are recognized as a major breakthrough in view of cost. PSCs showed a certified efficiency of 22.7% due to their high extinction coefficient [18], high mobility of charge carriers [19], small binding energy of excitons [20], long diffusion length of charge carriers [21], and solution processability [22]. Based on these considerations, semitransparent OSCs and PSCs have great potentials for power generation that promises flexibility, neutral coloring, and pleasant appearance when there are assembled on windows, foldable curtains, automotive windshields, and other architectural and fashionable items.

This review focuses on the recent studies of semitransparent OSCs (STOSCs) and semitransparent PSCs (STPSCs), including their transparent electrodes and harvesting materials. Unlike opaque cells, the performance characteristics of STOSCs and STPSCs strongly depend on the optical parameters affecting not only the PCE but also the visual comfort. Despite several review papers on this topic, most of them dealt with single-junction cells and some of them were biased to STOSCs or STPSCs [23–26], but this review covers a wide range of multi-junction cells as well as single-junction ones. Tables 2 and 3 provide overview for the progress of semitransparent photovoltaic cells.

Table 2. Recently-reported photovoltaic parameters of STOSCs.

Device Structure	V_{oc} (V)	J_{sc} (mA·cm²)	FF (%)	PCE (%)	AVT (%)	Reference
Indium tin oxide (ITO)/poly(3,4-ethylenedioxythiophene):poly(styrenesulfonate) (PEDOT:PSS)/PhanQ: phenyl-C_{61}-butyric acid methyl ester (PCBM)/buckminsterfullerene (C_{60})-surfactant/Ag	0.84	8.0	63.0	4.2	32	[16]
ITO/ZnO/poly(3-hexylthiophene) (P3HT):PCBM/PEDOT:PSS CPP 105D (CPP-PEDOT:PSS)	0.55	7.4	58.0	2.4	–	[27]
ITO/PEO/APFO₃:PCBM/PEDOT:PSS P VP Al 4083 (PEDOT-EL)/PEDOT:PSS	0.67	2.3	45.0	0.7	–	[28]
ITO/ZnO nanoparticles (NPs)/P3HT:PCBM/poly(allylamine hydrochloride) and dextran(PAH-D)/PEDOT:PSS	0.59	6.7	47.3	1.9	–	[29]
PEDOT:PSS/P3HT:IDT-2BR/polyethylenimine (PEIE)/PEDOT:PSS	0.84	5.9	57.9	2.9	50	[30]
ITO/n-C_{60}/C_{60}/zinc phthalocyanine (ZnPc):C_{60}//2,20-(perfluoronaphthalene-2,6-diylidene) dima-lononitrile(F6TCNNQ) dopedN,N0-((diphenyl-N,N0-bis)9,9,-dime-thyl-fluoren-2-yl)-benzidine (p-BF-DPB)/carbonnanotube (CNT)	0.57	4.5	58.2	1.5	–	[31]
ITO/ZnO/P3HT:PCBM/CNT	0.57	10.5	40.2	2.5	80 @ 670 nm	[32]
ITO/polymer thieno[3,4-b]thiophene/benzodithiophene (PTB7):PCBM/MoO$_x$/HNO₃-doped CNT	0.66	6.5	40.0	1.8	80 @ 400 nm	[33]
ITO/ZnO/P3HT:PCBM/graphene oxide (GO)/graphene	0.54	10.5	44.0	2.5	70 @ 650 nm	[34]
ITO/ZnO/P3HT:PCBM/PEDOT:PSS/ Au-doped Graphene	0.59	10.6	43.3	2.7	–	[35]
graphene/PEDOT:PSS/PTB7: PhanQ: phenyl-C_{71}-butyric acid methyl ester (PC₇₁BM)/ZnO-NP/PEDOT:PSS/graphene	0.67	12.1	41.4	3.4	40	[36]
ITO/TiO₂/P3HT:PCBM/PEDOT:PSS/graphene/Au grid/	0.62	8.2	55.0	2.8	–	[37]
Triethylene tetramine (TETA) -doped graphene/ZnO/P3HT:PCBM/ PEDOT:PSS/(trifluoromethanesulfonyl)-amide (TFSA)-doped graphene	0.62	9.8	54.5	3.3	70 @ 650 nm	[38]
ITO/copper phthalocyanine (CuPc):C_{60}/bathocuproine (BCP)/Ag/Ag NWs	0.44	1.9	55.0	0.6	26	[39]
AZO/P3HT:PCBM/PEDOT:PSS/Ag NWs	0.56	5.8	65.1	2.1	–	[40]
Ag NWs/aluminum doped ZnO (AZO)/P3HT: polymer poly[(4,40-bis(2-ethylhexyl) dithieno[3,2-b:2',3'-d]-silole)-2,6-diyl-alt-(4,7-bis(2-thienyl)-2,1,3-benzothiadiazole)-5,5'-diyl] (Si-PCPDTBT):PCBM/PEDOT:PSS/Ag NWs	0.57	6.4	60.0	2.2	33	[41]
PEDOT:PSS/GO/PEDOT:PSSP3HT:PCBM/ZnO/TiO$_x$/Ag NWs	0.58	8.2	49.0	2.3	60 @ 700 nm	[42]
ITO/PEDOT:PSS/poly(2,60-4,8-bis(5-ethylhexylthienyl)benzo-[1,2-b;3,4-b]dithiophene-alt-5-dibutyloctyl-3,6-bis(5-bromothiophen-2-yl)pyrrolo[3,4-c]pyrrole-1,4-dione) (PBDTT-DPP):PCBM/TiO₂/Ag NWs	0.77	9.3	56.2	4.0	66 @ 550 nm	[43]
Ag NWs/PEDOT:PSS/PV2000PCBM/ZnO/Ag NWs	0.76	10.7	52.8	4.3	–	[44]
ITO/PEDOT:PSS/poly[(4,4'-bis(2-ethylhexyl)dithieno[3,2-b:2',3'-d]silole)-2,6-diyl-alt-(2,1,3-benzothiadiazole)-4,7-diyl] (PSBTBT):PCBM/LiCoO₂/Al/ZnO:Al	0.61	10.7	42.0	2.8	–	[45]
ITO/ZnO/Poly[[4,8-bis[5-(2-ethylhexyl)-2-thienyl]benzo[1,2-b:4,5-b']dithiophene-2,6-diyl][2-(2-ethyl-1-oxohexyl)thieno[3,4-b]thiophenediyl]] (PBDTTT-C-T):PC₇₁BM/MoO₃/Ag	0.77	12.1	61.0	5.7	28	[46]
ITO/PEDOT:PSS/poly[N-900-hepta-dec-anyl-2,7-carbazole-alt-5,5-(40,70-di-2-thienyl-20,10,30-benzo-thiadiazole)] (PCDTBT):PC₇₁BM/LiF/Al/Ag	0.76	4.6	49.0	1.7	35	[47]

Table 2. Cont.

Device Structure	V_{oc} (V)	J_{sc} (mA·cm²)	FF (%)	PCE (%)	AVT (%)	Reference
ITO/ZnO/Poly[[2,6′-4,8-di(5-ethylhexylthienyl)benzo[1,2-b:3,3-b]dithiophene][3-fluoro-2[(2-ethylhexyl)carbonyl]thieno[3,4-b]thiophenediyl]] (PTB7-Th):ATT-2/MoO₃/Ag	0.71	16.0	55.0	6.3	45	[48]
ITO/ZnO/poly[(5,6-difluoro-2,1,3-benzothiadiazol-4,7-diyl)-alt-(3,3‴-di(2-octyldodecyl)-2,2′:5′,2″:5″,2‴-quaterthiophen-5,5‴-diyl)] (PffBT4T-2OD):PCBM:PC₇₁BM/MoO₃/Ag	0.76	13.7	56.0	5.8	6	[49]
Graphene/PEDOT:PSS/ZnO/poly[2,7-(5,5-bis-(3,7-dimethyl octyl)-5Hdithieno[3,2-b:20,30-d]pyran)-alt-4,7-(5,6-difluoro-2,1,3-benzothiadiazole) (PDTP-DFBT):PCBM/MoO₃/graphene	0.67	12.4	45.0	3.7	54 @ 500 nm	[50]
ITO/ZnO/poly[[2,6-(4,8-bis(5-(2-ethylhexyl)thiophen-2-yl)benzo[1,2-b:4,5-b′]-dithiophene)-alt-(5,5-(1′,3′-di-2-thienyl-5′,7′-bis(2-ethylhexyl)benzo[1′,2′-c:4′,5′-c′]dithiophene-4,8-dione)]] (PBDB-T): 9-bis(2-methylene-(3-(1,1-dicyanomethylene)indanone)-5,5,11,11-tetrakis(4-hexylphenyl)dithieno[2,3-d:2′,3′-d′]s-indaceno[1,2-b:5,6-b′]-dithiophene (ITIC)/MoO₃/Ag/MoO₃	0.88	13.8	59.8	7.3	25	[51]
ITO/ZnMgO/PTB7-Th/PC₇₁BM/MoO₃/Ag	0.77	14.84	59.5	6.8	22	[52]
ITO/PEDOT:PSS/PTB7-Th:2,2′-((2Z,2′Z)-(((4,4,9,9-trisi(4-hexylphenyl)-4,9-dihydro-s-indaceno[1,2-b:5,6-b dithiophene-2,7-diyl) bis(4-(2-ethylhexyl)oxythiophene-5,2-diyl)) bis(methanylylidene) bis(5,6-dichloro-3-oxo-2,3-dihydro-1H-indene-2,1-diylidene))dimalononitrile (IEICO-4Cl)/ poly[9,9-bis[6′-(N,N,Ntrimethylammonium) hexyl]fluorene-alt-co-1,4-phenylene]bromide (PFN-Br)/Au	0.73	19.6	59.0	8.4	26	[53]
FTO/TiO₂/((C₃₀H₃₈N₂S₉)Sil)n Poly[(4,4′-bis(2-ethylhexyl) dithieno[3,2-b:2′,3′-d]silole)-2,6-diyl-alt-(2,1,3-benzothiadiazole)-4,7-diyl] (PSBtBT):PCBM/WO₃/Ag/one-dimensional photonic crystals (1DPCs)	0.64	7.90	48.7	2.5	40	[54]
PEDOT:PSS/PTB7:PC₇₁BM/PEN/ITO/1DPC	0.77	10.03	55.6	4.3	39	[55]
ITO/ZnO/PTB7-Th:PC₇₁BM/MoO₃/Ag/6 pairs (MoO₃/LiF)	0.79	13.1	67.7	7.0	12	[56]
ITO/ZnO/P3HT:PCBM/MoO₃/Ag/6 pairs (MoO₃/LiF)	0.63	10.89	66.0	4.32	60 @ 650 nm	[57]
3 pairs (TiO₂/SiO₂)/ZnO/PTB7:PCBM/2 pairs (MoO₃/Ag)/LiF	0.73	10.7	67.9	5.3	21	[58]
ITO/C₆₀:NDN1/C₆₀/F₆-ZnPc:C₆₀/5 wt.% Di-NPB:NDP9/10 wt.% Di-NPB:NDP9/C₆₀:NDN1/C₆₀/DCV6T-C₆₀/BPAPF/BPAPF: NDP9/Di-NPB:NDP9/C₆₀:NDN1/Al/Ag/tris(8-hydroxyquinolinato)-aluminum	1.54	5.2	61.0	4.9	24	[59]
ITO/poly(3,3′-((9′,9′-dioctyl-9H,9′H-[2,2′-bifluorene]-9,9-diyl)bis(4,1-phenylene)]bis(oxy)bis(N,N-dimethylpropan-1-amine))(PFPA-1)/ poly[N,N′-bis(2-hexyldecyl)isoindigo-6,6′-diyl-alt-thiophene-2,5-diyl] P3TI:PC₇₁BM/PEDOT:PSS/ITO/ poly(3,3′-((9′,9′-dioctyl-9H,9′H- [2,2′-bifluorene]-9,9-diyl)bis(4,1-phenylene)]bis(oxy)bis(N,N-dimethylpropan-1-amine)) (PFPA-1)/ poly[2,3-bis-(3-octyloxyphenyl)quinoxaline-5,8-diyl-alt-thiophene-2,5-diyl] (TQ1)]PC₇₁BM/PEDOT:PSS	1.46	4.47	67.0	4.35	–	[60]
Graphene mesh/PEDOT:PSS/poly[(4,4′-bis(3-ethyl-hexyl)dithieno[3,2-b:′3′-d]silole)-2,6-diyl-alt-(2,5-(3-ethyl-hexyl)thiophen-2-yl)thiazolo[5,4-d]thiazole] (PSEtTT): indene-C₆₀-bisadduct (IC₆₀BA)/ZnO/PEDOT:PSS/PBDTT-DPP:PC₇₁BM/TiO₂/Ag NWs	1.62	7.62	64.2	8.02	45	[61]
ITO/ZnO/fullerene self-assembled monolayer (C₆₀-SAM)/PCBM/PEDOT:PSS/PEDOT:PSS (PH1000)/ZnO/C₆₀-SAM/PCBM/Ag	1.68	5.93	68.6	8.5	40	[62]
ITO/PEDOT:PSS/PBDTT-FDPP-C₁₂:PCBM/PFN/TiO₂/PEDOT:PSS/PBDTT-SeDPP:PC₇₁BM/TiO₂/Ag NWs	1.47	8.4	59.0	7.3	30	[63]

Table 3. Recently-reported photovoltaic parameters of STPSCs.

Device Structure	V_{oc} (V)	J_{sc} (mA·cm^2)	FF (%)	PCE (%)	AVT (%)	Reference
Fluorine-doped tin oxide (FTO)/TiO$_2$/CH$_3$NH$_3$PbI$_3$ (MAPbI$_3$) islands/ 2,20,70-tetrakis-(N,N-di-pmethoxyphenylamine)-9,90-spirobifluorene (spiro-OMeTAD))/Au	0.71	8.1	61.0	3.5	22	[64]
FTO/TiO$_2$/polystyrene (PS)/MAPbI$_3$/spiro-OMeTAD/Au	0.95	19.2	64.0	11.7	36	[65]
ITO/copper thiocyanate (CuSCN)/MAPbI$_3$/PCBM/Bis-C$_{60}$/Ag	1.06	13.0	73.0	10.7	37	[66]
ITO/PEDOT:PSS/poly[N,N0-bis(4-butylphenyl)-N,N0-bis(phenyl)benzidine] (poly-TPD)/MAPbI$_3$/PCBM/Au/LiF	1.04	5.66	57.7	3.39	35.4	[67]
FTO/TiO$_2$/MAPbI$_3$/spiro-OMeTAD/MoO$_3$/Au/MoO$_3$	0.72	9.7	66.0	5.3	31	[68]
Indium tin oxide (ITO)//PEDOT:PSS/MAPbI$_3$/PCBM/ C$_{60}$/11-amino-1-undecanethiol hydrochloride (AUH)/Ag	0.95	12.1	71.0	8.2	34	[69]
FTO/TiO$_2$/Al$_2$O$_3$/MAPbI$_{3-x}$Cl$_x$-NiO nanoparticles (NPs)/spiro-OMeTAD/Au	0.79	20.62	61.8	10.1	27	[70]
FTO/TiO$_2$/MAPbI$_3$/poly-triarylamine (PTAA)/ poly(3,4-ethylenedioxythiophene)-poly(styrenesulfonate) (PEDOT:PSS)/ITO	1.10	15.2	75.0	12.6	17	[71]
FTO/ZnO/ PhanQ/ phenyl-C$_{61}$-butyric acid methyl ester (PCBM)//MAPbI$_3$/spiro-OMeTAD/MoO$_3$/In$_2$O$_3$:H	1.10	17.4	73.6	14.2	72 @ 800 nm	[72]
FTO/TiO$_2$/MAPbI$_3$ islands/PTAA/PEDOT:PSS/ITO	0.96	15.5	71.2	10.6	20.9	[73]
FTO/TiO$_2$/MAPbI$_3$ islands/PTAA/PEDOT:PSS/ITO (sub-module)	0.98	16.52	73.8	12.0	20.1	[74]
ITO/PEDOT:PSS/MAPbI$_3$/ZnO/Ag NWs/Al$_2$O$_3$	0.96	15.87	69.7	10.8	25.5	[75]
ITO/PEDOT:PSS/MAPbI$_3$/PCBM/ZnO/Ag NWs	0.96	13.18	66.8	8.5	28.4	[76]
ITO/PEDOT:PSS/MAPbI$_3$/PCBM/AZO/SnO$_x$/Ag/SnO$_x$	0.86	18.3	72.3	11.4	70 @ 800 nm	[77]
FTO/TiO$_2$/mesoporous (mp)-TiO$_2$/MAPbI$_3$/ spiro-OMeTAD/PEDOT:PSS	0.94	11.4	27.3	2.9	7	[78]
PEDOT:PSS/ZnO/MAPbI$_3$/spiro-OMeTAD/PEDOT:PSS	1.06	19.3	68.0	13.9	60 @ 800 nm	[79]
FTO/TiO$_2$/mp-TiO$_2$/carbon mesoscopic/MAPbI$_3$/multiwall carbon nanotubes (MWCNT)	0.87	18.10	52.1	8.2	24	[80]
FTO/TiO$_2$/MAPbI$_3$/spiro-OMeTAD/PEDOT:PSS/graphene	0.96	19.17	67.2	12.4	–	[81]
ITO/MoO$_x$/spiro-OMeTAD/MAPbI$_3$/PCBM/indium zinc oxide (IZO)	1.03	17.5	77.7	14.0	–	[82]
ITO/PTAA/MAPbI$_3$/PCBM/ buckminsterfullerene (C$_{60}$)/ bathocuproine (BCP)/Cu/Au	1.08	20.6	74.1	16.5	–	[83]
FTO/TiO$_2$/MAPbI$_3$/spiro-OMeTAD/MoO$_x$/Au/Ag/MoO$_x$	1.05	14.6	75.1	11.5	–	[84]
FTO/TiO$_2$/MAPbI$_3$/spiro-OMeTAD/Ag NWs	1.03	17.5	71.0	12.7	–	[85]

2. Figure of Merits for Characterizing Semitransparent Solar Cells

Semitransparent photovoltaic windows are considered as an effective solution for balancing between energy generation and visual comfort by integrating photovoltaic cells into existing and new buildings. Visible light is only part of the total solar irradiation, whose power is spectrally distributed in the 380–780 nm range. The solar power distribution of the light transferred to buildings and houses through the transparent window has a big influence on visual comfort of humans. Other parameters in addition to PCE, such as AVT, transparency color perception, corresponding color temperature, and color rendering index (CRI) should be evaluated for optimizing semitransparent solar cells (STSCs). Semitransparency of photovoltaic cells provide wide varieties in the color selection from light source, which allows residents to enjoy outside scenes with natural colors.

2.1. Average Visible Light Transmittance

The AVT can be estimated from the averaged transparency of a cell in the visible wavelength range of 400 to 700 nm, considering the spectral response of the human eye. The transmittance required for semitransparent photovoltaic cells depends on where they are actually used, but it is generally accepted that 20–30% AVT is the minimum requirement for the window applications [86]. To further evaluate the sensed transmittance, the measured transmittance spectrum of STSCs should be calibrated with respect to human sensitivity for obtaining their lightness or visible light transmittance. The spectrum integral value of the visible light transmitted through the semitransparent photovoltaic devices corresponds to the light recognized by the human eye [87,88]. In addition, the color of STSCs is generally determined by the photoactive layers and electrodes.

2.2. Transparency Color Perception

The optical perception of STSCs by human eye was first taken into account in 2010 [87] because their transparency color perception by human eye is usually different from that disclosed by measurements. International Lighting Commission (CIE) 1931 xy chromaticity diagram, specially established for color perception by humans, can be used to characterize the transparency color of STSCs. The transparency color recognition of STSCs is determined by calculating the color coordinates (x, y, z) from their transmittance pattern, based on the CIE 1931 xyz standard [89]. In fact, the sum of x, y, and z is equal to 1 so that the color coordinates can be simplified to two-dimensional coordinates (x, y). Standard daylight illuminant D65 and air mass 1.5 global (AM 1.5G) solar spectrum are typically selected as reference light sources (incident light) for evaluating the color parameters of STSCs. Even if colorful STSCs can be useful for the decoration, neutral-color STSCs with color coordinates close to "white point" (0.3333, 0.3333), illuminant D65 (0.3128, 0.3290) and AM 1.5G light (0.3202, 0.3324) are usually taken for window applications because the environmental lighting condition is almost not changed [90].

2.3. Corresponding Color Temperature

The human eye can adjust white light perception to any virtual blackbody radiator. The absolute temperature of the black body (full radiator) having the chromaticity closest to that of the white light source color is defined as the correlated color temperature. Chromaticity point cloud group with the same color temperature exists on the black body locus. That is, even though the correlated color temperature is the same, the chromaticity is not the same. In this case, the degree and direction deviating from the blackbody locus are displayed together with the color temperature. The color temperature of the light source at the closest point on the Planckian orbit is referred to as the corresponding color temperature when the color coordinates are in CIE 1931 color space chromaticity.

2.4. Color Rendering Index

As another important parameter, the color rendering properties of STSCs cannot be simply determined from their transparency perception. Therefore, color rendering index (CRI) should be evaluated for presenting the degree of the variation between the transmitted and the incident lights for the cells. The CRI of a STSC is evaluated from the transmitted light, following the standard CIE procedure, and is scaled from 0 to 100. Higher CRI indicates better color rendering capacity, i.e., higher neutral color degree. In other words, the higher the CRI, the more the transmitted light that can adequately represent the color, and the lower the CRI, the less the ability to express color. The CRI of STSCs is influenced by a number of factors such as device architecture, active material, and electrode transmittance. Several methods have been employed to improve the CRI of STSCs. For example, the use of the materials with low band gap [45,46] or complementary optical absorption [63] and the combination of dyes for the active layer [90] have been proposed, but most CRI values must be compromised with the device compliance. Recently, a new strategy has been developed to improve the CRI of STSCs while maintaining the device performance. In this approach, a reflective layer was located on top of the semitransparent metal anode of OSCs, resulting in a CRI close to 100% [91].

3. Semitransparent Organic Solar Cells

Unlike inorganic solar cells, the OSCs have inherent advantages that can provide a wide range of functional abilities for semi-transparent PVMs. Especially, the color of STOSCs can be changed controlled by varying the components of the photoactive layer, useful for aesthetic architectural applications in electric windows of buildings and automobiles [16,92,93]. The transmittance of STOSCs is typically characterized by an AVT measured in the visible range of 370 to 740 nm. For semitransparent solar cells, the front and back electrodes must be transparent, and it is therefore important to develop high-transparency/low-sheet-resistance transparent conductive electrodes (TCEs) that can be easily prepared on the photoactive layer. The material/device design is also essential for achieving high-performance STSCs. For example, theoretical expectations for selecting the photoactive layer and determining the thickness of the electrode are required to obtain optimized current density (J_{sc}) [94]. Simulating the optical field distribution of the cell while varying the thickness of photoactive layer can help optimize the electrical and optical performance of the cell [95].

The TCEs of STOSCs serve to transmit the light, to transfer the carriers to the photoactive layer, or to provide a distributed electric field by which the carriers transport [96]. Recently, many studies have focused on the development of transparent conductive oxide (TCO)-free TCEs such as conductive polymers, carbon nanotubes (CNTs), graphene, ultra-thin metal films, and metal nanowires (NWs). The TCO-free TCE-based STOSCs are discussed in the following sections.

3.1. Transparent Conductive Electrodes for Semitransparent Organic Solar Cells

3.1.1. Transparent Conducting Polymers

Transparent conductive polymers are regarded as attractive electrodes for STOSCs because they can be prepared on a large area in an easy and simple way. In particular, poly(3,4-ethylenedioxythiophene):poly(styrenesulfonate) (PEDOT:PSS) among various transparent conductive polymer electrode materials have been intensively employed for STOSCs [97,98]. Recently, a structure of PEDOT:PSS CPP 105D (H.C.Starck GmbH, Goslar, Germany) (CPP-PEDOT:PSS)/poly(3-hexylthiophene) (P3HT):phenyl-C_{61}-butyric acid methyl ester ($PC_{61}BM$)/ZnO/indium tin oxide (ITO) for STOSCs has been studied by using ITO and PEDOT:PSS as the bottom cathode and the top anode, respectively, thereby showing a PCE of 2.4% [27]. In another approach, an ITO/polyethylene oxide (PEO)/poly[2,7-(9,9-dioctylfluorene)-alt-5,5-(4,7′-di-2-thienyl-2′,1′,-3-benzothiadiazole) (APFO$_3$): $PC_{61}BM$/low-conductive PEDOT:PSS P VP Al 4083 (PEDOT-EL)/PH500 PEDOT:PSS-type OSC employing PEO as a reflector has exhibited a PCE of 0.7% [28]. In the same structure, by inserting poly(allylamine hydrochloride)-dextran between the PEDOT:PSS electrode and the active layer and changing the active layer with P3HT:$PC_{61}BM$, the PCE increased to 1.86% [29]. When P3HT:IDT-2BR (non-fullerene acceptor) as a photoactive layer and PH1000 PEDOT:PSS as a top electrode were used for semitransparent, non-phenylene, and flexible OSCs, resulting PCE and AVT were 2.88% and ~50%, respectively, as shown in Figure 1 [30]. These results show the possibility of further progress in the performance of the STOSCs with transparent conducting polymer electrodes in near future, but the improvement may be limited due to the inherently-low conductivity (high sheet resistance) of the transparent conducting polymers.

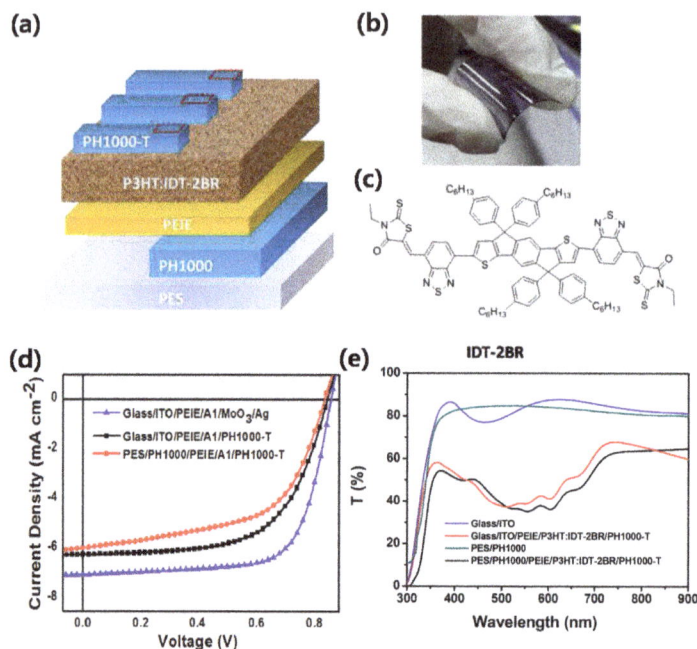

Figure 1. (**a**) Schematic diagram of a typical fullerene-free all-plastic solar cell; (**b**) photo image; (**c**) chemical structure of IDT-2BR; (**d**) *J-V* curves of all-plastic OSCs based on different acceptors; (**e**) Transmittance spectra of electrodes and corresponding cells. Reprinted with permission from [30]. Copyright 2016 Elsevier.

3.1.2. Carbon Nanotubes

CNTs are attractive for flexible electrodes due to the uniform transparency in the entire visible wavelength range, intrinsically excellent electrical characteristics, high flexibility, and simple solution process at room temperature [99,100]. The resistance of a CNT network is determined by the contact resistance between the CNTs and their individual resistance, depending on length, diameter, metallic/semiconducting volume ratio, synthesis method, and purity. In addition, the surface roughness of CNT films is another major factor as a TCE for OSCs because it can result in significant shunt resistance loss. Recently, free-standing multi-walled-CNTs (f-MWCNT)-based STOSCs have been fabricated by depositing f-MWCNTs on the active layer by room-temperature self-laminating process [31], as shown in Figure 2a. Figure 2b shows current density-voltage (*J-V*) curves of the zinc phthalocyanine (ZnPc):C_{60}-based STOSCs as a function of hole transfer layer (HTL) thickness, resulting in 60% FF and 1.5% PCE at a HTL thickness of 20 nm. In another approach, STOSCs with CNT film TCE and P3HT:$PC_{61}BM$ as the active layer showed 2.48% PCE and 60%–80% transmittance in the NIR region [32]. As a different structure, MoO_3/PTB7: [6,6]-Phenyl-C_{71}-butyric Acid Methyl Ester ($PC_{71}BM$)/ZnO/ITO-type STOSCs employing single-wall CNTs exhibited 1.8% PCE and ~80% transmittance in the visible light range of 400 to 800 nm [33]. Figure 2c shows fabrication processes of STOSCs using CNTs p-type-doped by two steps: HNO_3 doping through 'sandwich transfer' and MoO_x thermal doping through "bridge transfer". Resulting PCE of the STOSC using the thin-CNT with 90%

transmittance in the visible light region was 3.7%, but the PCE was about 50% lower compared to non-transparent metal-based solar cells (7.8%).

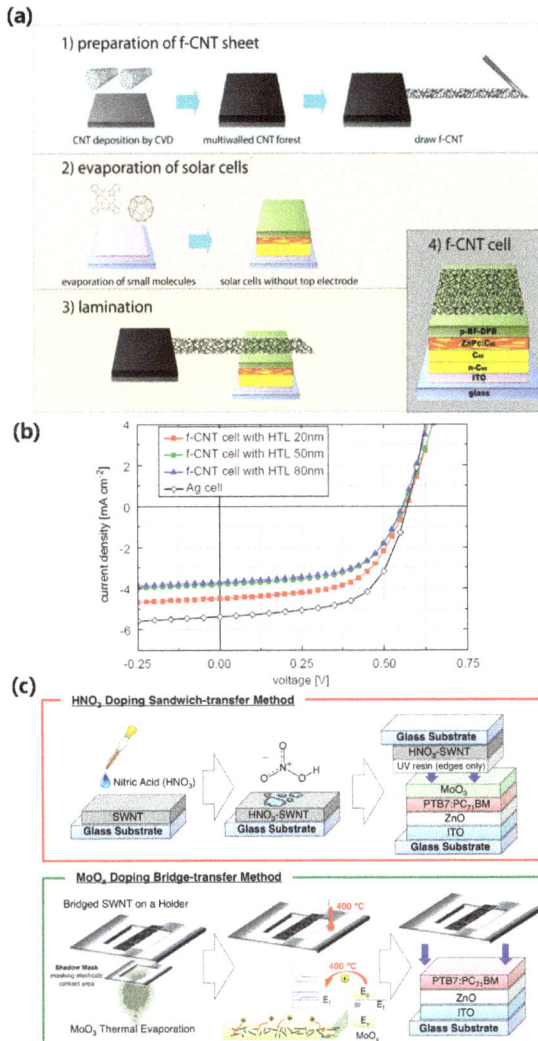

Figure 2. (a) Fabrication processes for f-CNT top electrode-based STOSC; (b) *J-V* characteristics of Ag reference cell and f-CNT-based cells with different HTL thicknesses of 20, 50, and 80 nm. Reprinted with permission from [31]. Copyright 2012 Elsevier. (c) Schematics of HNO_3 doping sandwich transfer process (above) and MoO_x thermal doping bridge transfer process (below). Reprinted with permission from [33]. Copyright 2016 Nature.

3.1.3. Graphene

Graphene is very useful as a TCE for OSCs thanks to its inherently-high chemical/thermal stabilities, excellent mechanical properties, and small contact resistance with organic materials. Graphene can be fabricated by several techniques such as mechanical exfoliation, laser-induced exfoliation, reduction of graphite oxide (rGO), unzipping of carbon nanotubes, chemical synthesis, and chemical vapor deposition (CVD) [101]. Among these, rGO and CVD are most promising in view of mass-production of flexible TCEs.

Recently, STOSCs using a top-laminated graphene electrode have been successfully fabricated without any damage in the underlying organic photoactive layer, as shown in Figure 3a [34]. The PCE of the STOSCs reached approximately 76% of what was obtained from the counterpart cells using opaque-type Ag metal electrodes. These results are promising in that low-cost STSCs can be fabricated by using graphene TCEs, and are possibly useful for power-generated windows and multi-junction or bifacial photovoltaic panels. As another approach, STOSCs with P3HT and $PC_{61}BM$ were fabricated by employing single-layer graphene and ITO as the top and bottom electrodes, respectively (Figure 3b) [35]. Here, the graphene electrodes were hybridized with Au NPs and PEDOT:PSS to enhance their conductance, resulting in a maximum PCE of 2.7% at a 20 mm^2 area.

Figure 3c shows a different type of STOSC structure: graphene/PEDOT:PSS/PTB7:$PC_{71}BM$/ZnO NP/PEDOT:PSS/graphene [36]. The AVT and maximum PCE of the STOSC are ~40% and 3.4%, respectively on both sides. The fabrication method for this cell is compatible with those of other photonic devices because almost all the procedures including transfer of film, spin coating, and lamination were done at low temperatures. Another OSC with a structure of ITO/TiO_2/P3HT:$PC_{61}BM$/PEDOT:PSS/graphene/Au grid/polyethylene terephthalate (PET) exhibited high light transmittance and small sheet resistance, resulting in 2.8% PCE on the graphene side [37].

Very recently, the use of p and n-type graphene TCEs for both electrodes of the STOSC improved conductivity while maintaining good transparency [63]. Figure 3d shows another STOSC consisting of triethylene tetramine (TETA)-doped graphene/ZnO/P3HT:$PC_{61}BM$/PEDOT:PSS/bis-(trifluoromethanesulfonyl)-amide (TFSA)-doped graphene prepared following a simple solution process. The PCEs of 3.30% and 3.12% were obtained by the irradiation through TFSA-doped and TETA-doped graphene surfaces, respectively, and the transmittance was 30%–40% at ~400 to 550 nm and 70% at ~650 nm. On the other hand, a flexible OSC on PET retained more than 99% of its initial PCE even after inner/outer bending at a curvature radius (*R*) of 6 mm [38].

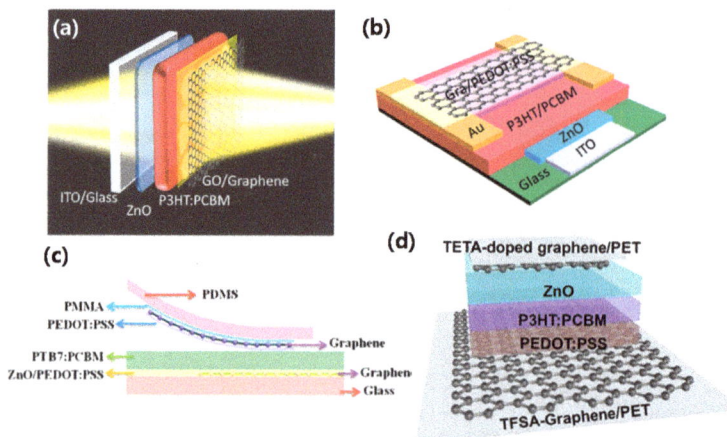

Figure 3. Schematic diagram of graphene-TCE-based various STOSC structures: (**a**) ITO/ZnO/P3HT:PCBM/graphene oxide (GO)/graphene top electrode. Reprinted with permission from [34]. Copyright 2011 ACS; (**b**) ITO/ZnO/P3HT:PCBM/PEDOT:PSS/graphene. Reprinted with permission from [35]. Copyright 2011 ACS. (**c**) PDMS/PMMA/graphene/PEDOT:PSS/PTB7:$PC_{71}BM$/ZnO-NP/PEDOT:PSS/graphene. Reprinted with permission from [36]. Copyright 2015 ACS. (**d**) TFSA-doped graphene/PEDOT:PSS/P3HT:PCBM/ZnO/TETA-doped graphene. Reprinted with permission from [38]. Copyright 2018 ACS.

3.1.4. Metal Nanowires

Metallic NWs were widely studied for the application as TCEs on flexible substrates due to their high transmittance, small sheet resistance, and excellent mechanical flexibility, especially useful for reducing the overall cost of the mass production [102]. For example, the Ag NWs electrode is a good alternative to ITO because the energy payback time of the modules is reduced almost 17% by use of the former [103]. Recently, metal NWs as a TCE have received much attention due to their outstanding electrical and optical properties. Especially, Ag NWs are very useful as solution-processable TCE for the application in the STOSCs [104–109] STOSCs were also fabricated using laminated Ag NWs meshes without damaging the photoactive layer in 2010 [39], thereby showing an AVT of 26% in the spectral range of 400 to 800 nm and a low PCE of 0.63%.

Recently, there has been a report on P3HT:PC$_{61}$BM-based STOSCs with Ag NWs as a top electrode [40], whose performance (PCE: 2.1%) was comparable to that of an OSC with thermally-evaporated Ag electrode (PCE: 2.6%). More interestingly, the STOSCs exhibited nearly same optical and device-performance characteristics under illumination on both sides. The OSCs with Ag NWs as both electrodes, completely processed by solution treatment [41], showed 2.2% PCE and 30% transmission at a 550 nm wavelength. In particular, they exhibited almost same transmittance in the ultraviolet (UV)-vis-NIR range, not available when ITO was used as one of the electrodes, demonstrating better optical characteristics of Ag NWs. These results indicate important technological progress for fully printed OSCs and further suggest that ITO-free STOSCs can be successfully fabricated.

In another approach, Ag NWs and PH1000 PEDOT:PSS film were employed as both electrodes for translucent and ITO-free OSCs fully solution-processed on P3HT:PC$_{61}$BM [42]. Figure 4a shows the real image of the OSC fabricated at low temperature, whose transmittance is approximately 55% at wavelengths longer than 650 nm without absorption of the P3HT:PC$_{61}$BM active layer, as shown in Figure 4b, resulting in 2.3% and 2.0% PCEs under the illumination at the cathode and anode sides, respectively (Figure 4c). This performance is comparable to that of STOSCs with ITO bottom electrodes [34]. To exclude possible damage of the underlying organic photoactive layer by the TCE, the OSCs were produced by using spray-coated Ag NWs-based composites [43]. Here, the used photoactive material was sensitive to the light in the NIR range, but highly transparent to visible light. Figure 4d shows a schematic and a photo image of the cell with a PCE of 4% and a maximum transparency of 66% at 550 nm, as shown in Figure 4e,f. Inkjet printing technique was employed to fabricate Ag NWs electrodes from Ag NWs solutions [44]. The printed Ag NWs mesh showed similar uniformity, conductivity, and transmittance with Ag NW films prepared by conventional methods such as slot die or spray coating. In this study, STOSCs using bulk heterojunction (donor (PV2000): acceptor (PC$_{71}$BM)) as the active layer and ink-jet-printed Ag NWs mesh as both electrodes showed 4.3% PCE at a 1 cm^2 area and about 10% transmittance in the visible range.

As mentioned above, TCEs play an important role in determining the photovoltaic performance of STOSCs. The TCEs such as conducting polymers, CNTs, graphene, and Ag NWs have greatly improved the performance of the STOSCs, but there still remain many issues for further optimization because all the TCEs have trade-off correlations between their sheet resistance and transmittance. For example, STOSCs with high-transmittance/low-conductivity TCEs exhibit excellent transmittance but poor device performance whist those with low-transmittance/high-conductivity TCEs are operated to the contrary. Especially, the conductivity of conducting polymers, CNTs, and graphene are still low despite their excellent transmittance, and therefore limits the performance of STOSCs. If the TCEs with sheet resistances similar or even better than those of TCOs while maintaining high transmittance are successfully developed, solar cells with excellent performance/transparency at the same time can be realized.

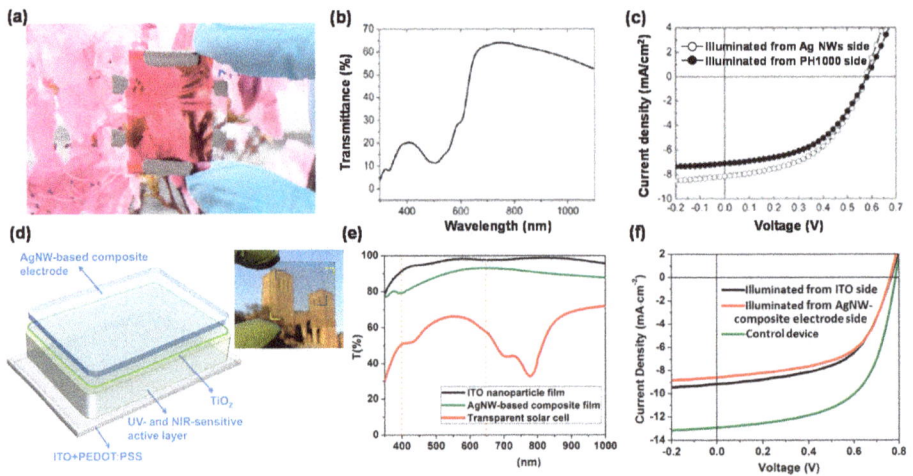

Figure 4. (**a**) A photograph of a typical P3HT:PCBM-based STOSC; (**b**) Optical transmittance spectrum of a STOSC with Ag NWs- and PH1000-based TCEs; (**c**) *J-V* characteristics of the STOSCs illuminated through the Ag NWs and the PH1000 electrodes. Reprinted with permission from [42]. Copyright 2014 ACS. (**d**) Schematic and photograph of the device architecture for solution-processed visibly-transparent OSCs; (**e**) Transmission spectra of the pristine ITO nanoparticle film, the Ag NW-based composite transparent electrode, and a STOSCs; (**f**) *J-V* characteristics of the transparent device (illuminated from ITO side or Ag NW composite electrode side) and the control device. Reprinted with permission from [43]. Copyright 2012 ACS.

3.2. High Color Neutral Perception

In addition to transparency and efficiency as important factors for STOSCs, color perception should be also considered. Color perception is generally determined by photoactive layer materials and electrodes. The transparency color perceptions of STOSCs are described based on the CIE 1931 chromaticity diagram. The coordinates of the white light are (0.33, 0.33), and the closer to (0.33, 0.33) the color coordinates of the STOSCs, the better their color neutrality [16]. The maximum photopic (vision in bright light) and scotopic (vision in dim light) sensitivities of human eye are at 555 and 507 nm respectively. Therefore, the key idea of designing STOSCs is to completely utilize the solar illumination in NIR region by transmitting the light in the range of 400 to 600 nm, most sensitive to human vision [110].

3.2.1. Photoactive Layer Materials

It is important to select photoactive layer materials with strong absorbance in the UV and NIR range to obtain high J_{sc} and PCE from STOSCs with reasonable visible transparency. Recently, efficient STOSCs were successfully fabricated using a blend of poly[(4,4′-bis(2-ethylhexyl)dithieno [3,2-*b*:2′,3′-*d*]silole)-2,6-diyl-alt-(2,1,3-benzothiadiazole)-4,7diyl] (PSBT BT) polymer donor and PC$_{71}$BM acceptor with long-term absorption power in the IR region, as shown in Figure 5a [45]. The transparent color recognition of the STOSCs was very close to white light (Figure 5b), as confirmed by the color coordinates obtained from the transmission spectrum. Moreover, the STOSCs exhibited remarkable rendering characteristics together with overall PCE of 3%. Another kind of STOSC was fabricated using a photoactive layer composed of low-bandgap polymers [46], thereby showing grayish or neutral color appearance due to the complementary and balanced absorption of the active layers [poly[2,6-(4,4-bis-(2-ethylhexyl)-4H-cyclopenta[2,1-*b*;3,4-*b*0] dithiophene)-alt-4,7-(2,1,3-benzothiadiazole)] (PBDTT-C-T) and PC$_{71}$BM]. As the reflective Ag electrode thickened from 6 to 60 nm, the average visible light transmittance and PCE varied from 36%

to 2% and 4.25% to 7.56%, respectively. Furthermore, the CRI of the STOSCs was shown to be close to 100, promising for their power-generating window applications.

As a different method, STOSCs were fabricated by employing a photoactive layer of poly (indacenodithiophene-co-phenanthro[9,10-*b*]quinoxaline) (PIDT-PhanQ):PC$_{71}$BM [16]. The STOSCs with a 30 nm Ag cathode exhibited 5.1% PCE, about 25% AVT, and good color recognition close to white light. Other groups used blade coating technology to fabricate a series of STOSCs using different polymer donors [47]. The STOSCs with 6-bis(trimethyltin)-4,8-bis(5-(2-ethylhexyl) thiophen-2-yl)benzo[1,2-*b*:4,5-*b'*]dithiophene (PBDTTT-CT):PC$_{71}$BM and poly[4,8-bis(5-(2-ethylhexyl) thiophen-2-yl)benzo[1,2-*b*;4,5-*b*0]dithiophene-2,6-diylalt-(4-(2-ethylhexyl)-3 fluorothieno [3,4-*b*]thiophene-)-2-carboxylate-2-6-diyl)] (PBDTTT-EFT):PC$_{71}$BM showed transparent color recognition close to white based on standard light source D65 with CRI of 95.4 and 87.1, respectively. The STOSCs using PBDTTT-CT:PC$_{71}$BM and PBDTTT-EFT:PC$_{71}$BM showed 5.2%/3.8% and 5.6%/5.3% PCEs at active area of 0.04/10.8 cm^2, respectively. More recently, a small-bandgap electron acceptor (ATT-2) was blended with a polymer donor poly[[2,6′-4,8-di(5-ethylhexylthienyl)benzo[1,2-*b*;3,3-*b*]dithiophene][3-fluoro-2[(2-ethylhexyl)carbonyl]thieno[3,4-*b*]thiophenediyl]] (PTB7-Th) to fabricate nonfullerene STOSCs [48]. Figure 5c shows complementary absorption spectrum of the photoactive blend film, useful for improving the light harvesting, resulting in the increase of J_{sc}. The STOSCs exhibited excellent color rendering with color coordinates of (0.2805, 0.3076), correlated color temperature of 9113 K, and excellent CRI of 94.1. In addition, the AVT and PCE increased to 37% and 7.74%, respectively due to absorption of light up to the NIR region, as shown in Figure 5d,e.

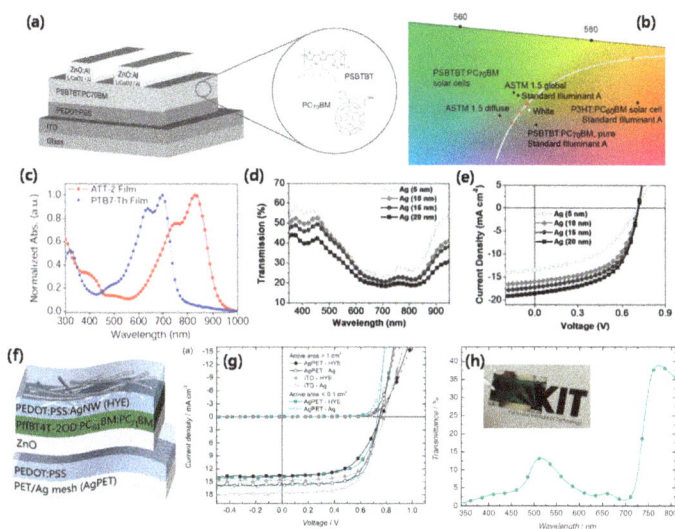

Figure 5. (**a**) Device architecture of STOSCs comprising a PSBTBT:PC$_{70}$BM active layer, a sputtered ZnO:Al transparent cathode, and a LiCoO$_2$/Al interfacial layer; (**b**) Determination of the color rendering index. The white dots represent the color coordinates of the TCS01–08 illuminated by the reference Standard Illuminant D. The black needles correspond to the TCS01–08 illuminated from the PSBTBT:PC$_{70}$BM solar cell transmitted light. Reprinted with permission from [45]. Copyright 2011 Wiley. (**c**) Normalized UV-vis-NIR absorption spectra of PTB7-Th and ATT-2 in thin films; (**d**) Transmission spectra of STOSCs as a function of Ag thickness; (**e**) *J*-*V* curves of STOSC devices. Reprinted with permission from [48]. Copyright 2011 Wiley; (**f**) Schematic diagram of a typical Ag mesh/PEDOT:PSS/ZnO/PffBT4T-2OD:PC$_{61}$BM:PC$_{71}$BM/PEDOT:PSS:Ag NWs STOSC; (**g**) *J*-*V* curves of STOSCs comprising different electrode combinations under AM1.5 illumination (solid lines) and in the dark (dashed lines); (**h**) Transmittance spectrum of STOSCs. Reprinted with permission from [49]. Copyright 2016 Wiley.

As a simple approach, ITO-free OSCs were fabricated by employing hybrid top and bottom electrodes and ternary polymer:fullerene photo-active layers [49]. Here, the polymer:fullerene absorber blend poly[(5,6-difluoro-2,1,3-benzothiadiazol-4,7-diyl)-alt-(3,3'''-di(2-octyldodecyl)-2,2';5', 2'';5'',2'''-quaterthiophen-5,5'''-diyl)] (PffBT4T-2OD):PC$_{61}$BM:PC$_{71}$BM and all other functional layers, which are high-efficiency tertiary polymers as active layers, were prepared using a doctor blade method from non-halogenated solvents to meet industrial device manufacturing requirements. Figure 5f shows a schematic diagram of the completed STOSC with maximum PCEs of 6.6% and 5.9% on active areas ≤0.1 cm^2 and >1 cm^2, respectively (Figure 5g) and 13% peak transparency at 515 nm (Figure 5h). Unlike conventional methods, flexible transparent OSCs using graphene TCEs as both anode and cathode were recently prepared based on a device structure of graphene/PEDOT:PSS/ZnO/poly[2,7-(5,5-bis-(3,7-dimethyl octyl)-5Hdithieno[3,2-*b*:20,30-*d*]pyran)-alt-4,7-(5,6-difluoro-2,1,3-benzothiadiazole)] (PDTP-DFBT):PC$_{61}$BM/MoO$_3$/graphene, thereby showing 3.7% PCE and 54% AVT in the visible regime (400–650 nm) [50]. The PCE was almost invariant after 100 cycles at $R = 1.2$ mm. As another approach, a highly efficient STOSC device was fabricated based on a novel polymer:non-fullerene (PBDB-T:ITIC) system [51], thereby showing optimized PCE > 7% with an AVT of 25%. These results suggest that optimizing the thickness of the active layer does not only allow images to be displayed accurately but also makes the STOSCs exhibit high efficiency/visible light transmittance. On the other hand, a combination of a cathode modified with transparent ZnMgO and a thin MoO$_3$/Ag anode yielded stable and efficient STOSCs with clear color recognition [52]. In this device, the PCE/AVT, controlled by the thickness of the Ag electrode, was 6.83%/21.6% and 8.58%/0.6% at Ag thicknesses of 10 and 50 nm, respectively. In addition, this method resulted in long-term device life due to the barrier effect on oxygen and water. More importantly, the STOSCs showed a very-long life span, i.e., high PCE of 7.02% even after two-year storage. These results may provide a useful way for developing STOSCs with high efficiency, long lifetime, and excellent color recognition for practical applications. Recently, another group designed and synthesized "IEICO-4Cl", a nonlelerene receptor with ultra-low band gap, and produced blend films with other polymer donors (poly[4-(4,8-bis(5-(2-ethylhexyl)thiophen-2-yl) benzo[1,2-b:4,5-b′]dithiophen-2-yl)-5,6-difluoro-2-(2-2hexyldecyl)-7-(thiophen-2-yl)-2H-benzo[d][1,2,3] triazole] (J52), poly[(2,6-(4,8-bis(5-(2-ethylhexyl)thiophen-2-yl) benzo [1,2-b:4,5-b′] dithiophene)-co-(1,3-di(5-thiophene-2-yl)-5,7-bis(2-ethylhexyl)benzo[1,2-c:4,5-c′]dithiophene-4,8-dione)] (PBDB-T), and PTB7-Th) [53]. The colors of the blend films such as J52:IEICO-4Cl, PBDB-T:IEICO-4Cl, and PTB7-Th:IEICO-4Cl could be adjusted from purple to blue and cyan, resulting in 10.1%/35.1%, 9.7%/35.7%, and 10.3%/33.5% PCE/AVT, respectively. These results suggest that design and use of ultralow-bandgap nonfullerene acceptor can be an efficient route for enhancing the photovoltaic parameters of STOSCs.

3.2.2. Photonic Crystals

The TCE absorbing only a fraction of the incident light and thinner photoactive layer should be used to maintain overall transmittance of STOSCs. The trade-off between the transmission and the absorption is the main factor for determining the PCE of STOSCs, in contrast with their opaque counterparts. A light-trapping approach should be therefore employed to improve the PCE of STOSCs. Chalesteric liquid crystals were developed in 2011 as an organic wavelength-dependent reflector, which can be fabricated by solution treatment for STOSCs [111]. The cholesteric liquid crystal can be properly adjusted to reflect the light only in the spectral range compatible with that absorbed by the photoactive layer and transmit the light in the other wavelength range. Then, the unabsorbed light that can produce photocurrent (PC) are reflected and reabsorbed by the photoactive layer, resulting in the increase of PC and PCE without reducing the transparency. The PCE was 6% increased when the spectral range of the light reflected by the cholesteric liquid crystal was matched to that absorbed by the cell.

Photonic crystals are nanostructured materials with the dielectric function periodically modulated, which can be employed as a distributed Bragg reflector in OSCs for enhancing the light capture [112–115]. To trap NIR and near-UV photons for the OSC, another group introduced a new approach for employing aperiodic photonic crystals in 2013 [88]. The photon manipulation by the nonperiodic 1D structure was very effective for enhancing the light harvesting of the STOSCs to 77% of what was achieved from the opaque counterpart cells. By the light trapping technique, the STOSCs exhibited 5.6% PCE and ~30% luminosity. In other study, one-dimensional photonic crystal (1DPC) was prepared on top of the Ag anode for STOSCs as a distribution reflector [54]. By this approach, the PCE has improved by 28% (absolutely from 1.92% to 2.46%) due to high reflectivity close to 100% at 580–780 nm. The AVT of 40% was maintained over the remaining visible wavelength range of 380–580 nm, and this was 33% improvement compared to the cells without 1DPCs. Subsequently, a series of translucent OSCs were fabricated to show further-enhanced PCE and transmittance using 1DPCs as reflectors [55,56,95,116–118]. In other report, the fullerene blends were successfully sandwiched between a bottom 1DPC and a top solution-processed highly-conductive PEDOT:PSS, as shown in Figure 6a [55]. Resulting PCE and J_{sc} were 5.20% and 12.25 mA·cm^{-2}, respectively, 37% and 38% rises compared to the OSCs without 1DPC and highest ever known for inverted OSCs with highly-conductive PEDOT:DOT layer at the illumination side.

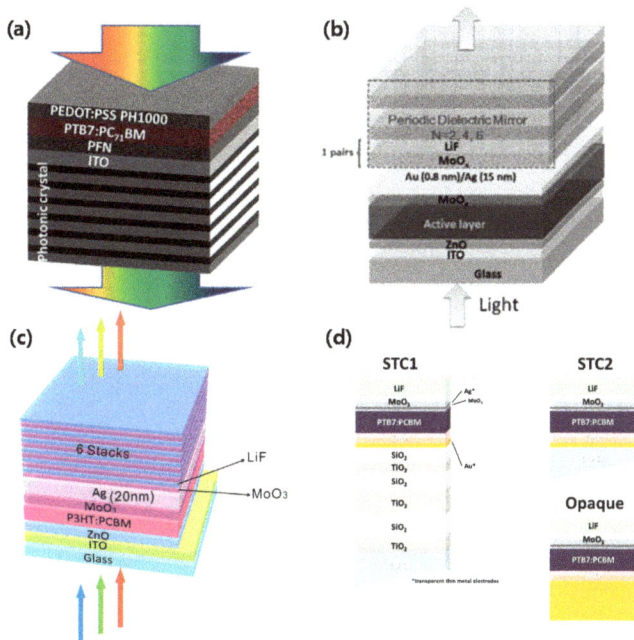

Figure 6. Device engineering of STOSCs for improving light harvesting: (**a**) Schematic of a STOSC with one-dimensional photonic crystal. Reprinted with permission from [55]. Copyright 2016 Royal Society of Chemistry. Schematic of STOSCs with (**b**) MoO$_x$/LiF and (**c**) MoO$_3$/LiF. Reprinted with permission from [56,57]. Copyright 2017 Wiley; Copyright 2013 ACS. (**d**) Schematic illustration of the semitransparent device cell architecture incorporating the multilayer dielectric structure (MLD) between the glass and the Au thin metal electrode and antireflection coating above the Ag thin metal electrode. Reprinted with permission from [58]. Copyright 2017 Wiley.

Recently, high-quality hybrid Au/Ag transparent top electrodes and 6 pairs of fine-tuned dielectric mirrors (DMs) were employed for inverted OSCs with PTB7-Th:PC$_{71}$BM, as shown in Figure 6b [56], showing a maximum PCE of 7.0% with an AVT of 12.2%. This PCE was highest ever reported for translucent PSCs as well as 81.4% of what was obtained from the opaque counterparts. Similar device design and processing methods were also successfully applied to flexible substrates, thereby achieving

6.4% PCE and 11.5% AVT. As a different method, 6 pairs of MoO_3/LiF layers (60 nm for MoO_3 and 90 nm for LiF, respectively) were used as reflectors to fabricate STOSCs, as shown in Figure 6c [57]. By the optimization of the reflector structure, enhanced light harvesting was obtained, resulting in an increase of J_{sc} from 8.1 to 10.9 mA·cm^{-2} (the increase of the PCE from 3.36% to 4.32%), caused by the enhanced reflection of the light in the wavelength range of 450 to 600 nm, corresponding to the wavelength range of the absorption by the active layer.

According to another approach, PTB7:PC$_{71}$BM blend active material were employed to enhance the light trapping inside the STOSC [58]. The multilayer dielectric structure (MLD) of the STOSCs was made to improve external quantum efficiency in the UV and NIR wavelength range with the transparency in the visible region being almost unchanged [54]. In this study, two types of STOSCs (STOSC1 and STOSC2) and opaque cells (reference) were fabricated to better evaluate the photovoltaic performance of STOSCs (Figure 6d). The J_{sc} of the STOSC1 containing this cavity configuration was 10.7 mA·cm^{-2}, 96.4% of what was obtained from the opaque counterpart cell, resulting in a PCE of 5.3%, comparable to that (5.9%) of the opaque cell. On the other hand, the J_{sc} of the STOSC2 was lower due to the lack of effective optical trapping effect.

3.2.3. Tandem Cells

Tandem OSCs have the advantage of maximizing the absorption of the solar energy by stacking two or more different cells that respond to different parts of the solar spectrum. The absorption, transparency, and efficiency of the cell can be also finely adjusted by constructing a series of stacks by introducing multiple of photoactive materials [59]. Two semitransparent sub-cell stacks were connected in parallel to enhance the PCE of OSCs in 2012 [60]. The total loss of light in the semitransparent tandem solar cells by the reflection, parasitic absorption, and transmission can be less than that in a single cell with the same photoactive materials. In this study, a new and efficient STOSC was constructed using polymer poly(3,3′-([(9′,9′-dioctyl-9H,9′H-[2,2′-bifluorene]-9,9-diyl)bis(4,1-phenylene)]bis(oxy))bis(N,N-dimethylpropan-1-amine)) (PFPA-1) modified ITO bottom cathode and high conductivity PEDOT: PSS PH1000 top anode, thereby showing PCE of 3.35% and 2.61% depending on the composition of the active layer with poly[N,N′-bis(2-hexyldecyl) isoindigo-6,6′-diyl-alt-thiophene-2,5-diyl] (P3TI):PC$_{71}$BM and poly[2,3-bis-(3-octyloxyphenyl) quinoxaline-5,8-diyl-alt-thiophene-2,5-diyl] (TQ1):PC$_{71}$BM, respectively. Furthermore, a series of tandem solar cells using two cells were also fabricated to exhibit an efficiency of 4.35%. In a different approach, semitransparent tandem OSCs were fabricated using graphene meshes and laminated Ag NWs, each of which was solution-treated for a transparent anode and a cathode, respectively (Figure 7a) [61]. In this study, the AVT was adjusted from 44.9% to 39.9% as the thickness of TiO$_2$ increased from 5 to 20 nm, as shown in Figure 7b. In the same range of TiO$_2$ thickness, the J_{sc} decreased from 7.62 to 5.90 mA·cm^{-2}, resulting in the decrease of the PCE from 8.02% to 5.52%.

As a final issue, when the light was incident at the side of the graphene-mesh electrode, the PCE was remarkably enhanced to 8.02% with an average transmittance of 44.9%, but the irradiation from the stacked Ag NW side resulted in a PCE of 6.47% (Figure 7c). Alternatively, for the production of high-efficiency tandem STOSC, the same polymer donor blended with PC$_{61}$BM and PC$_{71}$BM were used as the active layer, as shown in Figure 7d [62]. The tandem cells using PIDT-phanQ and same middle-bandgap polymer showed a high open circuit voltage (V_{oc}) of 1.71 V, thereby achieving a high PCE of 8.5%. Since the total active layer is very thin, the PIDT-phanQ can be used as a donor for very efficient semitransparent tandem cells. As shown in Figure 7e, around 40% AVT was obtained, promising for their use in power window applications. In addition, tandem architecture OSCs were configured to adjust the external appearance from visible transparency to semitransparency [63], resulting in 7.3% PCE and 30% AVT. Interestingly, the photovoltaic properties of these tandem OSCs did not differ significantly from those at the top and bottom lighting modes. As shown above, the trade-off relation between the transmittance and absorption of the photoactive layers affects the performance and transparency of STOSCs, and the related parameters should be therefore systematically adjusted through the improvement of the photoactive layer and the manipulation of the cell architecture.

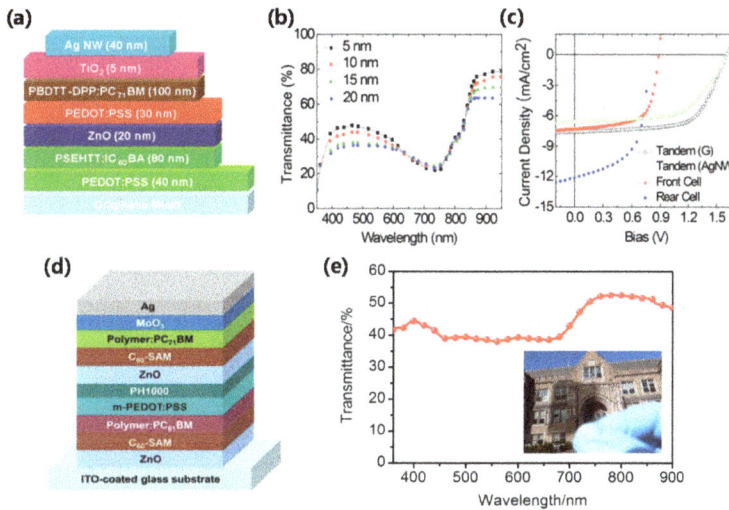

Figure 7. (**a**) Schematic diagram of a typica tandem STOSC; (**b**) Optical transmittance of the tandem cell with different TiO$_2$ thicknesses; (**c**) *J*-*V* curves of front, rear, and tandem cells (light projected from graphene mesh or Ag NWs) Reprinted with permission from [61]. Copyright 2014 Wiley. (**d**) Schematic representation of the tandem device architecture; (**e**) Optical transmittance of the tandem STOSC. The inset shows a photograph of the cell. Reprinted with permission from [62]. Copyright 2013 Wiley.

4. Semitransparent Perovskite Solar Cells

PSCs have received much attention in recent years due to their high efficiency and low-cost solution processibility [119–121]. The photoactive layer of PSCs, organic-inorganic metal halide perovskites, can be prepared on flexible substrates by low-temperature solution treatment, similar to OSCs. The bandgap and thickness of perovskite films can easily be adjusted to be semitransparent in the visible region, especially useful for flexible and translucent PSCs. Considering the close correlation and strong similarities between planar PSCs and OSCs, this section will review the progress of the semitransparent planar PSCs to provide the reader with an overall understanding for the state of the art of this research field. The STPSCs are also highly attractive due to their potential applications in building integrated photovoltaics and smart glasses. The principal difference between STPSCs and STOSCs is the optically active layer. In general, the transparency of the perovskite film in STPSCs is smaller than that of the organic photoactive layer in STOSCs because the perovskite film is thicker and more absorptive. The transparency of the perovskite film can be enhanced by decreasing the surface coverage of the island structure or by simply reducing its thickness. In addition, the performance of STPSCs depends on the electrode type of the cell, similar to STOSCs.

4.1. Single-Junction Cells

4.1.1. General Trend

Typically, the device architecture of PSCs is classified as n-i-p (normal) and p-i-n (inverted) types in terms of what type of material is placed on the TCO such as FTO or ITO. A metal electrode with proper work function was used as the top electrode. Recently, microstructure array of the "island" perovskite for STPSCs were formed by dewetting process, as shown in Figure 8a [64]. Figure 8b shows a scanning electron micrograph (SEM) of top surface of a typical film composed of perovskite islands (paler regions) on a TiO$_2$-coated FTO substrate. Figure 8c shows a photo image of a STPSC formed on glass. The correlation between the transmittance and the PCE of the cell was controlled by the surface coverage of the perovskite active layer. The average PCE is 7% when the perovskite surface coverage is 100%, but the AVT is 7% due to the low transmittance, as shown in Figure 8d. However, the cells with highest AVT of about 30% showed about 3.5% PCE. As another approach, a macroporous perovskite semitransparent film was aligned with FTO glass by a sacrificial polystyrene (PS) microsphere template [65] because the macropores inside the perovskite film was known to increase the light transmittance. Thus, the AVT in macroporous perovskite films was adjusted from 20% to 45% by simply changing the PS diameter and precursor concentration. The STPSCs optimized with the macroporous perovskite film showed 36.5% AVT and 11.7% PCE. All cells showed relatively-low V_{oc} due to the poor shape and crystallinity of the perovskite film.

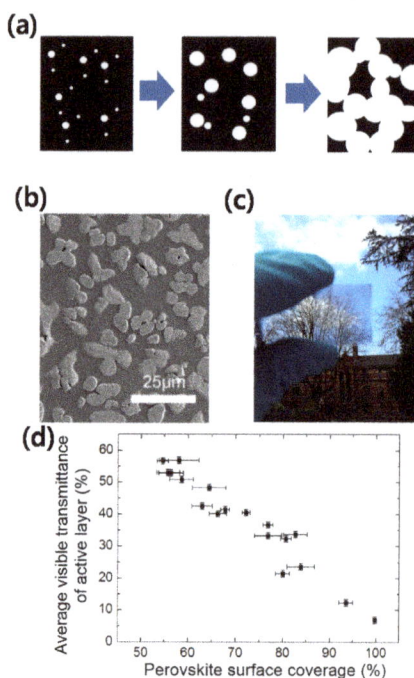

Figure 8. Controlled dewetting for varying transmittance of a perovskite film. (**a**) Schematic of the film dewetting process showing morphology change over time, from as-cast film to discrete islands. Perovskite material and pores are represented with black and white, respectively; (**b**) Scanning electron micrograph of the top surface of a representative film of perovskite islands (paler regions) on a TiO$_2$-coated FTO substrate; (**c**) Photograph of a STPSC; (**d**) Dependence of average visible transmittance of the active layer on perovskite surface coverage. Reprinted with permission from [64]. Copyright 2014 ACS.

It was also possible to produce STPSCs with excellent transparency in the visible region by using copper thiocyanate (CuSCN) HTL to reduce the thickness of the perovskite film [66]. Figure 9a,b shows a schematic of a typical CuSCN HTL-based high-efficiency STPSC and its corresponding energy band diagram. Figure 9c shows the transmittance for various thicknesses of perovskite films coated on CuSCN/ITO coated glass, resulting in the variation of the PCE and AVT. The PCE increases from 3.39% to 7.73% as the thickness increases from 40 to 280 nm, but the AVT decreases from 35% to 10% in the same range of thickness (Figure 9d,e). In addition, the use of an extremely thin Au layer as an electrode helped minimize parasitic absorption, and the LiF capping layer played a decisive role in decreasing the energy loss through the specular reflection, resulting in the improvement of the transparency [67]. Consequently, based on a simple cell architecture without need for high temperature processes, the STPSCs showed the AVT close to 29% and highest PCE of 6.4%. In another method using a $MoO_3/Au/MoO_3$ top electrode, a STPSC with orange-brown color was successfully fabricated through perovskite thickness control [68]. This cell showed 5.3% PCE/31% AVT at 55 nm and 13.6% PCE/7% AVT at 290 nm. Another group developed STPSCs based on a structure of $ITO/PEDOT:PSS/MAPbI_3/PC_{61}BM/C_{60}/11$-amino-1-undecanethiol hydrochloride/Ag [69]. Here, 110 nm thin perovskite film was used to show high performance of 8.2% PCE and 34% AVT. STPSCs were also fabricated using the island structure-$MAPbI_{3-x}Cl_x$-NiO composite and the interfacial modification with Al_2O_3/NiO layer [70]. Resulting performance was 10.06% PCE and 27% AVT together with excellent air stability over 270 days.

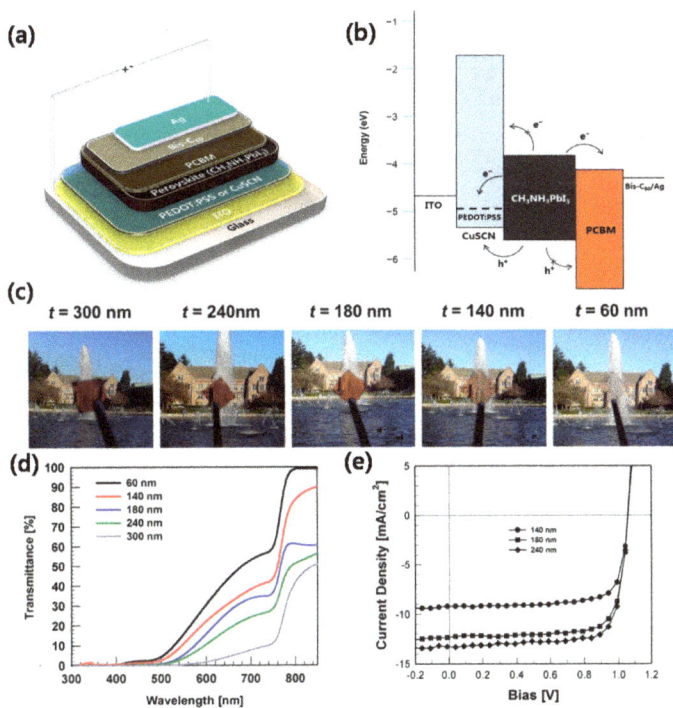

Figure 9. (**a**) Device architecture of a STPSC and (**b**) corresponding energy band diagram relative to vacuum level; (**c**) Photographs of $MAPbI_3$ films as a function of thickness and (**d**) UV-Vis transmittance of the corresponding films; (**e**) *J-V* curves of a STPSC with thin perovskite layers. Reprinted with permission from [66]. Copyright 2015 Wiley.

4.1.2. Use of TCO Electrodes

As mentioned above, the metal electrodes are limited in their use for improving the transmittance of STSCs. Since TCO has low resistivity and broadband transparency, it is expected that the efficiency and the transmittance of STPSCs can be improved if TCO electrodes are used at the both sides. Recently, STPSCs were fabricated based on a simple structure of $FTO/TiO_2/MAPbI_3$/hole transport materials and PEDOT:PSS/ITO [71]. The AVT and PCE of the cells were varied from 17.3% and 12.55% to 6.3% and 15.80% by adjusting the thickness of the $MAPbI_3$ layer. Another group fabricated high-efficiency semitransparent planar PSCs by high-frequency magnetron sputtering at room temperature [72], where hydrogenated indium oxide (In_2O_3:H) with high mobility was used as the transparent rear electrode, thereby showing 14.2% PCE. These cells also exhibited an additional 13.5% power improvement (absolutely, 20.5% PCE) when operated in a double-sided cell with a perovskite top cell and a copper indium gallium diselenide (CIGS) bottom cell in a 4-terminal tandem configuration. In another approach, a highly efficient translucent sandwich-type $MAPbI_3$ island PSCs were fabricated by employing a PS passivation interlayer to prevent the TiO_2 electron conductor and the poly[bis(4-phenyl)(2,4,6-trimethylphenyl)amine (PTAA) hole conductor from being directly contacted [73]. The selective deposition of the PS passivation interlayer on bare TiO_2 or between $MAPbI_3$ island crystals was done by a spin-washing process using toluene solvent. Resulting PCE and AVT were 10.2% and 20.9%, respectively. The same group reported semitransparent neutral-color sandwich-type $MAPbI_3$ perovskite submodules (active area = 25 cm^2) [74]. In this study, UV-curable monomer compounds coated on island-shaped $MAPbI_3$ were successfully employed to fabricate highly-efficient translucent solar cells by preventing the HTL and the TiO_2 blocking layer from being directly contacted. Here, 40 cells were measured to show 10.28% $\pm$ 0.96% PCE and 20.1% AVT.

4.1.3. Use of One or Both TCO—Free Electrodes

As described in the previous section, TCO-free TCEs such as PEDOT:PSS, Ag NWs, CNT, and graphene have many advantages for their use in STPSCs. Recently, STPSCs were fabricated based on a structure of Al_2O_3/Ag NWs/ZnO/$MAPbI_3$/PEDOT:PSS/ITO, as shown in Figure 10a [75]. Here, low temperature treatment (below 100 °C) and atomic layer deposition (ALD) of ZnO and Al_2O_3 films for the cathode buffer layer (CBL) and encapsulation layers were employed for very efficient and environmentally stable STPSCs. The cells with reflective opaque Ag electrodes showed remarkable PCE of up to 16.5% with high repeatability. In addition, the use of ALD ZnO CBL in STPSCs resulted in 25.5% AVT and 10% PCE. These results can provide a new route for developing highly-efficient, stable, low-temperature-processed semitransparent photovoltaic cells, indicating significant advancements in the practical application of PSCs. In another approach, efficient STPSCs were fabricated using solution-treated Ag NWs top electrodes made with careful selection of reasonable interface engineering and deposition techniques [76]. In this study, the ZnO layer inserted between the ETL $PC_{61}BM$ and the upper electrode Ag NWs did not only allow the Ag NW solution deposition without damaging the $PC_{61}BM$ layer, but also ensured the ohmic contact between $PC_{61}BM$/ZnO and ZnO/Ag NWs. The resulting translucent PSCs showed high PCE of 8.5% and an AVT of 28.4%. These results suggest that high-performance PSCs can be realized by integrating low-cost materials and following a completely liquid processing path, thereby greatly reducing the production cost. As a different approach, very robust PSCs were prepared by using a translucent top electrode sandwiched between an ultra-thin Ag layer and SnO_x grown by low-temperature ALD [77]. Figure 10b shows a SEM of the STPSC produced in this study. Here, SnO_x acts as an electrically conductive transmission barrier to protect the perovskite film and ultra-thin Ag electrodes from moisture. The STPSC showed an average transmittance of about 70% for wavelength >800 nm in the NIR region and an efficiency of 11% or more with an average transmittance of 29% at wavelength of 400–900 nm. In addition, the cells exhibited remarkable stability over 4500 h regardless of ambient atmosphere or to elevated temperatures. Another group reported STPSCs using PEDOT:PSS film as a TCE, as shown in Figure 10c [78]. The PEDOT:PSS electrode was deposited using transfer lamination technology to escape from the damage to the $MAPbI_3$ perovskite membrane by direct contact of PEDOT:PSS aqueous solution. The STPSCs showed approximately 10.1% PCE and 7.3% AVT for active area 0.06 cm^2. Recently, high-efficiency STPSCs without TCO was manufactured using PEDOT:PSS electrode which can be manufactured under low-temperature and vacuum-free environments [79]. These cells showed high PCEs of 13.9% and 10.3% on glass and PET substrates, respectively. In addition, the flexible substrate-based PSCs retained more than 90% of its initial value even after 1000-cycles bending at a small radius of 5 mm. In more recent years, other groups have successfully produced carbon-grid electrodes that are fully printable on STPSCs [80]. In this study, the AVT/PCE was varied from 6.5%/11.3% to 24.0%/5.1% by controlling the spacing between the carbon grid lines. Finally, by employing MWCNTs to the top electrode of the cell, the PCE was improved to 8.21% while maintaining 24.0% AVT. These results suggest that CNT-based carbon grids provide a low-cost route to fabricate highly-efficient STPSCs. CVD multilayer graphene was also employed for STPSCs (Figure 10d) [81]. By adjusting the preparation conditions, the cells exhibited the average PCE up to 12.02% and 11.65% under irradiation at the bottom and top electrode sides, respectively.

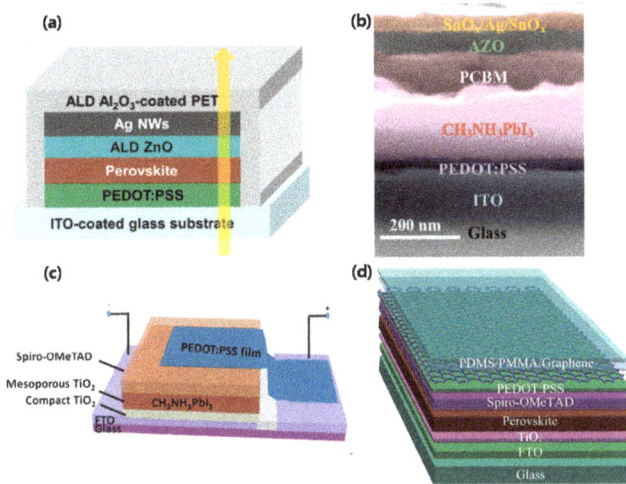

Figure 10. (**a**) Schematic illustration of a Ag NWs-TCE-based STPSC device architecture. Reprinted with permission from [75]. Copyright 2015 ACS. (**b**) Scanning electron microscope image of a STPSC with the layer sequence: ITO/PEDOT:PSS/MAPbI$_3$/PCBM/AZO/SnO$_x$/Ag/SnO$_x$. Reprinted with permission from [77]. Copyright 2017 Wiley. (**c**) Structure of a STPSC with PEDOT:PSS electrode. Reprinted with permission from [78]. Copyright 2015 ACS. (**d**) Schematic diagram of a graphene TCE-based STPSC. Reprinted with permission from [81]. Copyright 2015 Wiley.

4.2. Multi-Junction Cells

The photoactive layer used in a single-junction solar cell is limited in absorbing all of the sunlight due to its inherent bandgap properties. To solve this problem, multi-junction solar cells have been actively studied by employing two or more optically absorbing semiconductors complementary to each other, stacked vertically to minimize transmitted or heat-lost energy. Multi-junction or tandem solar cells were developed primarily in two schemes, mechanically-stacked 4-terminal and monolithic 2-terminal configurations. In case of perovskites, it is possible to control the bandgap easily by a solution process, thereby making the thin films with high extinction coefficient. High efficiency can be then achieved by reducing losses through the conventional tandemization with a crystalline Si (1.1 eV) or CIGS (1.1 eV) solar cells. Recently, perovskite/c-Si tandem solar cells were successfully produced to show the PCE up to 14.5% [82]. Figure 11a shows a schematic diagram of a typical perovskite/Si tandem solar cell. The PCE of the single-junction cells of perovskite (0.25 cm^2) and Si (1.22 cm^2) reached 14% and 16.8%, respectively. The PCE of the hybrid tandem solar cell was improved to 19.2% at a 1.22 cm^2 area, as shown in Figure 11b. Both PCEs were much better than those of the single-junction counterpart cells. This single-junction and tandem PSCs showed no hysteresis and were very stable under maximum power point tracking for several min (The inset of Figure 11b). Another group studied translucent perovskite solar cells and IR-enhanced Si heterojunction cells for high-efficiency serial devices [83]. The STPSCs were fabricated by thermally depositing a semitransparent metal electrode (7 nm Au/1 nm Cu seed layer) with excellent electrical conductivity and optical transparency.

The STPSC and Si cells showed 16.5% and 21.2% PCEs, respectively whilst the perovskite/Si tandem solar cells exhibited 23.0% PCE. In a different study, dielectric/metal/dielectric transparent top electrodes were successfully fabricated for hybrid halide PSC [84]. The Au-coated Ag film significantly enhanced the electrical conductivity and transparency of the electrodes compared to the separate Ag or Au film due to the change of the wetting properties of the metal deposition. The STPSC irradiated from the multilayer top electrode showed 11.5% PCE together without reduction of V_{oc} and FF compared to the conventional cells. Figure 11c shows the 4-terminal serial tandem cell with the PSC top single cell solution-processed on the bottom CIGS single cell. Figure 11d shows the *J-V* curve of single and tandem cells, resulting in 15.5% PCE for the tandem cell. These results possibly provide a useful route for further enhancement of the PCE by choosing tandem-type perovskite/CIGS solar cell. Another tandem solar cell was also obtained by mechanically placing an Ag NWs-TCE-based PSC on CIGS solar cell and low-quality polycrystalline Si [85], thereby realizing low-cost and high-efficiency (>25%) solar-energy source.

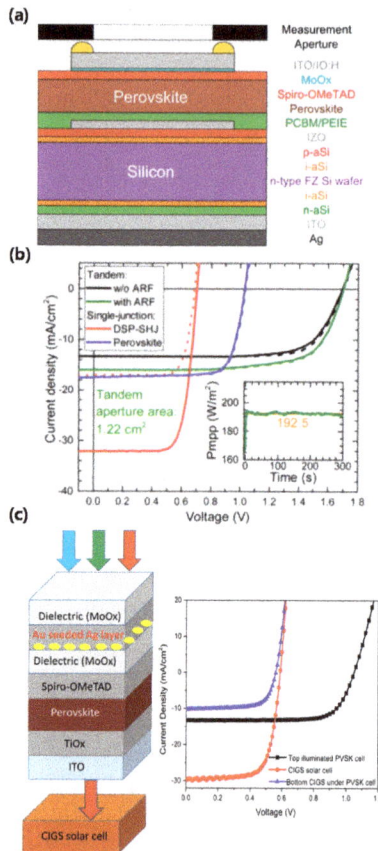

Figure 11. (**a**) Schematic drawing of a typical planar perovskite/Si heterojunction tandem cell layer stack; (**b**) *J-V* curves of the best perovskite/Si tandem and of the single junction perovskite and Si cell. Reprinted with permission from [81]. Copyright 2016 ACS. (**c**) Device architecture of a typical perovskite/GIGS tandem cell and *J-V* curves of top illuminated perovskite, CIGS, and perovskite/CIGS tandem cell. Reprinted with permission from [83]. Copyright 2015 ACS.

5. Conclusions and Outlook

In the last decade, there have been rapid progress in the studies of STOSCs and STPSCs by developing new flexible bottom/top TCEs, performing the processes at low temperatures, designing/synthesizing high-quality photoactive layers, and engineering device structures. To date, many researchers have made much efforts to achieve high PCE/AVT for the STOSCs and STPSCs, but there still remain several issues to overcome for their practical applications. The PCE and AVT of the semitransparent cells is currently much lower than those of their opaque counterpart cells, and their neutral transparency color perceptions can be hardly achieved. Therefore, the TCEs for STSCs are required to be highly transparent/conductive, a way to minimize the performance degradation while maintaining the device transparency. In addition, the TCEs should be structured for their good tolerance to defects and easy patterning, and should be available for mass production at low cost. To fundamentally improve the PCE/AVT of the STOSCs and STPSCs, new TCEs with the desired optical, electrical, and structural properties are ultimately necessary. Ideally, most of the photons in the UV and NIR range are absorbed by band gap engineering of the material because the composition of organic semiconductors and perovskite materials can be fully controlled. In addition, the color and transparency of the device can be adjusted by introducing periodic 1D dielectric mirrors with enhanced optical reflectivity in selected wavelength range to recapture unabsorbed photons. Although there still remain problems to be solved, STOSCs and STPSCs are expected to contribute greatly to harvesting of solar energy with high efficiency/semitransparency in the near future.

Funding: This research was funded by the Ministry of Science and ICT supported by Basic Science Research Program through the National Research Foundation of Korea (NRF) (NRF-2017R1A2B3006054).

Conflicts of Interest: The authors declare no conflicts interest.

References

1. Gaur, A.; Tiwari, G.N. Performance of photovoltaic modules of different solar cells. *Sol. Energy* **2013**, *2013*, 734581. [CrossRef]
2. Green, M.A.; Hishikawa, Y.; Dunlop, E.D.; Levi, D.H.; Hohl-Ebinger, J.; Ho-Baillie, A.W.Y. Solar cell efficiency tables. *Prog. Photovolt. Res. Appl.* **2018**, *26*, 427–436. [CrossRef]
3. Green, M.A.; Ho-Baillie, A.; Snaith, H.J. The emergence of perovskite solar cells. *Nat. Photonics* **2014**, *8*, 506–514. [CrossRef]
4. Brittman, S.; Adhyaksa, G.W.P.; Garnett, E.C. The expanding world of hybrid perovskites: Materials properties and emerging applications. *MRS Commun.* **2015**, *5*, 7–26. [CrossRef] [PubMed]
5. O'Regan, B.; Grätzel, M. A low-cost, high-efficiency solar cell based on dye-sensitized colloidal TiO$_2$ films. *Nature* **1991**, *353*, 737–740. [CrossRef]
6. Martinson, A.B.F.; Elam, J.W.; Liu, J.; Pellin, M.J.; Marks, T.J.; Hupp, J.T. Radial electron collection in dye-sensitized solar cells. *Nano Lett.* **2008**, *8*, 2862–2866. [CrossRef] [PubMed]
7. Bagnis, D.; Beverina, L.; Huang, H.; Silvestri, F.; Yao, Y.; Yan, H.; Pagani, G.A.; Marks, T.J.; Facchetti, A. Marked alkyl- vs. alkenyl-substitutent effects on squaraine dye solid-state structure, carrier mobility, and bulk-heterojunction solar cell efficiency. *J. Am. Chem. Soc.* **2010**, *132*, 4074–4075. [CrossRef] [PubMed]
8. Zhou, N.; Prabakaran, K.; Lee, B.; Chang, S.H.; Harutyunyan, B.; Guo, P.; Butler, M.R.; Timalsina, A.; Bedzyk, M.J.; Ratner, M.A.; et al. Metal-free tetrathienoacene sensitizers for high-performance dye sensitized solar cells. *J. Am. Chem. Soc.* **2015**, *137*, 4414–4423. [CrossRef] [PubMed]
9. Yella, A.; Lee, H.-W.; Tsao, H.N.; Yi, C.; Chandiran, A.K.; Nazeeruddin, M.K.; Diau, E.W.-G.; Yeh, C.-Y.; Zakeeruddin, S.M.; Grätzel, M. Porphyrinsensitized solar cells with cobalt (II/III)-based redox electrolyte exceed 12% efficiency. *Science* **2011**, *334*, 629–634. [CrossRef] [PubMed]
10. Mathew, S.; Yella, A.; Gao, P.; Humphry-Baker, R.; Curchod, B.F.E.; Ashari-Astani, N.; Tavernelli, I.; Rothlisberger, U.; Nazeeruddin, M.K.; Grätzel, M. Dye-sensitized solar cells with 13% efficiency achieved through the molecular engineering of porphyrin sensitizers. *Nat. Chem.* **2014**, *6*, 242–247. [CrossRef] [PubMed]

11. Kakiage, K.; Aoyama, Y.; Yano, T.; Oya, K.; Fujisawa, J.; Hanaya, M. Highly-efficient dye-sensitized solar cells with collaborative sensitization by silyl-anchor and carboxy-anchor dyes. *Chem. Commun.* **2015**, *51*, 15894–15897. [CrossRef] [PubMed]

12. Selvaraj, P.; Baig, H.; Mallick, T.K.; Siviter, J.; Montecucco, A.; Lib, W.; Paul, M.; Sweet, T.; Gaoc, M.; Knox, A.R.; et al. Enhancing the efficiency of transparent dye-sensitized solar cells using concentrated light. *Sol. Energy Mater. Sol. Cells* **2018**, *175*, 29–34. [CrossRef]

13. Yoon, S.; Tak, S.; Kim, J.; Jun, Y.; Kang, K.; Park, J. Application of transparent dye-sensitized solar cells to building integrated photovoltaic systems. *Build. Environ.* **2011**, *46*, 1899–1904. [CrossRef]

14. Chen, D.; Huang, F.; Cheng, Y.B.; Caruso, R.A. Mesoporous anatase TiO_2 beads with high surface areas and controllable pore sizes: A superior candidate for high-performance dye-sensitized solar cells. *Adv. Mater.* **2009**, *21*, 2206–2210. [CrossRef]

15. Li, D.; Qin, D.; Deng, M.; Luo, Y.; Meng, Q. Optimization the solid-state electrolytes for dye-sensitized solar cells. *Energy Environ. Sci.* **2009**, *2*, 283–291. [CrossRef]

16. Chueh, C.-C.; Chien, S.-C.; Yip, H.-L.; Salinas, J.F.; Li, C.-Z.; Chen, K.-S.; Chen, F.-C.; Chen, W.-C.; Jen, A.K.-Y. Toward high-performance semi-transparent polymer solar cells: Optimization of ultra-thin light absorbing layer and transparent cathode architecture. *Adv. Energy Mater.* **2013**, *3*, 417–423. [CrossRef]

17. Ye, L.; Zhang, S.; Zhao, W.; Yao, H.; Hou, J. Highly efficient 2D-conjugated benzodithiophene-based photovoltaic polymer with linear alkylthio side chain. *Chem. Mater.* **2014**, *26*, 3603–3605. [CrossRef]

18. Park, N.-G. Perovskite solar cells: An emerging photovoltaic technology. *Mater. Today* **2015**, *18*, 65–72. [CrossRef]

19. Stoumpos, C.C.; Malliakas, C.D.; Kanatzidis, M.G. Semiconducting tin and lead iodide perovskites with organic cations: Phase transitions, high mobilities, and near-infrared photoluminescent properties. *Inorg. Chem.* **2013**, *52*, 9019–9038. [CrossRef] [PubMed]

20. Miyata, A.; Mitioglu, A.; Plochocka, P.; Portugall, O.; Wang, J.T.-W.; Stranks, S.D.; Snaith, H.J.; Nicholas, R.J. Direct mesaurement of the exciton binding energy and effective masses for charge carriers in organic-inorganic tri-halide perovskites. *Nat. Phys.* **2015**, *11*, 582–587. [CrossRef]

21. Stranks, S.D.; Eperon, G.E.; Grancini, G.; Menelaou, C.; Alcocer, M.J.P.; Leijtens, T.; Herz, L.M.; Petrozza, A.; Snaith, H.J. Electron-hole diffusion lengths exceeding 1 micrometer in an organometal trihalide perovskite absorber. *Science* **2013**, *342*, 341–344. [CrossRef] [PubMed]

22. Zhao, Y.; Zhu, K. Organic–inorganic hybrid lead halide perovskites for optoelectronic and electronic applications. *Chem. Soc. Rev.* **2016**, *45*, 655–689. [CrossRef] [PubMed]

23. Sun, J.; Jasieniak, J.J. Semi-transparent solar cells. *J. Phys. D Appl. Phys.* **2017**, *50*, 093001. [CrossRef]

24. Xue, Q.; Xia, R.; Brabec, C.J.; Yip, H.-L. Recent advances in semi-transparent polymer and perovskite solar cells for power generating window applications. *Energy Environ. Sci.* **2018**, *11*, 1688–1709. [CrossRef]

25. Li, Y.; Xu, G.; Cui, C.; Li, Y. Flexible and semitransparent organic solar cells. *Adv. Energy Mater.* **2018**, *8*, 1701791. [CrossRef]

26. Lee, K.-T.; Guo, L.J.; Park, H.J. Neutral- and multi-colored semitransparent perovskite solar cells. *Molecules* **2016**, *21*, 475. [CrossRef] [PubMed]

27. Zhou, Y.; Cheun, H.; Choi, S.; Fuentes-Hernandez, C.; Kippelen, B. Optimization of a polymer top electrode for inverted semitransparent organic solar cells. *Org. Electron.* **2011**, *12*, 827–831. [CrossRef]

28. Zhou, Y.; Li, F.; Barrau, S.; Tian, W.; Inganäs, O.; Zhang, F. Inverted and transparent polymer solar cells prepared with vacuum-free processing. *Sol. Energy Mater. Sol. Cells* **2009**, *93*, 497–500. [CrossRef]

29. Dong, Q.; Zhou, Y.; Pei, J.; Liu, Z.; Li, Y.; Yao, S.; Zhang, J.; Tian, W. All-spin-coating vacuum-free processed semi-transparent inverted polymer solar cells with PEDOT:PSS anode and PAH-D interfacial layer. *Org. Electron.* **2010**, *11*, 1327–1331. [CrossRef]

30. Wang, Y.; Jia, B.; Qin, F.; Wu, Y.; Meng, W.; Dai, S.; Zhou, Y.; Zhan, X. Semitransparent, non-fullerene and flexible all-plastic solar cells. *Polymer* **2016**, *107*, 108–112. [CrossRef]

31. Kim, Y.H.; Müller-Meskamp, L.; Zakhidov, A.A.; Sachse, C.; Meiss, J.; Bikova, J.; Cook, A.; Zakhidov, A.A.; Leo, K. Semi-transparent small molecule organic solar cells with laminated free-standing carbon nanotube top electrodes. *Sol. Energy Mater. Sol. Cells* **2012**, *96*, 244–250. [CrossRef]

32. Xia, X.; Wang, S.; Jia, Y.; Bian, Z.; Wu, D.; Zhang, L.; Cao, A.; Huang, C. Infrared-transparent polymer solar cells. *J. Mater. Chem.* **2010**, *20*, 8478–8482. [CrossRef]

33. Jeon, I.; Delacou, C.; Kaskela, A.; Kauppinen, E.I.; Maruyama, S.; Matsuo, Y. Metal-electrode-free window-like organic solar cells with p-doped carbon nanotube thin-film electrodes. *Sci. Rep.* **2016**, *6*, 31348. [CrossRef] [PubMed]

34. Lee, Y.-Y.; Tu, K.-H.; Yu, C.-C.; Li, S.-S.; Hwang, J.-Y.; Lin, C.-C.; Chen, K.-H.; Chen, L.-C.; Chen, H.-L.; Chen, C.-W. Top laminated graphene electrode in a semitransparent polymer solar cell by simultaneous thermal annealing/releasing method. *ACS Nano* **2011**, *5*, 6564–6570. [CrossRef] [PubMed]

35. Liu, Z.; Li, J.; Sun, Z.-H.; Tai, G.; Lau, S.-P.; Yan, F. The application of highly doped single-layer graphene as the top electrodes of semitransparent organic solar cells. *ACS Nano* **2012**, *6*, 810–818. [CrossRef] [PubMed]

36. Liu, Z.; You, P.; Liu, S.; Yan, F. Neutral-color semitransparent organic solar cells with all-graphene electrodes. *ACS Nano* **2015**, *9*, 12026–12034. [CrossRef] [PubMed]

37. Lin, P.; Choy, W.C.H.; Zhang, D.; Xie, F.; Xin, J.; Leung, C.W. Semitransparent organic solar cells with hybrid monolayer graphene/metal grid as top electrodes. *Appl. Phys. Lett.* **2013**, *102*, 113303. [CrossRef]

38. Shin, D.H.; Jang, C.W.; Lee, H.S.; Seo, S.W.; Choi, S.-H. Semitransparent flexible organic solar cells employing doped-graphene layers as anode and cathode electrodes. *ACS Appl. Mater. Interfaces* **2018**, *10*, 3596–3601. [CrossRef] [PubMed]

39. Lee, J.-Y.; Connor, S.T.; Cui, Y.; Peumans, P. Semitransparent organic photovoltaic cells with laminated top electrode. *Nano Lett.* **2010**, *10*, 1276–1279. [CrossRef] [PubMed]

40. Krantz, J.; Stubhan, T.; Richter, M.; Spallek, S.; Litzov, I.; Matt, G.J.; Spiecker, E.; Brabec, C.J. Spray-coated silver nanowires as top electrode layer in semitransparent P3HT:PCBM-based organic solar cell devices. *Adv. Funct. Mater.* **2013**, *23*, 1711–1717. [CrossRef]

41. Guo, F.; Zhu, X.; Forberich, K.; Krantz, J.; Stubhan, T.; Salinas, M.; Halik, M.; Spallek, S.; Butz, B.; Spiecker, E.; et al. ITO-free and fully solution-processed semitransparent organic solar cells with high fill factors. *Adv. Energy Mater.* **2013**, *3*, 1062–1067. [CrossRef]

42. Yim, J.H.; Joe, S.-Y.; Pang, C.; Lee, K.M.; Jeong, H.; Park, J.-Y.; Ahn, Y.H.; de Mello, J.C.; Lee, S. Fully solution-processed semitransparent organic solar cells with a silver nanowire cathode and a conducting polymer anode. *ACS Nano* **2014**, *8*, 2857–2863. [CrossRef] [PubMed]

43. Chen, C.-C.; Dou, L.; Zhu, R.; Chung, C.-H.; Song, T.-B.; Zheng, Y.B.; Hawks, S.; Li, G.; Weiss, P.S.; Yang, Y. Visibly transparent polymer solar cells produced by solution processing. *ACS Nano* **2012**, *6*, 7185–7190. [CrossRef] [PubMed]

44. Maisch, P.; Tam, K.C.; Lucera, L.; Egelhaafa, H.-J.; Scheiber, H.; Maier, E.; Brabec, C.J. Inkjet printed silver nanowire percolation networks as electrodes for highly efficient semitransparent organic solar cells. *Org. Electron.* **2016**, *38*, 139–143. [CrossRef]

45. Colsmann, A.; Puetz, A.; Bauer, A.; Hanisch, J.; Ahlswede, E.; Lemmer, U. Efficient semi-transparent organic solar cells with good transparency color perception and rendering properties. *Adv. Energy Mater.* **2011**, *1*, 599–603. [CrossRef]

46. Chen, K.-S.; Salinas, J.-F.; Yip, H.-L.; Huo, L.; Hou, J.; Jen, A.K.Y. Semi-transparent polymer solar cells with 6% PCE, 25% average visible transmittance and a color rendering index close to 100 for power generating window applications. *Energy Environ. Sci.* **2012**, *5*, 9551–9557. [CrossRef]

47. Wong, Y.Q.; Meng, H.-F.; Wong, H.Y.; Tan, C.S.; Wu, C.-Y.; Tsai, P.-T.; Chang, C.-Y.; Horng, S.-F.; Zan, H.-W. Efficient semitransparent organic solar cells with good color perception and good color rendering by blade coating. *Org. Electron.* **2017**, *43*, 196–206. [CrossRef]

48. Liu, F.; Zhou, Z.; Zhang, C.; Zhang, J.; Hu, Q.; Vergote, T.; Liu, F.; Russell, T.P.; Zhu, X. Efficient semitransparent solar cells with high NIR responsiveness enabled by a small-bandgap electron acceptor. *Adv. Mater.* **2017**, *29*, 1606574. [CrossRef] [PubMed]

49. Czolk, J.; Landerer, D.; Koppitz, M.; Nass, D.; Colsmann, A. Highly efficient, mechanically flexible, semi-transparent organic solar cells doctor bladed from non-halogenated solvents. *Adv. Mater. Technol.* **2016**, *1*, 1600184. [CrossRef]

50. Song, Y.; Chang, S.; Gradecak, S.; Kong, J. Visibly-transparent organic solar cells on flexible substrates with all-graphene electrodes. *Adv. Energy Mater.* **2016**, *6*, 1600847. [CrossRef]

51. Upama, M.B.; Wright, M.; Elumalai, N.K.; Mahmud, M.A.; Wang, D.; Xu, C.; Uddin, A. High-efficiency semitransparent organic solar cells with non-fullerene acceptor for window application. *ACS Photonics* **2017**, *4*, 2327–2334. [CrossRef]

52. Yin, Z.; Wei, J.; Chen, S.-C.; Cai, D.; Ma, Y.; Wang, M.; Zheng, Q. Long lifetime stable and efficient semitransparent organic solar cells using a ZnMgO-modified cathode combined with a thin MoO_3/Ag anode. *J. Mater. Chem. A* **2017**, *5*, 3888–3899. [CrossRef]

53. Cui, Y.; Yang, C.; Yao, H.; Zhu, J.; Wang, Y.; Jia, G.; Gao, F.; Hou, J. Efficient semitransparent organic solar cells with tunable color enabled by an ultralow-bandgap nonfullerene acceptor. *Adv. Mater.* **2017**, *29*, 1703080. [CrossRef] [PubMed]

54. Yu, W.; Shen, L.; Shen, P.; Meng, F.; Long, Y.; Wang, Y.; Lv, T.; Ruan, S.; Chen, G. Simultaneous improvement in efficiency and transmittance of low bandgap semitransparent polymer solar cells with one-dimensional photonic crystals. *Sol. Energy Mater. Sol. Cells* **2013**, *117*, 198–202. [CrossRef]

55. Zhang, Y.; Peng, Z.; Cai, C.; Liu, Z.; Lin, Y.; Zheng, W.; Yang, J.; Hou, L.; Cao, Y. Colorful semitransparent polymer solar cells employing a bottom periodic one-dimensional photonic crystal and a top conductive PEDOT:PSS layer. *J. Mater. Chem. A* **2016**, *4*, 11821–11828. [CrossRef]

56. Xu, G.; Shen, L.; Cui, C.; Wen, S.; Xue, R.; Chen, W.; Chen, H.; Zhang, J.; Li, H.; Li, Y.; et al. High-performance colorful semitransparent polymer solar cells with ultrathin hybrid-metal electrodes and fine-tuned dielectric mirrors. *Adv. Funct. Mater.* **2017**, *27*, 1605908. [CrossRef]

57. Zhang, D.-D.; Jiang, X.-C.; Wang, R.; Xie, H.-J.; Ma, G.-F.; Ou, Q.-D.; Chen, Y.-L.; Li, Y.-Q.; Tang, J.-X. Enhanced performance of semitransparent inverted organic photovoltaic devices via a high reflector structure. *ACS Appl. Mater. Interfaces* **2013**, *5*, 10185–10190. [CrossRef] [PubMed]

58. Pastorelli, F.; Romero-Gomez, P.; Betancur, R.; Martinez-Otero, A.; Mantilla-Perez, P.; Bonod, N.; Martorell, J. Enhanced light harvesting in semitransparent organic solar cells using an optical metal cavity configuration. *Adv. Energy Mater.* **2014**, *5*, 1400614. [CrossRef]

59. Meiss, J.; Menke, T.; Leo, K.; Uhrich, C.; Gnehr, W.-M.; Sonntag, S.; Pfeiffer, M.; Riede, M. Highly efficient semitransparent tandem organic solar cells with complementary absorber materials. *Appl. Phys. Lett.* **2011**, *99*, 043301. [CrossRef]

60. Tang, Z.; George, Z.; Ma, Z.; Bergqvist, J.; Tvingstedt, K.; Vandewal, K.; Wang, E.; Andersson, L.M.; Andersson, M.R.; Zhang, F.; et al. Semi-transparent tandem organic solar cells with 90% internal quantum efficiency. *Adv. Energy Mater.* **2012**, *2*, 1467–1476. [CrossRef]

61. Yusoff, A.R.B.M.; Lee, S.J.; Shneider, F.K.; da Silva, W.J.; Jang, J. High-performance semitransparent tandem solar cell of 8.02% conversion efficiency with solution-processed graphene mesh and laminated Ag nanowire top electrodes. *Adv. Energy Mater.* **2014**, *4*, 1301989. [CrossRef]

62. Chang, C.-Y.; Zuo, L.; Yip, H.-L.; Li, C.-Z.; Li, Y.; Hsu, C.-S.; Cheng, Y.-J.; Chen, H.; Jen, A.K.-Y. Highly efficient polymer tandem cells and semitransparent cells for solar energy. *Adv. Energy Mater.* **2013**, *4*, 1301645. [CrossRef]

63. Chen, C.-C.; Dou, L.; Gao, J.; Chang, W.-H.; Li, G.; Yang, Y. High-performance semi-transparent polymer solar cells possessing tandem structures. *Energy Environ. Sci.* **2013**, *6*, 2714–2720. [CrossRef]

64. Eperon, G.E.; Burlakov, V.M.; Goriely, A.; Snaith, H.J. Neutral color semitransparent microstructured perovskite solar cells. *ACS Nano* **2014**, *8*, 591–598. [CrossRef] [PubMed]

65. Chen, B.-X.; Rao, H.-S.; Chen, H.-Y.; Li, W.-G.; Kuang, D.-B.; Su, C.-Y. Ordered macroporous $CH_3NH_3PbI_3$ perovskite semitransparent film for high-performance solar cells. *J. Mater. Chem. A* **2016**, *4*, 15662–15669. [CrossRef]

66. Jung, J.W.; Chueh, C.-C.; Jen, A.K.-Y. High-performance semitransparent perovskite solar cells with 10% power conversion efficiency and 25% average visible transmittance based on transparent CuSCN as the hole-transporting material. *Adv. Energy Mater.* **2015**, *5*, 1500486. [CrossRef]

67. Roldán-Carmona, C.; Malinkiewicz, O.; Betancur, R.; Longo, G.; Momblona, C.; Jaramillo, F.; Camacho, L.; Bolink, H.J. High efficiency single-junction semitransparent perovskite solar cells. *Energy Environ. Sci.* **2014**, *7*, 2968–2973. [CrossRef]

68. Gaspera, E.D.; Peng, Y.; Hou, Q.; Spiccia, L.; Bach, U.; Jasieniak, J.J.; Cheng, Y.-B. Ultra-thin high efficiency semitransparent perovskitesolarcells. *Nano Energy* **2015**, *13*, 249–257. [CrossRef]

69. Bag, S.; Durstock, M.F. Efficient semi-transparent planar perovskite solar cells using a 'molecular glue'. *Nano Energy* **2016**, *30*, 542–548. [CrossRef]

70. Wang, Y.; Mahmoudi, T.; Yang, H.-Y.; Bhat, K.S.; Yoo, J.-Y.; Hahn, Y.-B. Fully-ambient-processed mesoscopic semitransparent perovskite solar cells by islands-structure-$MAPbI_{3-x}Cl_x$-NiO composite and Al_2O_3/NiO interface engineering. *Nano Energy* **2018**, *49*, 59–66. [CrossRef]

71. Heo, J.H.; Han, H.J.; Lee, M.; Song, M.; Kim, D.H.; Im, S.H. Stable semi-transparent CH₃NH₃PbI₃ planar sandwich solar cells. *Energy Environ. Sci.* **2015**, *8*, 2922–2927. [CrossRef]

72. Fu, F.; Feurer, T.; Jäger, T.; Avancini, E.; Bissig, B.; Yoon, S.; Buecheler, S.; Tiwari, A.N. Low-temperature-processed efficient semi-transparent planar perovskite solar cells for bifacial and tandem applications. *Nat. Commun.* **2015**, *6*, 8932. [CrossRef] [PubMed]

73. Heo, J.H.; Jang, M.H.; Lee, M.H.; Han, H.J.; Kang, M.G.; Lee, M.L.; Im, S.H. Efficiency enhancement of semi-transparent sandwich type CH₃NH₃PbI₃ perovskite solar cells with island morphology perovskite film by introduction of polystyrene passivation layer. *J. Mater. Chem. A* **2016**, *4*, 16324–16329. [CrossRef]

74. Heo, J.H.; Han, J.; Shin, D.H.; Im, S.H. Highly stable semi-transparent CH₃NH₃PbI₃ sandwich type perovskite solar sub-module with neutral color. *Mater. Today Energy* **2017**, *5*, 280–286. [CrossRef]

75. Chang, C.-Y.; Lee, K.-T.; Huang, W.-K.; Siao, H.-Y.; Chang, Y.-C. High-performance, air-stable, low-temperature processed semitransparent perovskite solar cells enabled by atomic layer deposition. *Chem. Mater.* **2015**, *27*, 5122–5130. [CrossRef]

76. Guo, F.; Azimi, H.; Hou, Y.; Przybilla, T.; Hu, M.; Bronnbauer, C.; Langner, S.; Spiecker, E.; Forberich, K.; Brabec, C.J. High-performance semitransparent perovskite solar cells with solution-processed silver nanowires as top electrodes. *Nanoscale* **2015**, *7*, 1642–1649. [CrossRef] [PubMed]

77. Zhao, J.; Brinkmann, K.O.; Hu, T.; Pourdavoud, N.; Becker, T.; Gahlmann, T.; Heiderhoff, R.; Polywka, A.; Görrn, P.; Chen, Y.; et al. Self-encapsulating thermostable and air-resilient semitransparent perovskite solar cells. *Adv. Energy Mater.* **2017**, *7*, 1602599. [CrossRef]

78. Bu, L.; Liu, Z.; Zhang, M.; Li, W.; Zhu, A.; Cai, F.; Zhao, Z.; Zhou, Y. Semitransparent fully air processed perovskite solar cells. *ACS Appl. Mater. Interfaces* **2015**, *7*, 17776–17781. [CrossRef] [PubMed]

79. Zhang, Y.; Wu, Z.; Li, P.; Ono, L.K.; Qi, Y.; Zhou, J.; Shen, H.; Surya, C.; Zheng, Z. Fully solution-processed TCO-free semitransparent perovskite solar cells for tandem and flexible applications. *Adv. Energy Mater.* **2018**, *8*, 1701569. [CrossRef]

80. Li, F.R.; Xu, Y.; Chen, W.; Xie, S.H.; Li, J.Y. Nanotube enhanced carbon grids as top electrodes for fully printable mesoscopic semitransparent perovskite solar cells. *J. Mater. Chem. A* **2017**, *5*, 10374–10379. [CrossRef]

81. You, P.; Liu, Z.; Tai, Q.; Liu, S.; Yan, F. Efficient semitransparent perovskite solar cells with graphene electrodes. *Adv. Mater.* **2015**, *27*, 3632–3638. [CrossRef] [PubMed]

82. Werner, J.; Weng, C.-H.; Walter, A.; Fesquet, L.; Seif, J.P.; Wolf, S.D.; Niesen, B.; Ballif, C. Efficient monolithic perovskite/silicon tandem solar cell with cell area >1 cm². *J. Phys. Chem. Lett.* **2016**, *7*, 161–166. [CrossRef] [PubMed]

83. Chen, B.; Bai, Y.; Yu, Z.; Li, T.; Zheng, X.; Dong, Q.; Shen, L.; Boccard, M.; Gruverman, A.; Holman, Z.; et al. Efficient semitransparent perovskite solar cells for 23.0%-efficiency perovskite/silicon four-terminal tandem cells. *Adv. Energy Mater.* **2016**, *6*, 1601128. [CrossRef]

84. Yang, Y.; Chen, Q.; Hsieh, Y.-T.; Song, T.-B.; Marco, N.D.; Zhou, H.; Yang, Y. Multilayer transparent top electrode for solution processed perovskite/Cu(In,Ga)(Se,S)₂ four terminal tandem solar cells. *ACS Nano* **2015**, *9*, 7714–7721. [CrossRef] [PubMed]

85. Bailie, C.D.; Christoforo, M.G.; Mailoa, J.P.; Bowring, A.R.; Unger, E.L.; Nguyen, W.H.; Burschka, J.; Pellet, N.; Lee, J.Z.; Grätzel, M.; et al. Semi-transparent perovskite solar cells for tandems with silicon and CIGS. *Energy Environ. Sci.* **2015**, *8*, 956–963. [CrossRef]

86. Drolet, N. Organic Photovoltaic: Efficiency and lifetime challenges for commercial viability. In Proceedings of the 2012 MRS Spring Meeting & Exhibit, San Francisco, CA, USA, 12 April 2012.

87. Ameri, T.; Dennler, G.; Waldauf, C.; Azimi, H.; Seemann, A.; Forberich, K.; Hauch, J.; Scharber, M.; Hingerl, K.; Brabec, C.J. Fabrication, optical modeling, and color characterization of semitransparent bulk-heterojunction organic solar cells in an inverted structure. *Adv. Funct. Mater.* **2010**, *20*, 1592–1598. [CrossRef]

88. Betancur, R.; Romero-Gomez, P.; Martinez-Otero, A.; Elias, X.; Maymó, M.; Martorell, J. Transparent polymer solar cells employing a layered light-trapping architecture. *Nat. Photonics* **2013**, *7*, 995–1000. [CrossRef]

89. Amara, M.; Mandorlo, F.; Couderc, R.; Gérenton, F.; Lemiti, M. Temperature and color management of silicon solar cells for building integrated photovoltaic. *EPJ Photovolt.* **2018**, *9*, 1. [CrossRef]

90. Czolk, J.; Puetz, A.; Kutsarov, D.; Reinhard, M.; Lemmer, U.; Colsmann, A. Inverted semi-transparent polymer solar cells with transparency color rendering indices approaching 100. *Adv. Energy Mater.* **2013**, *3*, 386–390. [CrossRef]

91. Yu, W.; Jia, X.; Yao, M.; Zhu, L.; Long, Y.; Shen, L. Semitransparent polymer solar cells with simultaneously improved efficiency and color rendering index. *Phys. Chem. Chem. Phys.* **2015**, *17*, 23732–23740. [CrossRef] [PubMed]

92. Bailey-Salzman, R.F.; Rand, B.P.; Forrest, S.R. Semitransparent organic photovoltaic cells. *Appl. Phys. Lett.* **2006**, *88*, 233502. [CrossRef]

93. Huang, J.; Li, G.; Yang, Y. A semi-transparent plastic solar cell fabricated by a lamination process. *Adv. Mater.* **2008**, *20*, 415–419. [CrossRef]

94. Upama, M.B.; Wright, M.; Elumalai, N.K.; Mahmud, M.A.; Wang, D.; Chan, K.H.; Xu, C.; Haque, F.; Uddin, A. High performance semitransparent organic solar cells with 5% PCE using non-patterned $MoO_3/Ag/MoO_3$ anode. *Curr. Appl. Phys.* **2017**, *17*, 298–305.

95. Yu, W.; Shen, L.; Long, Y.; Shen, P.; Guo, W.; Chen, W.; Ruan, S. Highly efficient and high transmittance semitransparent polymer solar cells with one-dimensional photonic crystals as distributed Bragg reflectors. *Org. Electron.* **2014**, *15*, 470–477. [CrossRef]

96. Granqvist, C.G. Transparent conductive electrodes for electrochromic devices: A review. *Appl. Phys. A* **1993**, *57*, 19–24. [CrossRef]

97. Alemu, D.; Wei, H.-Y.; Ho, K.-C.; Chu, C.-W. Highly conductive PEDOT: PSS electrode by simple film treatment with methanol for ITO-free polymer solar cells. *Energy Environ. Sci.* **2012**, *5*, 9662–9671. [CrossRef]

98. Kim, Y.H.; Sachse, C.; Machala, M.L.; May, C.; Müller-Meskamp, L.; Leo, K. Highly conductive PEDOT:PSS electrode with optimized solvent and thermal post-treatment for ITO-free organic solar cells. *Adv. Funct. Mater.* **2011**, *21*, 1076–1081. [CrossRef]

99. Park, S.; Vosguerichian, M.; Bao, Z. A review of fabrication and applications of carbon nanotube film-based flexible electronics. *Nanoscale* **2013**, *5*, 1727–1752. [CrossRef] [PubMed]

100. Zhang, M.; Fang, S.; Zakhidov, A.A.; Lee, S.B.; Aliev, A.E.; Williams, C.D.; Atkinson, K.R.; Baughman, R.H. Strong, transparent, multifunctional, carbon nanotube sheets. *Science* **2005**, *309*, 1215–1219. [CrossRef] [PubMed]

101. Pang, S.; Hernandez, Y.; Feng, X.; Müllen, K. Graphene as transparent electrode material for organic electronics. *Adv. Mater.* **2011**, *23*, 2779–2795. [CrossRef] [PubMed]

102. Kim, T.; Canlier, A.; Kim, G.H.; Choi, J.; Park, M.; Han, S.M. Electrostatic spray deposition of highly transparent silver nanowire electrode on flexible substrate. *ACS Appl. Mater. Interfaces* **2013**, *5*, 788–794. [CrossRef] [PubMed]

103. Lim, J.-W.; Cho, D.-Y.; Kim, J.; Na, S.-I.; Kim, H.-K. Simple brush-painting of flexible and transparent Ag nanowire network electrodes as an alternative ITO anode for cost-efficient flexible organic solar cells. *Sol. Energy Mater. Sol. Cells* **2012**, *107*, 348–354. [CrossRef]

104. Emmott, C.J.M.; Urbina, A.; Nelson, J. Environmental and economic assessment of ITO-free electrodes for organic solar cells. *Sol. Energy Mater. Sol. Cells* **2012**, *97*, 14–21. [CrossRef]

105. Gaynor, W.; Lee, J.-Y.; Peumans, P. Fully solution-processed inverted polymer solar cells with laminated nanowire electrodes. *ACS Nano* **2010**, *4*, 30–34. [CrossRef] [PubMed]

106. Gaynor, W.; Burkhard, G.F.; McGehee, M.D.; Peumans, P. Smooth nanowire/polymer composite transparent electrodes. *Adv. Mater.* **2011**, *23*, 2905–2910. [CrossRef] [PubMed]

107. Park, Y.; Bormann, L.; Müller-Meskamp, L.; Vandewal, K.; Leo, K. Efficient flexible organic photovoltaics using silver nanowires and polymer based transparent electrodes. *Org. Electron.* **2016**, *36*, 68–72. [CrossRef]

108. Hong, K.; Ham, J.; Kim, B.-J.; Park, J.Y.; Lim, D.C.; Lee, J.Y.; Lee, J.-L. Continuous 1D-metallic microfibers web for flexible organic solar cells. *ACS Appl. Mater. Interfaces* **2015**, *7*, 27397–27404. [CrossRef] [PubMed]

109. Angmo, D.; Andersen, T.R.; Bentzen, J.J.; Helgesen, M.; Søndergaard, R.R.; Jørgensen, M.; Carlé, J.E.; Bundgaard, E.; Krebs, F.C. Roll-to-roll printed silver nanowire semitransparent electrodes for fully ambient solution-processed tandem polymer solar cells. *Adv. Funct. Mater.* **2015**, *25*, 4539–4547. [CrossRef]

110. Geusebroek, J.-M.; van den Boomgaard, R.; Smeulders, A.W.; Gevers, T. Color constancy from physical principles. *Pattern Recognit. Lett.* **2003**, *24*, 1653–1662. [CrossRef]

111. Galagan, Y.; Debije, M.G.; Blom, P.W.M. Semitransparent organic solar cells with organic wavelength dependent reflectors. *Appl. Phys. Lett.* **2011**, *98*, 043302. [CrossRef]

112. Van der Mee, C.; Contu, P.; Pintus, P. One-dimensional photonic crystal design. *J. Quant. Spectrosc. Radiat. Transf.* **2010**, *111*, 214–225. [CrossRef]

113. Colodrero, S.; Mihi, A.; Häggman, L.; Ocaña, M.; Boschloo, G.; Hagfeldt, A.; Míguez, H. Porous one-dimensional photonic crystals improve the power-conversion efficiency of dye-sensitized solar cells. *Adv. Mater.* **2009**, *21*, 764–770. [CrossRef]

114. Fink, Y.; Winn, J.N.; Fan, S.; Chen, C.; Michel, J.; Joannopoulos, J.D.; Thomas, E.L. A dielectric omnidirectional refector. *Science* **1998**, *282*, 1679–1682. [CrossRef] [PubMed]

115. Yu, W.; Shen, L.; Long, Y.; Guo, W.; Meng, F.; Ruan, S.; Jia, X.; Ma, H.; Chen, W. Semitransparent polymer solar cells with one-dimensional (WO$_3$/LiF)N photonic crystals. *Appl. Phys. Lett.* **2012**, *101*, 153307. [CrossRef]

116. Yu, W.; Shen, L.; Shen, P.; Long, Y.; Sun, H.; Chen, W.; Ruan, S. Semitransparent polymer solar cells with 5% power conversion efficiency using photonic crystal reflector. *ACS Appl. Mater. Interfaces* **2014**, *6*, 599–605. [CrossRef] [PubMed]

117. Yu, W.; Jia, X.; Long, Y.; Shen, L.; Liu, Y.; Guo, W.; Ruan, S. Highly efficient semitransparent polymer solar cells with color rendering index approaching 100 using one-dimensional photonic crystal. *ACS Appl. Mater. Interfaces* **2015**, *7*, 9920–9928. [CrossRef] [PubMed]

118. Yu, W.; Ruan, S.; Long, Y.; Shen, L.; Guo, W.; Chen, W. Light harvesting enhancement toward low IPCE region of semitransparent polymer solar cells via one-dimensional photonic crystal reflectors. *Sol. Energy Mater. Sol. Cells* **2014**, *127*, 27–32. [CrossRef]

119. Yang, W.S.; Noh, J.H.; Jeon, N.J.; Kim, Y.C.; Ryu, S.; Seo, J.; Seok, S.I. High-performance photovoltaic perovskite layers fabricated through intramolecular exchange. *Science* **2015**, *348*, 1234–1237. [CrossRef] [PubMed]

120. Lee, M.M.; Teuscher, J.; Miyasaka, T.; Murakami, T.N.; Snaith, H.J. Efficient hybrid solar cells based on meso-superstructured organometal halide perovskites. *Science* **2012**, *338*, 643–647. [CrossRef] [PubMed]

121. Kim, H.-S.; Lee, C.-R.; Im, J.-H.; Lee, K.-B.; Moehl, T.; Marchioro, A.; Moon, S.-J.; Humphry-Baker, R.; Yum, J.-H.; Moser, J.E.; et al. Lead iodide perovskite sensitized all-solid-state submicron thin film mesoscopic solar cell with efficiency exceeding 9%. *Sci. Rep.* **2012**, *2*, 591. [CrossRef] [PubMed]

MDPI

St. Alban-Anlage 66

4052 Basel

Switzerland

Tel. +41 61 683 77 34

Fax +41 61 302 89 18

www.mdpi.com

Coatings Editorial Office

E-mail: coatings@mdpi.com

www.mdpi.com/journal/coatings

www.ingramcontent.com/pod-product-compliance
Lightning Source LLC
Chambersburg PA
CBHW041215220326
41597CB00033BA/5972